METHODS IN MOLECULAR BIOLOGY

Series Editor
John M. Walker
School of Life and Medical Sciences
University of Hertfordshire
Hatfield, Hertfordshire, AL10 9AB, UK

For further volumes:
http://www.springer.com/series/7651

Optogenetics

Methods and Protocols

Edited by

Arash Kianianmomeni

Department of Cellular and Developmental Biology of Plants,
University of Bielefeld, Bielefeld, Germany

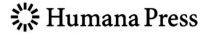

Editor
Arash Kianianmomeni
Department of Cellular
  and Developmental Biology of Plants
University of Bielefeld
Bielefeld, Germany

ISSN 1064-3745   ISSN 1940-6029 (electronic)
Methods in Molecular Biology
ISBN 978-1-4939-3510-9   ISBN 978-1-4939-3512-3   (eBook)
DOI 10.1007/978-1-4939-3512-3

Library of Congress Control Number: 2016931444

Springer New York Heidelberg Dordrecht London

Printed on acid-free paper

Humana Press is a brand of Springer
Springer Science+Business Media LLC New York is part of Springer Science+Business Media (www.springer.com)

Preface

During the past decade, specific light-sensitive modules have paved the way for the development of "Optogenetics," a technology using light switches for precise and spatial control of signaling pathways in individual cells and even in complex biological systems. Although the idea behind controlling cellular, physiological, and behavioral processes using external light was an old scientific desire, crucial factor for rapid development of optogenetics was the discovery of a handful of algal sensory photoreceptors, i.e., light-gated ion channels (channelrhodopsins), which can be easily used to control neural spiking with light. Concurrently, other light-sensitive modules were engineered to generate new photoswitches to control protein activity, protein localization, and gene expression. Fast progress in genome and transcriptome sequencing towards identification of new photoswitchable proteins as well as engineering of new variants with modified absorption and activity properties enriched the optogenetic toolkit and allowed fine-tuned regulation of multiple signaling pathways. In the light of current advances and growing diversity, future application of optogenetic tools for modulation of distinct cellular signaling pathways, even in complex biological systems, without need for chemical additives, seems to be more favorable in comparison to chemical systems.

In this book leading experts on optogenetics, synthetic biology, and neurobiology provide their state-of-the-art protocols and take a close look at current research and its promising applications. This volume provides a collection of the most recently developed technical protocols on optogenetic applications in neuroscience, brain mapping, treatment of neurological disorders, and restoration of visual function. Moreover, several introductory and discussion chapters give a deep, wide overview about sources and diversity of optogenetic tools, design strategies, and potential application in other fields like plant research. I assembled the volume to stimulate an interdisciplinary view of optogenetic applications and its great potential to develop as a fantastic molecular tool for basic research as well as biomedical and biotechnological applications.

Bielefeld, Germany   Arash Kianianmomeni

Contents

Contributors

CARMEN BARTIC • *Department of Physics and Astronomy, KU Leuven, Heverlee, Belgium Imec, Heverlee, Belgium*

FRANCESCO P. BATTAGLIA • *Donders Centre for Neuroscience, Radboud Universiteit, Nijmegen, The Netherlands*

KEN BERGLUND • *Department of Neurosurgery, Emory University, Atlanta, GA, USA*

SIMON M. BRANDL • *Faculty of Biology, University of Freiburg, Freiburg, Germany*

HENRIQUE CABRAL • *Donders Centre for Neuroscience, Radboud Universiteit, Nijmegen, The Netherlands*

CHRISTIAN CASANOVA • *Laboratoire des Neurosciences de la Vision, École d'optométrie, Université de Montréal, Montréal, QC, Canada*

ALEXANDRE CASTONGUAY • *École Polytechnique de Montréal, Montréal, QC, Canada*

ANN-SHYN CHIANG • *Institute of Biotechnology, National Tsing Hua University, Hsinchu, Taiwan; Brain Research Center, National Tsing Hua University, Hsinchu, Taiwan; Genomics Research Center, Academia Sinica, Taipei, Taiwan; Kavli Institute for Brain and Mind, University of California, San Diego, La Jolla, CA, USA*

YONG KU CHO • *Department of Chemical and Biomolecular Engineering, University of Connecticut, Storrs, CT, USA; Institute for Systems Genomics, University of Connecticut, Storrs, CT, USA*

JAN CHRISTOPH • *Research Group Biomedical Physics, Max Planck Institute for Dynamics and Self-Organization, Göttingen, Germany; DZHK German Center for Cardiovascular Research, Partnersite Göttingen, Göttingen, Germany*

ROMAIN DURAND-DE CUTTOLI • *Sorbonne Universités, UPMC Univ Paris 06, Paris, France; Neuroscience Paris Seine, CNRS, UMR 8246, Paris, France; Neuroscience Paris Seine, INSERM, U1130, Paris, France*

EDGAR DEYOE • *Radiology Department, Medical College of Wisconsin, Milwaukee, WI, USA*

EMILIA ENTCHEVA • *Department of Biomedical Engineering, Institute for Molecular Cardiology, Stony Brook University, Stony Brook, NY, USA*

IOLANDA FEOLA • *Laboratory of Experimental Cardiology, Department of Cardiology, Heart Lung Center Leiden, Leiden University Medical Center, Leiden, The Netherlands*

ANDRÉ FIALA • *Department of Molecular Neurobiology of Behavior, Johann-Friedrich-Blumenbach-Institute for Zoology and Anthropology, Georg-August-Universität Göttingen, Göttingen, Germany*

CHEIN-CHUNG FU • *Institute of Photonics Technologies, National Tsing Hua University, Hsinchu, Taiwan; Department of Power Mechanical Engineering, National Tsing Hua University, Hsinchu, Taiwan; Institute of Nanotechnology and Microsystems Engineering, National Tsing Hua University, Hsinchu, Taiwan*

CARLOS GASSER • *Institut für Biologie, Biophysikalische Chemie, Humboldt-Universität zu Berlin, Berlin, Germany*

ROBERT E. GROSS • *Department of Neurosurgery, Emory University, Atlanta, GA, USA; Coulter Department of Biomedical Engineering, Georgia Institute of Technology, Atlanta, GA, USA*

SONGCHAO GUO • *Qiushi Academy for Advanced Studies (QAAS), Zhejiang University, Hangzhou, China; Department of Biomedical Engineering, Key Laboratory of Biomedical Engineering of Education Ministry, Zhejiang University, Hangzhou, China; Zhejiang Provincial Key Laboratory of Cardio-Cerebral Vascular Detection Technology and Medicinal Effectiveness Appraisal, Hangzhou, China*

ARMIN HALLMANN • *Department of Cellular and Developmental Biology of Plants, University of Bielefeld, Bielefeld, Germany*

WON DO HEO • *Center for Cognition and Sociality, Institute for Basic Science, Seoul, Republic of Korea; Department of Biological Sciences, Korea Advanced Institute of Science and Technology, Daejeon, Republic of Korea*

BRYAN HIGASHIKUBO • *Department of Neuroscience, Brown University, Providence, RI, USA*

UTE HOCHGESCHWENDER • *Neuroscience Program and College of Medicine, Central Michigan University, Mt. Pleasant, MI, USA*

PO-YEN HSIAO • *Institute of Biotechnology, National Tsing Hua University, Hsinchu, Taiwan*

ARASH KIANIANMOMENI • *Department of Cellular and Developmental Biology of Plants, University of Bielefeld, Bielefeld, Germany*

JIN MAN KIM • *Graduate School of Medical Science and Engineering, Korea Advanced Institute of Science and Technology, Daejeon, Republic of Korea*

NURY KIM • *Center for Cognition and Sociality, Institute for Basic Science, Seoul, Republic of Korea*

ROBERT J. KITTEL • *Department of Neurophysiology, Institute of Physiology, Julius-Maximilians-Universität Würzburg, Würzburg, Germany*

HEINZ G. KÖRSCHEN • *Department of Molecular Sensory Systems, Research Center Caesar, Bonn, Germany*

MICHAEL KYWERIGA • *Department of Neuroscience, Canadian Centre for Behavioural Neuroscience, University of Lethbridge at Lethbridge, Lethbridge, AB, Canada*

SANGKYU LEE • *Center for Cognition and Sociality, Institute for Basic Science (IBS), Daejeon, Republic of Korea*

STEPHAN E. LEHNART • *DZHK German Center for Cardiovascular Research, Partnersite Göttingen, Göttingen, Germany; Heart Research Center Göttingen, Clinic of Cardiology and Pulmonology, University Medical Center Göttingen, Göttingen, Germany*

DAMIEN LEMOINE • *Sorbonne Universités, UPMC Univ Paris 06, Paris, France; Neuroscience Paris Seine, CNRS, UMR 8246, Paris, France; Neuroscience Paris Seine, INSERM, U1130, Paris, France*

FRÉDÉRIC LESAGE • *École Polytechnique de Montréal, Montréal, QC, Canada; Research Center, Montreal Heart Institute, Montréal, QC, Canada*

DAN LI • *Department of Chemical and Biomolecular Engineering, University of Connecticut, Storrs, CT, USA*

YEN-YIN LIN • *Brain Research Center, National Tsing Hua University, Hsinchu, Taiwan; Institute of Photonics Technologies, National Tsing Hua University,*

Hsinchu, Taiwan; Department of Electrical Engineering, National Tsing Hua University, Hsinchu, Taiwan

STEFAN LUTHER • Research Group Biomedical Physics, Max Planck Institute for Dynamics and Self-Organization, Göttingen, Germany; DZHK German Center for Cardiovascular Research, Partnersite Göttingen, Göttingen, Germany; Institute for Nonlinear Dynamics, Georg-August-Universität Göttingen, Göttingen, Germany

ANNE P. LUTZ • Department of Biology/Genetics, Philipps-University Marburg, Marburg, Germany

FRANCISCO MARTÍN-SAAVEDRA • CIBER de Bioingeniería, Biomateriales y Nanomedicina (CIBER-BBN), Madrid, Spain; Edificio de Investigación, Hospital Universitario La Paz-IdiPAZ, Madrid, Spain

TILO MATHES • Department of Exact Sciences/Biophysics, Vrije Universiteit, Amsterdam, The Netherlands

ANDREAS MÖGLICH • Institut für Biologie, Biophysikalische Chemie, Humboldt-Universität zu Berlin, Berlin, Germany; Faculty of Biology, Chemistry and Earth Sciences, Lehrstuhl für Biochemie, Universität Bayreuth, Bayreuth, Germany

MAJID H. MOHAJERANI • Department of Neuroscience, Canadian Centre for Behavioural Neuroscience, University of Lethbridge at Lethbridge, Lethbridge, AB, Canada

INÈS G. MOLLET • Department of Clinical Sciences, Malmö, Lund University Diabetes Centre, Lund University, Malmö, Sweden

CHRISTOPHER I. MOORE • Department of Neuroscience, Brown University, Providence, RI, USA

ALEXANDRE MOUROT • Sorbonne Universités, UPMC Univ Paris 06, Paris, France; Neuroscience Paris Seine, CNRS, UMR 8246, Paris, France; Neuroscience Paris Seine, INSERM, U1130, Paris, France

KONRAD MÜLLER • Faculty of Biology, University of Freiburg, Freiburg, Germany; Novartis Pharma AG, Biologics Process R&D, Basel, Switzerland

ZANETA NAVRATILOVA • Donders Centre for Neuroscience, Radboud Universiteit, Nijmegen, The Netherlands

THOA T. NGUYEN • Department of Physics and Astronomy, KU Leuven, Heverlee, Belgium

CHRISTOPHER NICOL • Institut für Biologie, Neurobiologie, Freie Universität Berlin, Berlin, Germany

ROCIO OCHOA-FERNANDEZ • Institute of Synthetic Biology, University of Düsseldorf, Düsseldorf, Germany; iGRAD Plant International Graduate Program for Plant Science, University of Düsseldorf, Düsseldorf, Germany

SHINZI OGASAWARA • Creative Research Institution Sousei (CRIS), Hokkaido University, Sapporo, Hokkaido, Japan

HYERIM PARK • Department of Biological Sciences, Korea Advanced Institute of Science and Technology (KAIST), Daejeon, Republic of Korea

RAMIN PASHAIE • Electrical Engineering Department, University of Wisconsin-Milwaukee, Milwaukee, WI, USA

CHRISTOPHER PAWELA • Biophysics Department, Medical College of Wisconsin, Milwaukee, WI, USA

DANIËL A. PIJNAPPELS • *Laboratory of Experimental Cardiology, Department of Cardiology, Heart Lung Center Leiden, Leiden University Medical Center, Leiden, The Netherlands*

MATTHIAS PRIGGE • *Department of Neurobiology, Weizmann Institute of Science, Rehovot, Israel*

THOMAS M. REINBOTHE • *Department of Physiology, University of Gothenburg, Gothenburg, Sweden*

JAKEB M. REIS • *Department of Chemistry, University of Toronto, Toronto, ON, Canada*

CHRISTIAN RENICKE • *Department of Biology/Genetics, Philipps-University Marburg, Marburg, Germany*

CLAUDIA RICHTER • *Research Group Biomedical Physics, Max Planck Institute for Dynamics and Self-Organization, Göttingen, Germany; DZHK German Center for Cardiovascular Research, Partnersite Göttingen, Göttingen, Germany*

THOMAS RIEMENSPERGER • *Department of Molecular Neurobiology of Behavior, Johann-Friedrich-Blumenbach-Institute for Zoology and Anthropology, Georg-August-Universität Göttingen, Göttingen, Germany*

SOPHIA L. SAMODELOV • *Faculty of Biology, University of Freiburg, Freiburg, Germany; Spemann Graduate School of Biology and Medicine (SGBM), University of Freiburg, Freiburg, Germany*

CHARLOTTE HELENE SCHUMACHER • *Institut für Biologie, Biophysikalische Chemie, Humboldt-Universität zu Berlin, Berlin, Germany*

MARTIN SCHWÄRZEL • *Institut für Biologie, Neurobiologie, Freie Universität Berlin, Berlin, Germany*

REINHARD SEIFERT • *Department of Molecular Sensory Systems, Research Center Caesar, Bonn, Germany*

CHRISTOF TAXIS • *Department of Biology/Genetics, Philipps-University Marburg, Marburg, Germany*

ALEXANDER TEPLENIN • *Laboratory of Experimental Cardiology, Department of Cardiology, Heart Lung Center Leiden, Leiden University Medical Center, Leiden, The Netherlands*

SÉBASTIEN THOMAS • *Laboratoire des Neurosciences de la Vision, École d'optométrie, Université de Montréal, Montréal, QC, Canada*

JACK K. TUNG • *Department of Neurosurgery, Emory University, Atlanta, GA, USA; Coulter Department of Biomedical Engineering, Georgia Institute of Technology, Atlanta, GA, USA*

NURIA VILABOA • *Edificio de Investigación, Hospital Universitario La Paz-IdiPAZ, Madrid, Spain; CIBER de Bioingeniería, Biomateriales y Nanomedicina (CIBER-BBN), Madrid, Spain*

ANTOINE A.F. DE VRIES • *Laboratory of Experimental Cardiology, Department of Cardiology, Heart Lung Center Leiden, Leiden University Medical Center, Leiden, The Netherlands; ICIN-Netherlands Heart Institute, The Netherlands*

LING WANG • *Department of Physics and Astronomy, KU Leuven, Heverlee, Belgium*

SHIYAO WANG • *Department of Chemical and Biomolecular Engineering, University of Connecticut, Storrs, CT, USA*

WILFRIED WEBER • *Spemann Graduate School of Biology and Medicine (SGBM), University of Freiburg, Freiburg, Germany; Faculty of Biology, University of Freiburg, Freiburg, Germany; BIOSS-Centre for Biological Signalling Studies, University of Freiburg, Freiburg, Germany*

ELKE WEHINGER • *Faculty of Biology, University of Freiburg, Freiburg, Germany*

JONAS WIETEK • *Experimental Biophysics, Humboldt University Berlin, Berlin, Germany*

G. ANDREW WOOLLEY • *Department of Chemistry, University of Toronto, Toronto, ON, Canada*

MING-CHIN WU • *Brain Research Center, National Tsing Hua University, Hsinchu, Taiwan*

KEDI XU • *Qiushi Academy for Advanced Studies (QAAS), Zhejiang University, Hangzhou, China; Department of Biomedical Engineering, Key Laboratory of Biomedical Engineering of Education Ministry, Zhejiang University, Hangzhou, China; Zhejiang Provincial Key Laboratory of Cardio-Cerebral Vascular Detection Technology and Medicinal Effectiveness Appraisal, Hangzhou, China*

JINZHU YU • *Department of Biomedical Engineering, Stony Brook University, Stony Brook, NY, USA*

JIACHENG ZHANG • *Qiushi Academy for Advanced Studies (QAAS), Zhejiang University, Hangzhou, China; Department of Biomedical Engineering, Key Laboratory of Biomedical Engineering of Education Ministry, Zhejiang University, Hangzhou, China; Zhejiang Provincial Key Laboratory of Cardio-Cerebral Vascular Detection Technology and Medicinal Effectiveness Appraisal, Hangzhou, China*

XIAOXIANG ZHENG • *Qiushi Academy for Advanced Studies (QAAS), Zhejiang University, Hangzhou, China; Department of Biomedical Engineering, Key Laboratory of Biomedical Engineering of Education Ministry, Zhejiang University, Hangzhou, China; Zhejiang Provincial Key Laboratory of Cardio-Cerebral Vascular Detection Technology and Medicinal Effectiveness Appraisal, Hangzhou, China*

THEA ZIEGLER • *Institut für Biologie, Biophysikalische Chemie, Humboldt-Universität zu Berlin, Berlin, Germany; Lehrstuhl für Biochemie, Universität Bayreuth, Bayreuth, Germany*

MATIAS D. ZURBRIGGEN • *Institute of Synthetic Biology, University of Düsseldorf, Düsseldorf, Germany; Faculty of Biology, University of Freiburg, Freiburg, Germany*

Chapter 1

Optogenetics: Basic Concepts and Their Development

Yong Ku Cho and Dan Li

Abstract

The discovery of light-gated ion channels and their application to controlling neural activities have had a transformative impact on the field of neuroscience. In recent years, the concept of using light-activated proteins to control biological processes has greatly diversified into other fields, driven by the natural diversity of photoreceptors and decades of knowledge obtained from their biophysical characterization. In this chapter, we will briefly discuss the origin and development of optogenetics and highlight the basic concepts that make it such a powerful technology. We will review how these enabling concepts have developed over the past decade, and discuss future perspectives.

Key words Optogenetics, Light-activated protein, Photoreceptor, Channelrhodopsin, Protein engineering

1 Introduction

The word 'optogenetics' was initially used within the context of neuroscience [1], to describe the approach of using light to image and control neuronal activity in the intact, living brain. The idea of controlling biological function using light has been around for a long time, but it is in the field of neuroscience that its need as a truly enabling tool has been envisioned early on [2, 3], and remarkably to the realization of such potential [4–6] with wide acceptance over the past decade. The success of optogenetics in neuroscience has sparked interest of many scientists and engineers in other fields, and now the definition of optogenetics has broadened to encompass the general field of biotechnology that combines genetic engineering and optics to enable gain or loss of well-defined function, often in the intact animal [7, 8]. In this chapter, we will briefly go over the historical background of the origin of optogenetics and highlight the basic concepts of optogenetics that make it such a powerful technology. We will review how these enabling concepts have developed over the past decade, and discuss future perspectives.

Arash Kianianmomeni (ed.), *Optogenetics: Methods and Protocols*, Methods in Molecular Biology, vol. 1408,
DOI 10.1007/978-1-4939-3512-3_1, © Springer Science+Business Media New York 2016

2   Original Concepts: A Brief Historical Perspective

The seed for the birth of optogenetics in neuroscience may have been planted a long time ago by the pioneering work of Ramón y Cajal, who provided foundational evidence that neurons are the signaling units of the nervous system, and that they exist in many distinctive morphologies [9]. Since then, studies followed to show that indeed many different types of neurons exist, classified based on their physiological characteristics, anatomical location, morphology, and gene expression profile [10–12]. It is still unclear how many cell types exist in the human brain, but it is speculated that there are about 1000 neuronal cell types just within the mammalian cortex [13, 14]. In a general sense, it is agreed that a specific cell type carries out the same function within a neural circuit [15]. Therefore, identification of all cell types and mapping out their connectivity is essential in order to understand how the nervous system works. Identification of different cell types must be accompanied with functional characterization within their normal context, the nervous system. For this reason, neuroscientists have been looking for a way to perturb individual cell types within the intact brain. This was clearly expressed by Francis Crick in his insightful discussion published in 1979, in which he proposed the need for a 'method by which all neurons of just one type could be inactivated, leaving the others more or less unaltered' [2].

Optogenetics is arguably the first technological breakthrough that enables such experiments. It makes the crucial connection between cell-type information and the ability to perform gain or loss of function experiments. Prior to the development of optogenetics, experimental approaches existed in neuroscience to use light as a way to control neural activity. For example, 'caged' compounds such as secondary messenger molecules, ions, and neurotransmitters have been developed that are initially inert, but become active upon light illumination [16, 17]. Even though these photochemical approaches did not provide ways to control specific cell types, they laid the groundwork for the use of millisecond timescale illumination in intact cells and tissue [17]. Cell-type specific activation of neurons was first achieved in a series of pioneering work by the Miesenböck group, by heterologously expressing an invertebrate rhodopsin with other interacting proteins [18], and ligand-gated ion channels that can be activated by synthetic photocaged precursors [19, 20]. Around this time, microbial rhodopsins that function as single component light-gated ion channels were discovered [21–23]. Remarkably, these microbial photoreceptors could be heterologously expressed in mammalian neurons to optically trigger action potentials with millisecond timescale precision [4]. These findings, along with the fact that these rhodopsins contain a covalently bound all-*trans* retinal chromophore naturally produced in nearly all cell types including mammalian cells and tissues [24, 25], catalyzed the wide adoption of these molecules in neuroscience.

3 Genetic Targeting: Cell-Type Specificity and Beyond

As discussed above, cell-type specificity is a defining feature of optogenetic experiments. In some ways, gain or loss of function experiments on distinct cell types are analogous to genetic knockin or knockout experiments in molecular biology. However, while the definition of gene is clear, the definition of cell type can be ambiguous and variable. This is particularly true in highly complex systems such as the mammalian neocortex [15, 26]. To account for such high complexity, many parameters have been used to describe each cell type, such as developmental lineage, anatomical location, dendritic and axonal morphology, electrophysiology, and gene expression. Nonetheless, recent advances in single cell analysis methods have provided evidence that gene expression patterns can capture many of these diverse cellular phenotypes [10, 27]. Therefore, the use of genetically encoded tools is well justified for probing the function of each cell type, and is likely to expand as we gather more gene expression data.

In principle, optogenetic tools can be driven by cell-type specific promoters or enhancers [28, 29] to achieve cell-type specific expression. However, this approach is not suitable in many cases for several reasons. First, it is rare to find a single genetic regulatory element exclusive for a given cell type. Cell identity in most cases seems to be defined by a combination of multiple gene expression patterns [8, 10, 15, 30]. Even when cell-type specific genetic regulatory elements are accessible, they may not drive high enough expression of optogenetic tools for efficient control [31]. Moreover, reproducing endogenous expression pattern of genes requires the entire transcription unit and all associated regulatory factors [11], which are undefined in many cases, and can be hard to transport. A widely used strategy to achieve cell-type specific expression that circumvents these problems is using viral vectors in transgenic animals expressing recombinase in genetically defined population of cells. In these approaches, viral vectors such as lentivirus or adeno-associated virus (AAV) carry an optogenetic tool under relatively strong promoters such as EF-1α promoter (Fig. 1a). The optogenetic tool encoding gene is initially present in reverse frame to prevent expression, and requires recombinase activity for inversion and subsequent expression (Fig. 1a). The recombinase gene expression is driven in genetically defined cell types, for which the expression level is relatively unimportant. Cre-recombinase has been widely used for this purpose, due to the availability of increasing number of cell-type specific Cre-transgenic

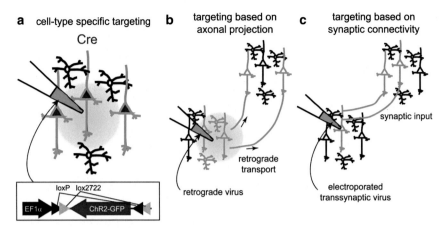

Fig. 1 Genetic targeting strategies used in optogenetics. (**a**) Cell-type specific targeting using a viral vector carrying an optogenetic construct (e.g., ChR2-GFP, indicated in *green*) in reverse frame. The optogenetic construct is expressed only in cells that express Cre-recombinase (indicated in *red*). (**b**) Targeting neurons that project in defined brain regions. Modified rabies virus can retrogradely transduce neurons. (**c**) Targeting neurons based on synaptic connectivity. Modified rabies virus can be electroporated into a single neuron, and spread transsynaptically to label cells making synaptic connections

lines [32]. Recently, this approach has been further developed to enable targeting of neural cell types defined by multiple genetic markers. Transgenic animals that express multiple recombinases driven by independent promoters are infected with viral vectors carrying a combination of these recombinase sites, resulting in intersectional targeting [33, 34].

Using viral vectors also enable other useful modes of targeting. One approach is to target neurons based on their axonal projections or synaptic connectivity (Fig. 1b, c). For example, modified rabies virus can retrogradely transduce neurons [35, 36], while herpes simplex virus and vesicular stomatitis virus can anterogradely transduce across axon terminals [37, 38]. These viruses can be used to deliver Cre-recombinase transsynaptically [37, 39], allowing the targeting of neurons projecting to a genetically defined populations. Recently, it was also shown that certain serotypes of AAV also can mediate efficient retrograde transduction [40]. These tools enable targeting optogenetic tools based on their axonal projection [41–43] and potentially synaptic connectivity.

Another targeting strategy is to express genes using promoters and enhancers induced by neuronal activity. These genetic elements were identified from immediate early genes that are known to rapidly induce expression upon neuronal activity [44, 45]. In many cases, these elements are triggered by calcium influx that activates calcium-dependent kinases [46]. These promoters have been used to specifically express optogenetic tools in neurons that gain activity during specific behavioral tasks in rodents, such as fear conditioning [47]. In a set of impressive demonstrations, Liu and others showed that these neurons could be specifically reactivated using the light-gated ion channel channelrhodopsin-2 (ChR2) driven by the activity-dependent promoter of c-Fos [47, 48]. This reactivation recreated the original behavior with only light activation, indicating optically controlled memory recall [47]. Ramirez and others later used the same approach to trigger fear response by optically activating a set of neurons activated during fear conditioning, demonstrating the creation of false memory [49]. Conceptually, these demonstrations show that optogenetics can be used to reactivate cell populations specifically activated during a behavioral task, enabling activity-based targeting. Considering the diverse functionality of transcription factors that can sense a wide range of input signals [50–52], such responsive transcription may be a generalizable approach to express optogenetic tools to control cell populations that drive a specific functional output.

4   Customizing Photoreceptors for Designed Control: Modes of Action

The use of microbial rhodopsins in mammalian neural cells is an inspiring demonstration of customizing naturally existing photoreceptors for controlling physiological properties in a completely new context. Owing to the diverse modes of action found in photoreceptors (Fig. 2), the concept of repurposing naturally existing parts to control new biological functions has been implemented in many applications, and is expected to further develop as new photoreceptors continue to be discovered.

Microbial rhodopsins (type 1 rhodopsins) mediate either transmembrane ion transport (Fig. 2a, b) or light sensing through signal transduction (Fig. 2c). Microbial rhodopsins that mediate light-driven ion transport have been widely used to control membrane potential in mammalian neurons [5]. They can be categorized into two mechanistically distinct forms: ion pumps (Fig. 2a) that can transport ions against their gradient and channels (Fig. 2b) that passively conduct ions along the gradient established by other active transporters. Interestingly, all known light-driven ion pumps result in hyperpolarization of membrane potential through outward transport of proton or inward transport of chloride ion [53]. When heterologously expressed in mammalian neurons, they generate enough current to enable optical silencing [5, 54–56]. Recently, a light-driven chloride pump with high sensitivity for far-red light named Jaws has been described that enabled noninvasive neural silencing through the intact mouse skull [57]. A light-driven outward sodium pump (KR2) has also been found recently [58], that may be used for optical silencing of neurons. Unlike ion pumps, microbial rhodopsin ion channels cannot transport ions against their gradient, but enable control of membrane potential through selective conduction of specific ions. First described

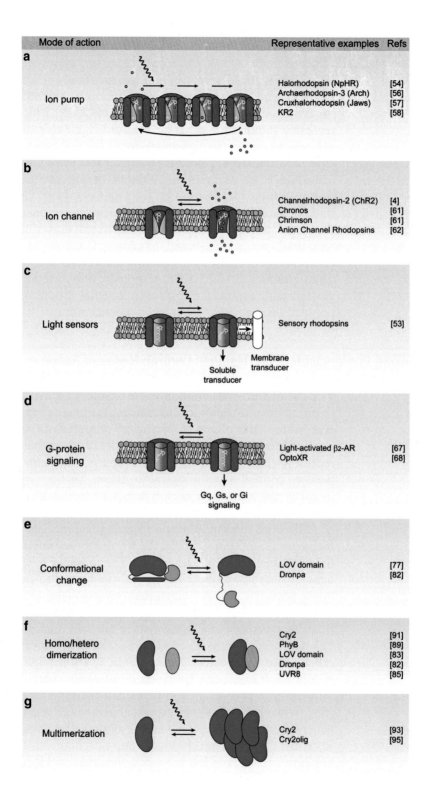

Fig. 2 Modes of action found in photoreceptors. (**a**) Light-activated ion pumps. (**b**) Light-gated ion channels. (**c**) Microbial sensory rhodopsins for light-activated signal transduction. (**d**) Animal rhodopsins that function as light-activated G-protein coupled receptors. (**e**) Light-triggered conformational change, leading to regulation of protein domain function through allosteric coupling or uncaging. (**f**) Homo/heterodimerization controlled by light. Dronpa and UVR8 monomerize in the lit-state. (**g**) Multimerization upon light activation. Figure panels modified from [63] with permission

examples are channelrhodopsins found in the green alga *Chlamydomonas reinhardtii* [21, 22], which conduct cations including proton, sodium, potassium, and calcium ions [22]. Even though the conductance of these channels are relatively low (about 100-fold lower than high-conductance ion channels) [59, 60], their current depolarize mammalian neurons above their threshold to initiate action potentials [4]. Recently, more than 60 channelrhodopsin homologues were identified by conducting a systematic search of transcriptome from 127 species of alga [61]. This study resulted in a high sensitivity channelrhodopsin with extremely fast channel kinetics named Chronos, and a red-sensitive channelrhodopsin named Chrimson. As a pair, these channelrhodopsins enabled independent multicolor activation of two distinct neural populations [61]. Another recent study identified anion-conducting light-gated channels in the cryptophyte *Guillardia theta* that enabled rapid and reversible optical silencing of rat neurons [62]. So far, microbial sensory rhodopsins have not been applied in optogenetic experiments, perhaps due to the requirement of transmembrane or soluble transducers that are not readily compatible with other cell types. In depth discussion of biophysical properties of microbial rhodopsins and their impact as optogenetic tools have been discussed elsewhere [63].

Animal rhodopsins (type II rhodopsins) share structural homology to microbial rhodopsins, but act as G-protein coupled receptors (GPCRs) that upon light illumination, catalyze GDP to GTP exchange in heterotrimeric G proteins (Fig. 2d). In fact, rhodopsins constitute the largest GPCR family—over 700 rhodopsin-like GPCRs have been found in humans [64]. Based on the extensive structure–function relationship studies of GPCRs [65, 66], Khorana and colleagues demonstrated that the cytoplasmic domains of a bovine rhodopsin can be replaced by analogous sequences from a non-light sensitive GPCR, such as β_2-adrenergic receptor (β_2-AR) to create a chimeric light-sensitive β_2-AR [67]. This strategy has been extended to create light-driven rhodopsins coupled to Gq and Gs signaling pathways [68], Gi/o coupled pathways [69, 70]. It was also found that heterologous expression of unmodified forms of rhodopsins enables light-driven activation of endogenous G proteins of the host cell [71]. Unlike microbial rhodopsins that keep the retinal chromophore throughout the photocycle, certain animal rhodopsins such as bovine rhodopsin lose the chromophore after light activation, requiring consistent supply of retinal cofactors [72]. Fortunately, in mammalian cells and brain tissue, sufficient concentration of retinal cofactors are present to generate enough functional rhodopsins. However, even in cell types where the retinal chromophore is less abundant, the retinal cofactor supplementation may be avoided by using 'non-bleaching' G-protein coupled opsins that remain associated with the chromophore [73].

Other than rhodopsin based photoreceptors, light-sensing proteins found in plants and microbes have been applied in controlling cell signaling in diverse cell types, including mammalian cells, yeast, and bacteria. These proteins function by inducing light-triggered conformational change coupled to other protein domains (Fig. 2e), homo/heterodimerization (Fig. 2f), or multimerization (Fig. 2g). A prominent example that undergoes conformational change upon light activation is the Light-Oxygen-Voltage (LOV) domains found in photoreceptor systems of plants, fungi, and bacteria [74, 75]. In these systems, LOV domains may control the activity of an effector domain directly fused to it through allosteric coupling or steric inhibition. For example, the bacterial chemosensor FixL was made light-activatable by replacing its Per-ARNT-Sims (PAS) motif (which is structurally homologous to LOV domains) with the LOV domain Ytva [76]. In this example, an α-helical coiled coil linker of the LOV domain undergoes a rotational movement upon light activation, which activates a histidine kinase domain fused to it. In other examples, the conformational change in the LOV domain leads to unfolding of an α-helix in the lit-state, resulting in 'uncaging' of the effector domain fused to it (Fig. 2e). Such light-activated uncaging approach has been successfully applied in several synthetic constructs including a photoactivatable GTPase Rac1 and Cdc42 [77], peptide binding motifs [78], and tethered toxins [79]. Even though such photo-uncaging approach seems to be a generally applicable design, in many cases it requires several levels of optimization customized to each molecule for effective control. For instance, in the photoactivatable Rac1, Ca^{2+}-mediated interaction between residues at the interface of the LOV domain and Rac1 turned out to be crucial for effective light control [80]. Since such conformation-dependent interactions are hard to be predicted, de novo design of a photoactivatable construct still remains a challenge. Such difficulties may be overcome by using Dronpa, which is a photoactivated fluorescent protein that seems to allow modular design of optical control. Upon light activation, Dronpa not only changes fluorescence, but also monomerizes through the unfolding of a β-sheet [81]. This feature has been used to design photoactivatable GTPases and proteases [82].

Another mode of action found in photoreceptor domains is light-induced interaction. Certain LOV domains, such as the fungal LOV domain Vivid (VVD) [83] and bacterial LOV domain EL222 [84], homodimerize upon light activation (Fig. 2f). Plant photoreceptors such as *Arabidopsis thaliana* UVR8 exist as dimers in the dark but monomerize upon UV illumination [85]. Other photoreceptor domains such as *Arabidopsis thaliana* Cryptochrome-2 (CRY2) and Phytochrome B (PhyB) heterodimerize with a specific partner (Fig. 2f), cryptochrome-interacting basic helix–loop–helix 1 (CIB1) and phytochrome interacting factor (PIF), respectively [74]. Such light-induced interactions have been used primarily to control intracellular signaling, by recruiting signaling proteins to or away from a specific intracellular site of action, resulting in signal activation or inhibition through sequestration. Activities of various kinases, including Ras/ERK [86], phosphoinositide-3 kinase (PI3K) [87], and receptor tyrosine kinase (RTK) [88] have been controlled using this approach. Several studies also demonstrated optical control of DNA transcription using light-induced binders that mediate the recruitment of transcription activation domains [89, 90] or activation of Cre-recombinase [91, 92]. Interestingly, CRY2 has been shown to undergo multimerization upon illumination (Fig. 2g), which has been used to mediate light-dependent activation of Rho GTPase [93], RTK fibroblast growth factor receptor [94], and actin polymerization [95]. The use of photoreceptors for optogenetic control of cellular signaling pathways has been extensively reviewed recently [63, 96, 97].

5   Engineering Photoreceptors: A Multidimensional Problem

Even though it seems that natural photoreceptor systems rely on modularity of light-activated protein domains for controlling diverse signaling pathways [74, 98], development of a new optogenetic tool in general requires significant engineering efforts for effective control. One major challenge is achieving adequate expression of optogenetic tools. As seen in the case of microbial opsins, the expression level of an optogenetic construct may be one of the limiting factors in achieving effective control. Heterologous expression of photoreceptor domains depends on the host cell type, and may yield low levels of functionally active form of the optogenetic tool. For example, a study that systematically compared channelrhodopsin homologues found that less than half of them showed detectable ion conduction when expressed in a mammalian cell line, and even less were functionally active in neurons [61]. One strategy to improve functional expression is to enhance the protein trafficking to the desired compartment using targeting sequences [56, 99]. However, this approach is not generalizable [100], and achieving adequate expression of optogenetic tools in the desired cell type remains a challenge.

Another major challenge is the fact that optogenetic tools require multiple properties to be optimized. For example, critical properties of channelrhodopsins that impact their effectiveness in controlling neural activity include ion conductance, light sensitivity, spectral sensitivity, channel kinetics, and ion selectivity. Structure-guided mutagenesis studies have resulted in improved channelrhodopsins with fast kinetics [101], enhanced photocurrent [102, 103], red-shifted spectral sensitivity [104], and altered ion selectivity [105, 106]. These studies demonstrated that key

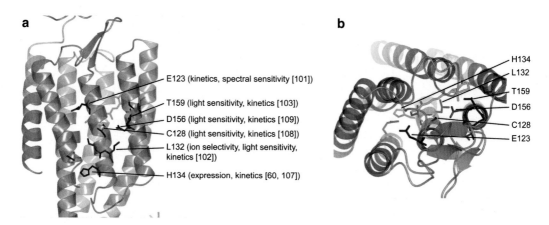

Fig. 3 ChR2 residues that affect multiple properties when mutated. Residue numbers based on ChR2 sequence are labeled on the X-ray structure obtained from a chimera between ChR1 and ChR2 (PDB code 3UG9). (**a**) Side-view. (**b**) Top-view

properties can be tuned using mutagenesis, but also revealed that a single mutation often affects multiple properties (Fig. 3). For example, mutations at E123 in ChR2 (corresponding to D85, which is the Schiff base counter-ion in bacteriorhodopsin), affect channel kinetics and red-shifts the action spectra [101]. Mutation H134R in ChR2 enhances photocurrent perhaps due to improved membrane expression, but slows channel kinetics [60, 107]. Mutations C128 and D156 slow down the channel kinetics drastically, which helps to keep open channels for longer periods of time, effectively enhancing the light sensitivity [108, 109]. The trend of multiple property modulation by a single mutation is also found in LOV domains. For example, in the *Avena sativa* phototropin 1 LOV2, mutations in the highly conserved residue Q513 reduce the structural changes between the dark and the lit-state and slow down the dark state return rate [110]. In addition, in the LOV domain VVD, mutations in M135 and M165 slow recovery kinetics and enhance the affinity of the lit-state VVD dimer [111]. Therefore, optimization of an optogenetic tool requires multidimensional characterizations to measure all essential parameters [109], and in certain cases individual properties may not be independently optimized. In other words, the fitness landscape of photoreceptors is a multidimensional space composed of potentially dependent parameters. Since typical screens used for protein engineering rely on rapid measurement of one or two parameters, strategies such as directed evolution to optimize one parameter may result in de-optimization of others that are critical for tool performance. Currently, the general strategy is to combine random mutagenesis-based methods with structure-guided approaches and rely on additive beneficial effects of multiple mutations to yield an optimal construct [105, 106, 112, 113]. In the future, screens that can characterize multiple parameters may be developed to perform multidimensional optimizations.

6   Analogies to Optical Imaging: Multicolor Imaging Versus Optogenetics

Another major direction in optogenetics has been to use multiple colors of light to control two independent cell types or processes using photoreceptor domains that have distinctive spectral sensitivities. This approach has been extensively explored using microbial opsins. For instance, ChR2, which has maximal sensitivity for blue light has been combined with a halorhodopsin which can be activated with yellow light to enable bidirectional control of neural activity [54, 55]. Proton pumps with separation in their spectral sensitivity have been used to achieve two-color neural silencing [56]. Even though these studies demonstrated that triggering or inhibiting action potentials in neurons can be controlled by leveraging the difference in spectral sensitivity of microbial rhodopsins, a recent study showed that achieving multicolor control relying on spectral separation alone may result in cross talk, due to the inherent blue-light sensitivity found in rhodopsins [61]. Even the red-sensitive channelrhodopsin Chrimson was shown to trigger action potentials under strong blue light in neurons with high expression levels [61, 63]. However, detailed biophysical characterization of Chrimson revealed that its channel opening rate under blue light is substantially slower than that under red light, and is also dependent on light intensity [61]. Therefore, cross-activation of neurons under blue light was minimized by pairing Chrimson with the fast and light-sensitive channelrhodopsin Chronos and limiting the intensity and duration of blue light [61, 63]. The blue-light sensitivity of microbial rhodopsins may not be completely abolished unless the chromophore itself is altered. Many photoreceptors, including LOV domains and cryptochromes are also maximally sensitive to blue light, which may be hard to eliminate as in the case for microbial rhodopsins. Therefore, strategies other than relying on spectral separation alone, such as leveraging differences in light sensitivity and kinetics under blue light may be necessary to reduce cross talk for multicolor control using other photoreceptors.

Although the concept of multicolor control in optogenetics may seem analogous to multicolor fluorescence imaging, practical differences exist that require caution in designing optogenetic experiments with multiple light sources. In multicolor fluorescence imaging, cross-activation of fluorophores occurs quite often, but the emission from multiple fluorophores can be filtered to obtain cross talk free images. In experiments using light-activated proteins, any cross-activation causes change that cannot be filtered or eliminated. Depending on the biological question asked, small changes induced by cross-activation may become important, such as in experiments that focus on subthreshold changes in membrane potential. Therefore, when designing experiments that require multiple sources of light, such as using light-gated ion channels with calcium reporters [114], the wavelength and intensity of light used for imaging should be tested for cross-activation of optical actuators. It is notable that blue-sensitive channelrhodopsins that seem to have negligible red-sensitivity can depolarize membrane potential under intense red light [113]. As described above, strategies that capitalize on all aspects of biophysical properties of photoreceptors may be used to prevent cross-activation, but may be difficult to completely eliminate small cross talks induced.

7   Conclusion and Future Perspectives

Over the past decade, optogenetics has made a major impact on neuroscience research, by enabling the control of specific cell types in the intact brain. A major advancement enabled using these tools was the ability to control specific cell types in intact systems with high spatial and temporal resolution. One potentially transformative future direction would be the integration of optical control methods with various detection methods to enable closed-loop control of biological systems, as achieved by several studies that demonstrated closed-loop control of neural circuit [115–117] and feedback-regulated control of cell signaling [118] and gene expression levels [119]. These studies suggest that optogenetic approaches will enable us to reveal the underlying mechanisms behind complex intracellular signaling systems or multicellular dynamics and precisely tune it to achieve a desired outcome. They will also allow us to develop and test models of intact, complex biological systems, opening an era of real-time 'process engineering' in biological systems. Optogenetics may enable the application of advanced process engineering approaches widely used in chemical and electrical systems to biological systems.

As discussed in this chapter, successful development of new optogenetic tools and their implementation is an interdisciplinary effort that requires tools tailored to the specific biological process studied. Targeting and achieving optimal expression in the desired cell type need to be tested, and the biophysical properties of photoreceptors used need to be optimized to the spatial and temporal properties to be controlled. Therefore, users of optogenetic tools need to pay close attention to their biophysical characteristics, and tool developers need to identify ways to tune them.

Acknowledgements

This work was funded by the University of Connecticut and the Brain and Behavior Research Foundation (NARSAD Young Investigator grant).

References

1. Deisseroth K, Feng G, Majewska AK et al (2006) Next-generation optical technologies for illuminating genetically targeted brain circuits. J Neurosci 26(41):10380–10386

2. Crick FH (1979) Thinking about the brain. Sci Am 241(3):219–232

3. Crick F (1999) The impact of molecular biology on neuroscience. Philos Trans R Soc Lond B Biol Sci 354(1392):2021–2025

4. Boyden ES, Zhang F, Bamberg E et al (2005) Millisecond-timescale, genetically targeted optical control of neural activity. Nat Neurosci 8(9):1263–1268

5. Boyden ES (2011) A history of optogenetics: the development of tools for controlling brain circuits with light. F1000 Biol Rep 3:11

6. Deisseroth K (2010) Controlling the brain with light. Sci Am 303(5):48–55

7. Deisseroth K (2011) Optogenetics. Nat Methods 8(1):26–29

8. Miesenbock G (2009) The optogenetic catechism. Science 326(5951):395–399

9. Ramón y Cajal S (1911) Histology of the nervous system of man and vertebrates. Oxford University Press, 1995 translation

10. Fishell G, Heintz N (2013) The neuron identity problem: form meets function. Neuron 80(3):602–612

11. Gong S, Zheng C, Doughty ML et al (2003) A gene expression atlas of the central nervous system based on bacterial artificial chromosomes. Nature 425(6961):917–925

12. Hawrylycz MJ, Lein ES, Guillozet-Bongaarts AL et al (2012) An anatomically comprehensive atlas of the adult human brain transcriptome. Nature 489(7416):391–399

13. Koch C (2004) The quest for consciousness: a neurobiological approach. Roberts and Co., Denver, CO

14. Stevens CF (1998) Neuronal diversity: too many cell types for comfort? Curr Biol 8(20): R708–R710

15. Luo L, Callaway EM, Svoboda K (2008) Genetic dissection of neural circuits. Neuron 57(5):634–660

16. Ellis-Davies GC (2007) Caged compounds: photorelease technology for control of cellular chemistry and physiology. Nat Methods 4(8):619–628

17. Kaplan JH, Somlyo AP (1989) Flash photolysis of caged compounds: new tools for cellular physiology. Trends Neurosci 12(2):54–59

18. Zemelman BV, Lee GA, Ng M et al (2002) Selective photostimulation of genetically chARGed neurons. Neuron 33(1):15–22

19. Zemelman BV, Nesnas N, Lee GA et al (2003) Photochemical gating of heterologous ion channels: remote control over genetically designated populations of neurons. Proc Natl Acad Sci U S A 100(3):1352–1357

20. Lima SQ, Miesenbock G (2005) Remote control of behavior through genetically targeted photostimulation of neurons. Cell 121(1):141–152

21. Nagel G, Ollig D, Fuhrmann M et al (2002) Channelrhodopsin-1: a light-gated proton channel in green algae. Science 296(5577): 2395–2398

22. Nagel G, Szellas T, Huhn W et al (2003) Channelrhodopsin-2, a directly light-gated cation-selective membrane channel. Proc Natl Acad Sci U S A 100(24):13940–13945

23. Sineshchekov OA, Jung KH, Spudich JL (2002) Two rhodopsins mediate phototaxis to low- and high-intensity light in Chlamydomonas reinhardtii. Proc Natl Acad Sci U S A 99(13):8689–8694

24. Blomhoff R, Blomhoff HK (2006) Overview of retinoid metabolism and function. J Neurobiol 66(7):606–630

25. Zhang F, Wang LP, Boyden ES et al (2006) Channelrhodopsin-2 and optical control of excitable cells. Nat Methods 3(10):785–792

26. Markram H, Toledo-Rodriguez M, Wang Y et al (2004) Interneurons of the neocortical inhibitory system. Nat Rev Neurosci 5(10):793–807

27. Sugino K, Hempel CM, Miller MN et al (2006) Molecular taxonomy of major neuronal classes in the adult mouse forebrain. Nat Neurosci 9(1):99–107

28. Spergel DJ, Kruth U, Hanley DF et al (1999) GABA- and glutamate-activated channels in green fluorescent protein-tagged gonadotropin-releasing hormone neurons in transgenic mice. J Neurosci 19(6):2037–2050

29. Oliva AA Jr, Jiang M, Lam T et al (2000) Novel hippocampal interneuronal subtypes identified using transgenic mice that express green fluorescent protein in GABAergic interneurons. J Neurosci 20(9):3354–3368

30. McGarry LM, Packer AM, Fino E et al (2010) Quantitative classification of somatostatin-positive neocortical interneurons identifies three interneuron subtypes. Front Neural Circuits 4:12

31. Zeng H, Madisen L (2012) Mouse transgenic approaches in optogenetics. Prog Brain Res 196:193–213

32. Gong S, Doughty M, Harbaugh CR et al (2007) Targeting Cre recombinase to specific neuron populations with bacterial artificial chromosome constructs. J Neurosci 27(37): 9817–9823

33. Fenno LE, Mattis J, Ramakrishnan C et al (2014) Targeting cells with single vectors using multiple-feature Boolean logic. Nat Methods 11(7):763–772

34. Madisen L, Garner AR, Shimaoka D et al (2015) Transgenic mice for intersectional targeting of neural sensors and effectors with high specificity and performance. Neuron 85(5):942–958

35. Wickersham IR, Finke S, Conzelmann KK et al (2007) Retrograde neuronal tracing with a deletion-mutant rabies virus. Nat Methods 4(1):47–49

36. Wickersham IR, Lyon DC, Barnard RJ et al (2007) Monosynaptic restriction of transsynaptic tracing from single, genetically targeted neurons. Neuron 53(5):639–647

37. Lo L, Anderson DJ (2011) A Cre-dependent, anterograde transsynaptic viral tracer for mapping output pathways of genetically marked neurons. Neuron 72(6):938–950

38. Beier KT, Saunders A, Oldenburg IA et al (2011) Anterograde or retrograde transsynaptic labeling of CNS neurons with vesicular stomatitis virus vectors. Proc Natl Acad Sci U S A 108(37):15414–15419

39. Wall NR, Wickersham IR, Cetin A et al (2010) Monosynaptic circuit tracing in vivo through Cre-dependent targeting and complementation of modified rabies virus. Proc Natl Acad Sci U S A 107(50):21848–21853

40. Rothermel M, Brunert D, Zabawa C et al (2013) Transgene expression in target-defined neuron populations mediated by retrograde infection with adeno-associated viral vectors. J Neurosci 33(38):15195–15206

41. Osakada F, Mori T, Cetin AH et al (2011) New rabies virus variants for monitoring and manipulating activity and gene expression in defined neural circuits. Neuron 71(4):617–631

42. Apicella AJ, Wickersham IR, Seung HS et al (2012) Laminarly orthogonal excitation of fast-spiking and low-threshold-spiking interneurons in mouse motor cortex. J Neurosci 32(20):7021–7033

43. Kress GJ, Yamawaki N, Wokosin DL et al (2013) Convergent cortical innervation of striatal projection neurons. Nat Neurosci 16(6):665–667

44. Smeyne RJ, Schilling K, Robertson L et al (1992) fos-lacZ transgenic mice: mapping sites of gene induction in the central nervous system. Neuron 8(1):13–23

45. Kawashima T, Okuno H, Nonaka M et al (2009) Synaptic activity-responsive element in the Arc/Arg3.1 promoter essential for synapse-to-nucleus signaling in activated neurons. Proc Natl Acad Sci U S A 106(1):316–321

46. Bito H, Deisseroth K, Tsien RW (1996) CREB phosphorylation and dephosphorylation: a Ca(2+)- and stimulus duration-dependent switch for hippocampal gene expression. Cell 87(7):1203–1214

47. Liu X, Ramirez S, Pang PT et al (2012) Optogenetic stimulation of a hippocampal engram activates fear memory recall. Nature 484(7394):381–385

48. Kubik S, Miyashita T, Guzowski JF (2007) Using immediate-early genes to map hippocampal subregional functions. Learn Mem 14(11):758–770

49. Ramirez S, Liu X, Lin PA et al (2013) Creating a false memory in the hippocampus. Science 341(6144):387–391

50. Sellick CA, Reece RJ (2005) Eukaryotic transcription factors as direct nutrient sensors. Trends Biochem Sci 30(7):405–412

51. Chandel NS, Maltepe E, Goldwasser E et al (1998) Mitochondrial reactive oxygen species trigger hypoxia-induced transcription. Proc Natl Acad Sci U S A 95(20):11715–11720

52. Tamura T, Yanai H, Savitsky D et al (2008) The IRF family transcription factors in immunity and oncogenesis. Annu Rev Immunol 26:535–584

53. Ernst OP, Lodowski DT, Elstner M et al (2014) Microbial and animal rhodopsins: structures, functions, and molecular mechanisms. Chem Rev 114(1):126–163

54. Han X, Boyden ES (2007) Multiple-color optical activation, silencing, and desynchronization of neural activity, with single-spike temporal resolution. PLoS One 2(3):e299

55. Zhang F, Wang LP, Brauner M et al (2007) Multimodal fast optical interrogation of neural circuitry. Nature 446(7136):633–639

56. Chow BY, Han X, Dobry AS et al (2010) High-performance genetically targetable optical neural silencing by light-driven proton pumps. Nature 463(7277):98–102

57. Chuong AS, Miri ML, Busskamp V et al (2014) Noninvasive optical inhibition with a red-shifted microbial rhodopsin. Nat Neurosci 17(8):1123–1129

58. Inoue K, Ono H, Abe-Yoshizumi R et al (2013) A light-driven sodium ion pump in marine bacteria. Nat Commun 4:1678

59. Feldbauer K, Zimmermann D, Pintschovius V et al (2009) Channelrhodopsin-2 is a leaky proton pump. Proc Natl Acad Sci U S A 106(30):12317–12322

60. Lin JY, Lin MZ, Steinbach P et al (2009) Characterization of engineered channelrhodopsin variants with improved properties and kinetics. Biophys J 96(5):1803–1814

61. Klapoetke NC, Murata Y, Kim SS et al (2014) Independent optical excitation of distinct neural populations. Nat Methods 11(3):338–346

62. Govorunova EG, Sineshchekov OA, Janz R et al (2015) Natural light-gated anion channels: a family of microbial rhodopsins for advanced optogenetics. Science 349(6248):647–650

63. Schmidt D, Cho YK (2015) Natural photoreceptors and their application to synthetic biology. Trends Biotechnol 33(2):80–91

64. Fredriksson R, Lagerstrom MC, Lundin LG et al (2003) The G-protein-coupled receptors in the human genome form five main families. Phylogenetic analysis, paralogon groups, and fingerprints. Mol Pharmacol 63(6):1256–1272

65. Farrens DL, Altenbach C, Yang K et al (1996) Requirement of rigid-body motion of transmembrane helices for light activation of rhodopsin. Science 274(5288):768–770

66. Rosenbaum DM, Rasmussen SG, Kobilka BK (2009) The structure and function of G-protein-coupled receptors. Nature 459(7245):356–363

67. Kim JM, Hwa J, Garriga P et al (2005) Light-driven activation of beta 2-adrenergic receptor signaling by a chimeric rhodopsin containing the beta 2-adrenergic receptor cytoplasmic loops. Biochemistry 44(7):2284–2292

68. Airan RD, Thompson KR, Fenno LE et al (2009) Temporally precise in vivo control of intracellular signalling. Nature 458(7241):1025–1029

69. Oh E, Maejima T, Liu C et al (2010) Substitution of 5-HT1A receptor signaling by a light-activated G protein-coupled receptor. J Biol Chem 285(40):30825–30836

70. Siuda ER, Copits BA, Schmidt MJ et al (2015) Spatiotemporal control of opioid signaling and behavior. Neuron 86(4):923–935

71. Cao P, Sun W, Kramp K et al (2012) Light-sensitive coupling of rhodopsin and melanopsin to G(i/o) and G(q) signal transduction in Caenorhabditis elegans. FASEB J 26(2):480–491

72. Terakita A, Koyanagi M, Tsukamoto H et al (2004) Counterion displacement in the molecular evolution of the rhodopsin family. Nat Struct Mol Biol 11(3):284–289

73. Koyanagi M, Terakita A (2014) Diversity of animal opsin-based pigments and their optogenetic potential. Biochim Biophys Acta 1837(5):710–716

74. Moglich A, Yang X, Ayers RA et al (2010) Structure and function of plant photoreceptors. Annu Rev Plant Biol 61:21–47

75. Herrou J, Crosson S (2011) Function, structure and mechanism of bacterial photosensory LOV proteins. Nat Rev Microbiol 9(10):713–723

76. Moglich A, Ayers RA, Moffat K (2009) Design and signaling mechanism of light-regulated histidine kinases. J Mol Biol 385(5):1433–1444

77. Wu YI, Frey D, Lungu OI et al (2009) A genetically encoded photoactivatable Rac controls the motility of living cells. Nature 461(7260):104–108

78. Strickland D, Lin Y, Wagner E et al (2012) TULIPs: tunable, light-controlled interacting protein tags for cell biology. Nat Methods 9(4):379–384

79. Schmidt D, Tillberg PW, Chen F et al (2014) A fully genetically encoded protein architecture for optical control of peptide ligand concentration. Nat Commun 5:3019

80. Winkler A, Barends TR, Udvarhelyi A et al (2015) Structural details of light activation of the LOV2-based photoswitch PA-Rac1. ACS Chem Biol 10(2):502–509

81. Mizuno H, Mal TK, Walchli M et al (2008) Light-dependent regulation of structural flexibility in a photochromic fluorescent protein. Proc Natl Acad Sci U S A 105(27):9227–9232

82. Zhou XX, Chung HK, Lam AJ et al (2012) Optical control of protein activity by fluorescent protein domains. Science 338(6108):810–814

83. Zoltowski BD, Schwerdtfeger C, Widom J et al (2007) Conformational switching in the fungal light sensor Vivid. Science 316(5827):1054–1057

84. Nash AI, McNulty R, Shillito ME et al (2011) Structural basis of photosensitivity in a bacterial light-oxygen-voltage/helix-turn-helix (LOV-HTH) DNA-binding protein. Proc Natl Acad Sci U S A 108(23):9449–9454

85. Muller K, Engesser R, Schulz S et al (2013) Multi-chromatic control of mammalian gene expression and signaling. Nucleic Acids Res 41(12):e124

86. Toettcher JE, Weiner OD, Lim WA (2013) Using optogenetics to interrogate the dynamic control of signal transmission by the Ras/Erk module. Cell 155(6):1422–1434

87. Idevall-Hagren O, Dickson EJ, Hille B et al (2012) Optogenetic control of phosphoinositide metabolism. Proc Natl Acad Sci U S A 109(35):E2316–E2323

88. Grusch M, Schelch K, Riedler R et al (2014) Spatio-temporally precise activation of engineered receptor tyrosine kinases by light. EMBO J 33(15):1713–1726

89. Shimizu-Sato S, Huq E, Tepperman JM et al (2002) A light-switchable gene promoter system. Nat Biotechnol 20(10):1041–1044

90. Liu H, Gomez G, Lin S et al (2012) Optogenetic control of transcription in zebrafish. PLoS One 7(11):e50738

91. Kennedy MJ, Hughes RM, Peteya LA et al (2010) Rapid blue-light-mediated induction of protein interactions in living cells. Nat Methods 7(12):973–975

92. Boulina M, Samarajeewa H, Baker JD et al (2013) Live imaging of multicolor-labeled cells in Drosophila. Development 140(7):1605–1613

93. Bugaj LJ, Choksi AT, Mesuda CK et al (2013) Optogenetic protein clustering and signaling activation in mammalian cells. Nat Methods 10(3):249–252

94. Kim N, Kim JM, Lee M et al (2014) Spatiotemporal control of fibroblast growth factor receptor signals by blue light. Chem Biol 21(7):903–912

95. Taslimi A, Vrana JD, Chen D et al (2014) An optimized optogenetic clustering tool for probing protein interaction and function. Nat Commun 5:4925

96. Tischer D, Weiner OD (2014) Illuminating cell signalling with optogenetic tools. Nat Rev Mol Cell Biol 15(8):551–558

97. Zhang K, Cui B (2015) Optogenetic control of intracellular signaling pathways. Trends Biotechnol 33(2):92–100

98. Losi A, Gartner W (2012) The evolution of flavin-binding photoreceptors: an ancient chromophore serving trendy blue-light sensors. Annu Rev Plant Biol 63:49–72

99. Gradinaru V, Zhang F, Ramakrishnan C et al (2010) Molecular and cellular approaches for diversifying and extending optogenetics. Cell 141(1):154–165

100. Kralj JM, Douglass AD, Hochbaum DR et al (2012) Optical recording of action potentials in mammalian neurons using a microbial rhodopsin. Nat Methods 9(1):90–95

101. Gunaydin LA, Yizhar O, Berndt A et al (2010) Ultrafast optogenetic control. Nat Neurosci 13(3):387–392

102. Kleinlogel S, Feldbauer K, Dempski RE et al (2011) Ultra light-sensitive and fast neuronal activation with the Ca(2)+-permeable channelrhodopsin CatCh. Nat Neurosci 14(4):513–518

103. Berndt A, Schoenenberger P, Mattis J et al (2011) High-efficiency channelrhodopsins for fast neuronal stimulation at low light levels. Proc Natl Acad Sci U S A 108(18):7595–7600

104. Prigge M, Schneider F, Tsunoda SP et al (2012) Color-tuned channelrhodopsins for multiwavelength optogenetics. J Biol Chem 287(38):31804–31812

105. Berndt A, Lee SY, Ramakrishnan C et al (2014) Structure-guided transformation of channelrhodopsin into a light-activated chloride channel. Science 344(6182):420–424

106. Wietek J, Wiegert JS, Adeishvili N et al (2014) Conversion of channelrhodopsin into a light-gated chloride channel. Science 344(6182):409–412

107. Nagel G, Brauner M, Liewald JF et al (2005) Light activation of channelrhodopsin-2 in excitable cells of Caenorhabditis elegans triggers rapid behavioral responses. Curr Biol 15(24):2279–2284

108. Berndt A, Yizhar O, Gunaydin LA et al (2009) Bi-stable neural state switches. Nat Neurosci 12(2):229–234

109. Mattis J, Tye KM, Ferenczi EA et al (2012) Principles for applying optogenetic tools derived from direct comparative analysis of microbial opsins. Nat Methods 9(2):159–172

110. Nash AI, Ko WH, Harper SM et al (2008) A conserved glutamine plays a central role in LOV domain signal transmission and its duration. Biochemistry 47(52):13842–13849

111. Zoltowski BD, Vaccaro B, Crane BR (2009) Mechanism-based tuning of a LOV domain photoreceptor. Nat Chem Biol 5(11):827–834

112. Gleichmann T, Diensthuber RP, Moglich A (2013) Charting the signal trajectory in a light-oxygen-voltage photoreceptor by random mutagenesis and covariance analysis. J Biol Chem 288(41):29345–29355

113. Hochbaum DR, Zhao Y, Farhi SL et al (2014) All-optical electrophysiology in mammalian neurons using engineered microbial rhodopsins. Nat Methods 11(8):825–833

114. Akerboom J, Carreras Calderon N, Tian L et al (2013) Genetically encoded calcium indicators for multi-color neural activity imaging and combination with optogenetics. Front Mol Neurosci 6:2

115. Krook-Magnuson E, Szabo GG, Armstrong C et al (2014) Cerebellar directed optoge-

netic intervention inhibits spontaneous hippocampal seizures in a mouse model of temporal lobe epilepsy. eNeuro 1(1):pii: e.2014

116. Paz JT, Davidson TJ, Frechette ES et al (2013) Closed-loop optogenetic control of thalamus as a tool for interrupting seizures after cortical injury. Nat Neurosci 16(1):64–70

117. Sohal VS, Zhang F, Yizhar O et al (2009) Parvalbumin neurons and gamma rhythms enhance cortical circuit performance. Nature 459(7247):698–702

118. Toettcher JE, Gong D, Lim WA et al (2011) Light-based feedback for controlling intracellular signaling dynamics. Nat Methods 8(10):837–839

119. Milias-Argeitis A, Summers S, Stewart-Ornstein J et al (2011) In silico feedback for in vivo regulation of a gene expression circuit. Nat Biotechnol 29(12):1114–1116

Chapter 2

Natural Resources for Optogenetic Tools

Tilo Mathes

Abstract

Photoreceptors are found in all kingdoms of life and mediate crucial responses to environmental challenges. Nature has evolved various types of photoresponsive protein structures with different chromophores and signaling concepts for their given purpose. The abundance of these signaling proteins as found nowadays by (meta-)genomic screens enriched the palette of optogenetic tools significantly. In addition, molecular insights into signal transduction mechanisms and design principles from biophysical studies and from structural and mechanistic comparison of homologous proteins opened seemingly unlimited possibilities for customizing the naturally occurring proteins for a given optogenetic task. Here, a brief overview on the photoreceptor concepts already established as optogenetic tools in natural or engineered form, their photochemistry and their signaling/design principles is given. Finally, so far not regarded photosensitive modules and protein architectures with potential for optogenetic application are described.

Key words Photoreceptor, Signal transduction, Optogenetics, Synthetic biology

1 Natural Photoreceptors

1.1 Chromophores

Naturally occurring photoreceptors cover the spectral range from the UV-B to the near-IR (Fig. 1). For this purpose, different chromophores derived from metabolic compounds and enzyme cofactors are utilized and their electronic properties are fine-tuned by the protein environment. Until lately the retinal cofactor of visual and nonvisual opsins has been considered the most versatile chromophore spanning from the near UV regions (~360 nm) in short wavelength sensitive rhodopsins (SWS) [1] to ~700 nm in one of the mantis shrimp rhodopsins [2]. With the discovery algal phytochromes [3] and especially of a novel type of linear tetrapyrrole binding photoreceptor, cyanobacteriochrome, which expand the range of tetrapyrrole chromophores to cover all of the visible spectrum and the near UV, tetrapyrroles outperform retinal in spectral coverage [3, 4]. In the blue to near-UV region, flavin binding photoreceptors—LOV (light–oxygen–voltage), BLUF (blue light sensors using FAD), and cryptochrome photoreceptors—are

Arash Kianianmomeni (ed.), *Optogenetics: Methods and Protocols*, Methods in Molecular Biology, vol. 1408,
DOI 10.1007/978-1-4939-3512-3_2, © Springer Science+Business Media New York 2016

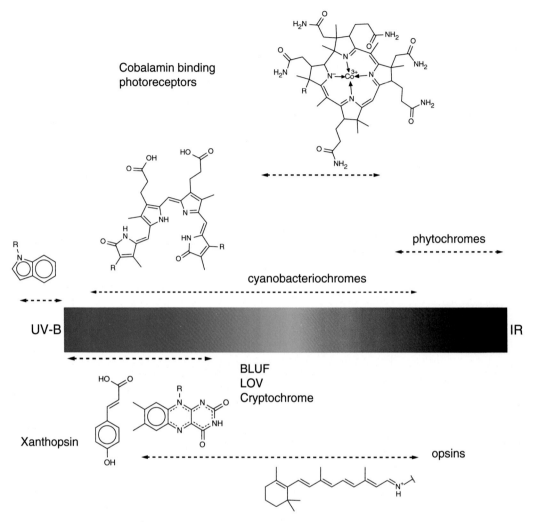

Fig. 1 Spectral coverage (not to scale) of naturally occurring photoreceptor chromophores

prominent and cover a large spectral region with their first two electronic transitions in the oxidized form but are rather limited in color tuning [5]. However, in different reduced states, semiquinone or fully reduced form, flavins have been described to respond to green/yellow or UV-A, respectively [6–8]. In the UV-B region, a recently described photoreceptor concept, UVR8, uses no cofactor at all and employs the UV absorbing nature of tryptophan side chains as a proteinogenic chromophore [9]. In addition, a few special cases exist like the Xanthopsins that employ *p*-coumaric acid as a blue light absorbing chromophore [10] or the just recently discovered cobalamin binding photoreceptors absorbing in the green to red wavelength region [11].

The photochemical properties of most of these chromophores in isolated and protein-bound form are well known. The key biophysical determinants for photoactivation are the quantum efficiency of signaling state generation and its deactivation reaction in the dark (Table 1). These properties directly affect the net biological signal/effect. The observed photoinduced reactions that lead to a signaling active conformation are isomerizations (retinal, tetrapyrrole, *p*-coumaric acid), proton coupled electron transfer (tryptophan, flavin), and chemical/redox reactions (flavin). The general photochemistry of the chromophores can be readily probed with state-of-the art time-resolved spectroscopic techniques [12, 13]; however, the interplay with the protein and the relevant structural changes in signal transduction are in most cases more difficult to address. The latter are essential for the feasibility of a potential optogenetic tool, since the light-dependent modulation of the biological signal is not only determined by the quantum yield of the phototransformation of the chromophore itself but rather by the stability of the resulting signaling state and its impact on the protein structure/the extent of modulation of effector activity. In addition to the photoactivation properties, thermal/dark noise of the given photosensor as discussed in GPCR signaling is a relevant aspect [14].

Secondary aspects are of course phototoxicity by photoinduced generation of reactive oxygen species (ROS) byproducts in the photoactivation process itself or by absorption in the cell/tissue that have to be considered for application and naturally depend on the energy required for sufficient optical stimulation. The latter is especially crucial in tissue or animals, where penetration depth is limiting. In such cases, long-wavelength excitation is desirable to minimize diffusion/scattering of the stimulating light. Moreover, energy dissipation from the receptor protein and the tissue itself are crucial parameters for design and evaluation of optogenetic experiments [15].

The ultimate requirement, however, is the bioavailability of the cofactor in the investigated cell type. While proteinogenic cofactors and flavins are ubiquitous in nature, the availability retinal and tetrapyrrole based cofactors may be limiting for the given application. Moreover, spectral orthogonality may have to be considered for applications where multiple selective stimuli are desired.

1.2 Photosensitive Modules/Chromophore Binding Domains

Representative crystal structural models of the binding domains for the above-described chromophores are shown in Fig. 2. While different architectures are found to bind flavins (PAS: Per–Arnt–Sim, BLUF: Blue Light sensors Using FAD, Cryptochrome/Photolyase), also some domain classes (e.g., PAS) are versatile in binding different chromophores (flavins, *p*-coumaric acid, tetrapyrroles). As mentioned above, the potential for optogenetics of these classes naturally depends on the given biophysical properties (Table 1). The small modular photoreceptors are usually more convenient to deliver into the given cell or organism, but most of these small molecule binding receptors feature a low molar

Table 1
Biophysical parameters of different photosensory modules

Module	Size (kDa)	Absorption cross section ε (M^{-1} cm^{-1})	Quantum yield of signaling state formation (%)	Lifetime of signaling state (s)	Apparent modulation of biological activity in native or engineered form (as reported)
BLUF	~14	~11,000–12,000 (FAD) (~445 nm) [85]	~20–90 [86–88]	1–300	Up to 300-fold (bPAC) [27]
LOV	~14	~11,000–12,000 (FMN) (~445 nm) [85]	~30–80 [89, 90]	1–10,000	Up to 1000-fold (YF1) [49]
Cryptochrome	~60	~11,000–12,000 (FAD) (~445 nm, oxidized) ~4000 (~600 nm, semiquinone) [85]	2–14 (oxidized to neutral radical) [91, 92]	180–1000	Up to 200-fold [45]
PYP	~14	~45,500 (~445 nm) [93]	14–18 [94]	100–200	Up to 12-fold [95]
Phytochrome (cyanobacteriochrome)	~60 [40]	40,000 (Pfr) to 90,000 (Pr) [96]	~10–17 phytochrome [97] (~8–30 cyanobacteriochrome [98])	~Bistable	Up to sevenfold (LaPD) [50]
Rhodopsin	~30	~40,000 [99]	~50 [100]	1–100	Up to 180-fold: Rh-GC [66, 67]
UVR8	~47	~154,000 calc.: $28 \cdot \varepsilon_{280nm}$ (Trp) [101]	~20 [17]	$>d$	Up to 800-fold [102, 103]

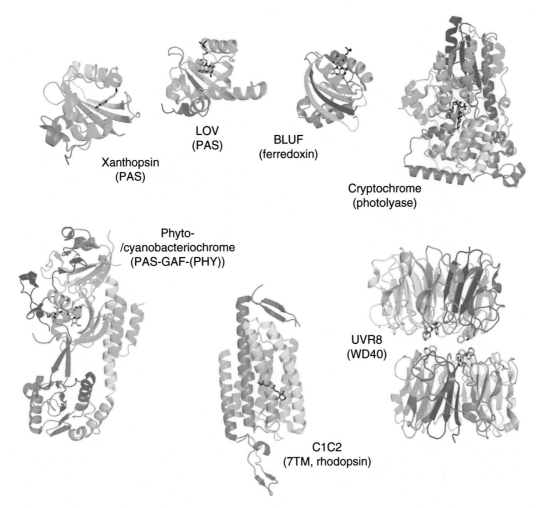

Fig. 2 Photoresponsive modules found in nature: Xanthopsins, LOV domains and phyto/-cyanobacteriochromes employ the PAS fold to assemble the light-sensitive module. Cyanobacteriochromes lack the PHY domain in contrast to phytochromes. BLUF domains use a ferredoxin-like fold to bind a flavin chromophore. Cryptochromes are structurally related to photolyases and bind FAD and an antenna cofactor (not shown). UVR8 belongs to the family of WD40 repeat proteins with a characteristic β-propeller structure. Pumping and channelrhodopsins belong to the greater family of seven transmembrane proteins and more specifically to the microbial rhodopsins. Structural data (PDB ID) used for illustration: Xanthopsin (1NWZ), AsLOV2 (2V1A), SyPixD (2HFO chain A), AtCry (4GU5), UVR8 (4D9S), C1C2 (3UG9), bacteriophytochrome (4OOP)

absorption cross section, which makes them generally less attractive for efficient photostimulation. Another important parameter is the quantum yield of phototransformation that illustrates the efficiency of the formation of the metastable signaling state. This is especially crucial in the case of the rhodopsin pumps where the phototransformation is stochiometrically related to the number of translocated ions. However, as described above the apparent modulation of the biological activity in most other photoreceptors depends strongly on how well regulated the cognate effector

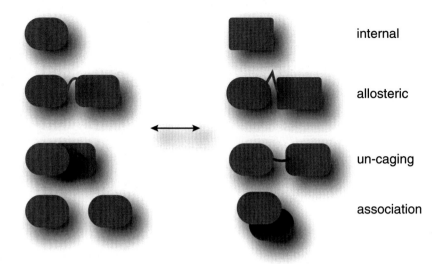

internal

allosteric

un-caging

association

Fig. 3 Signal transduction principles in naturally occurring photoreceptors

domain is and how strongly its activity becomes enhanced in the activated state. Therefore, even a photoreceptor with apparent low photochemical transformation efficiency (e.g., phytochrome) can be a potent optogenetic tool. Finally, the interaction with the given natural metabolism has to be considered as it might strongly lower the apparent activation compared to in vitro studies.

1.3 Signal Transduction Concepts

The above described photoreceptor classes use various concepts to transform the local structural changes of the chromophore and its protein surrounding into a biological signal and can be roughly categorized in four categories (Fig. 3). In case of the channelrhodopsins and pumping rhodopsins, these processes are intricately entangled as their activity, channeling, or ion pumping, directly involves the chromophore. Similarly, the UV-B induced monomerization of UVR8 is accomplished by direct interaction of the excited chromophore with the cross dimer salt-bridges [16, 17]. In the remaining photoreceptor classes and the sensory rhodopsins, the chromophore is in most cases located distal from the biological effector part, which is especially obvious for modular photoreceptor designs. The signaling concepts of such receptors can be classified into directly allosteric or allosterically induced uncaging and association/dissociation mechanisms [18].

The so-to-speak integral and modular photoreceptors provide different opportunities and limitations for optogenetic application and customization, accordingly. Integral photoreceptors are limited for optogenetic application solely due to their characteristic dark/thermal noise which results in more or less pronounced background signaling activity [14]. In this sense, most modular photoreceptor designs are more prone to background activity since the photosensory domain only modulates the activity of the effector domain and the dark activity thus depends strongly on the tightness of regulation by the photosensory domain [18]. The possibility to alter the principal functionality of an integral receptor is, however, very limited while modular photoreceptors are found in different sensor/effector combinations covering a wide range of biological activities already in their natural form. In addition, many photosensory modules elicit rather large scale structural changes like unfolding or (de-)oligomerization, which can be easily exploited in order to recruit or release other biomolecules.

1.4 Natural Resources of Light-Responsive Modules and Signaling Architectures

In contrast to the early times of genome analysis, the technological development in sequence analysis in combination with computational resources for automatic annotation available nowadays provides access to a vast and everyday increasing amount of information on the primary structure and, thus, on the function of proteins with known homologies. Especially the rise of metagenomics that does not rely on monoclonal samples provides access to diverse ensembles of organisms and their genomes and further increased the abundance of sequence information [19–21].

Currently, there are more than 3000 sequences available for the bacteriorhodopsin-like protein family PF01036 (http://pfam.xfam.org/family/PF01036) that feature ion pumps, ion channels, and sensory proteins. The greater family of seven transmembrane receptors (http://pfam.xfam.org/family/PF00001) with more than 58,000 sequences, covers about 8000 protein sequences of visual and nonvisual rhodopsins (http://www.ebi.ac.uk/interpro/entry/IPR001760). Even more remarkable is the greater family of PAS domains with more than 26,000 sequences (http://pfam.xfam.org/family/PF13426) that includes the LOV photosensory domains, Xanthopsins and phytochrome architectures. Less than 5 % of these sequences are found as single PAS domain containing proteins while the majority is arranged in a modular fashion in various architectures and with different effector domains. Given the high homology of PAS domains this abundance of structures provides a large number of naturally optimized architectures that can possibly be rendered light sensitive by replacing the given PAS domain with a suitable photosensitive domain, e.g., LOV. Similarly impressive are the phytochrome family (http://pfam.xfam.org/family/PF00360) that currently encompasses about 6300 sequences containing PAS–GAF(cGMP phosphodiesterase, adenylate cyclase, FhlA)–PHY(phytochrome-specific GAF related) modules and the BLUF family with about 3900 sequences (http://pfam.xfam.org/family/PF04940). Both classes feature a similar diversity in attached effector domains. The family of the cryptochromes (http://pfam.xfam.org/family/PF12546) currently features only 157 sequences, probably owing to their occurrence mainly in higher eukaryotic organisms for which less sequence information is available. The structurally closely related photolyase family (http://pfam.xfam.org/family/PF00875), however, features about 15,000 sequences.

The current knowledge on photosensitive modules and receptor architecture thus allows us to extract homologous proteins from genomic or metagenomic data, predict to some extent their properties, screen for desired traits and learn about modular architecture that can be exploited for the rational design of artificial photoreceptors. However, proteins with novel photo-activation concepts cannot be easily discovered in this way and still require thorough photo- and cell biological research. A few recent examples are given below.

2   Engineering of Photoreceptors and Artificial Signal Transduction Pathways

2.1   Photoreceptor Engineering

Besides directly harnessing the possibilities provided by nature itself, many photoreceptor functionalities have been extensively engineered. We only focus on a few selected examples, as concise review articles have been published recently (e.g., [18] and chapters 27 in this book). One of the most prominent examples is channelrhodopsin, which has been altered in its activation and deactivation properties, spectral sensitivity, and even ion selectivity [22–25]. Except for PACs, which have also been applied successfully in their natural form [26–29], most modular photoreceptors used in optogenetics have been cut down to the photosensitive module and fused with different interaction partners [18, 30]. The reason for this may be due to the related dark activity or unsuitable signaling activity of the given natural system.

For example, the light-modulated transcription factor Aureochrome found in yellow algae and diatoms at first glance seems like a perfect minimal system to regulate transcription in heterologous systems [31]. However, the short recognition sequence of the bZIP (Basic Leucine Zipper) domain and the ratio of light/dark activity prevent an efficient transfer of these systems into heterologous environments [32]. In its natural context probably other transcription effectors or regulatory elements are engaged to facilitate a physiologically sufficient light-induced change in activity. Therefore, a more efficient approach was to employ a hybrid approach using a (heterologous) LOV domain that undergoes light-induced dimerization, fused to a homologous DNA binding and a transactivator domain of the target cell type. Accordingly, light-induced binding of the homologous fusion protein to a specific sequence and a subsequent expression induction is facilitated [33]. Similarly, the naturally occurring LOV–HTH (helix–turn–helix) protein EL222 that naturally uses light-induced dimerization was fused with a transactivator domain, most likely to enhance optogenetic activation in the heterologous system [34]. Both systems have been applied successfully but differ in their dynamic range, mainly due to the lifetime of the signaling state of the given LOV domain.

In other artificial fusion constructs several modes of activation, uncaging, (de-)oligomerization, and even allostery directly have been efficiently employed. The most prominent uncaging scenario originates from the plant phototropins. Photoactivation of the LOV2 domain induces dissociation and unfolding of a short C-terminal helix [35] attached to the LOV-domain core and can be efficiently used, e.g., for uncaging of a RAC GTPase [36], a peptide toxin [37] or a cation channel [38]. Partial light-induced unfolding of a PAS domain is also witnessed in the photoactive yellow protein (PYP) and has been exploited for controlling protein/protein interactions [39].

Some naturally occurring LOV domains show light-induced dimerization that was successfully employed by fusion to DNA-binding motifs as described above [40, 41]. PYP in contrast can be engineered to feature such a photoinduced transformation [42]. Especially powerful also proved to be light-dependent heterodimerization of the Arabidopsis photoreceptors cryptochrome 2 (Arabidopsis) and CIB1 (cryptochrome-interacting basic-helix–loop–helix) [43] or phytochrome B and PIF (phytochrome-interacting factors) proteins [44], that was efficiently used to recruit and release proteins. Cryptochrome 2 (CRY2) also features homodimerization/clustering that was optimized and applied for spatial accumulation of proteins [45–47]. Similarly powerful proved to be the UV-B induced dissociation reaction of UVR8, that was used to drive expression or nuclear localization using the natural interaction partner COP1 (CONSTITUTIVELY PHOTOMORPHOGENIC 1) [48]. Despite its large size and slow dark reversion, the specific excitation with UV-B light, which is unbiased by the observation light source in light microscopic experiments, and the missing need for an exogenous cofactor makes UVR8 especially attractive as a fully genetically encoded photoswitch.

The above-described engineering approaches employ secondary effects of light-induced allosteric changes. Engineering allostery directly, however, is significantly more difficult due to the lack of knowledge on the molecular prerequisites in most cases. In the successful cases, careful homologies between different sensor/effector constructs have been investigated and applied in the construction of the artificial protein [49–53].

2.2 Orthogonal Signal Engineering

Besides changing the photosensory properties of a given photoreceptor or de- and reconstructing of sensor/effector combinations, nature also provides us with elegant ways of introducing complete artificial signaling cascades into a heterologous system. These approaches are especially interesting since highly orthogonal systems can be designed, that ideally prevent interference with the natural metabolism in a given cell. In this regard, especially modular photoreceptor architectures are appealing as an effector domain with a suitable activity can be chosen or engineered.

The most versatile approach in this regard is the optogenetic regulation of expression. An elegant example is the hybrid light-activated histidine kinase YF1, that has been engineered by exchanging a redox sensitive PAS domain from the oxygen activated histidine kinase FixL from *Bradyrhizobium japonicum* for a LOV domain from *Bacillus subtilis* [49, 54]. Together with its downstream phosphorylation target FixK and the well-described activation of its cognate FixK$_2$ promotor an orthogonal light-activated expression systems for *E. coli* was established. Due to the high specificity of such bacterial, two component systems, a cross talk with endogenous signaling cascades is unlikely.

Similarly, it is conceivable that non-native second messenger modules may be used for a given optogenetic task, e.g., to regulate the activity of a protein. At first glance, promising candidates seemed to be represented by dicyclic nucleotides, which regulate various processes in bacteria and corresponding (light-activated) cyclases and phosphodiesterases have been already described und functionally investigated [55–57]. However, only recently receptor proteins for c-di-GMP, c-di-AMP, and cGAMP have been discovered and functionally described in eukaryotes and even in mammals [58]. Accordingly, dicyclic nucleotides cannot be considered orthogonal second messengers in eukaryotic systems.

A more promising approach for orthogonality may therefore be the integration with chemical biology. Using so-called optochemical genetics it is possible to render an endogenous compound selectively light responsive with only minimal perturbance of the natural system [59, 60]. Moreover, it is possible to investigate the effect of light on the given cell independently of optical stimulation as in most cases the chromophore/photoactive ligand can be added independently of the expressed modified receptor. Similarly, it should be possible to generate non-natural enzyme activities that produce, in a light-responsive manner, artificial second messengers that are recognized by a cognate, non-natural designed receptor.

3   Potential Future Optogenetic Tools Found in Nature

The recent advances in optogenetics and optochemical genetics demonstrate that the combination of natural resources with knowledge driven molecular engineering provides seemingly endless opportunities for synthetic biology and biotechnology. Still, completely new photoreceptor principles are likely to be discovered and will add to the palette of the optogenetic construction kit in the future.

Apart from the so far described and successfully applied opsin based tools several new microbial and also some longer known opsins represent very promising candidates for optogenetics. The histidine kinase rhodopsins discovered in green algae, for example, show a unique multidomain architecture [61]. Directly fused to the transmembrane rhodopsin part are several soluble domains that constitute a complete phosphotransfer cascade on a single polypeptide. The terminal effector domain is a GTP cyclase and thus a promising natural light-activated cGMP synthase. Its function has yet to be demonstrated but the signaling network given by the protein architecture suggests an intricately regulated cyclase activity that offers many levers for optimization/adaptation in an optogenetic context. Especially intriguing is furthermore the bimodal nature of the rhodopsin photosensory domain of HKR1 that can be switched permanently between two states by UV and blue light, respectively. In that sense, also the long known non-visual rhodopsins like melanopsins are highly appealing as they also represent bimodal switches [62] and can be used to drive various G-protein signaling cascades. Melanopsins have recently been applied in modified forms in optogenetic assays for vision restoration [63] and controlling cardiomyocyte activity [64].

Another interesting modular photoreceptor candidate is a recently discovered directly rhodopsin coupled guanylyl cyclase identified in the fungus *Blastocladiella emersonii* [65]. The protein may be considered as a short-cut $G\alpha_s$GPCR that usually requires G-proteins to stimulate cyclase activity. Its potential for optogenetics has been recently demonstrated [66, 67]. In contrast to the BLUF regulated photoactivated cyclases, the protein is intrinsically membrane localized and thus enables subcellularly localized stimulation. It is very likely that the cGMP synthesis activity can be adapted to cAMP synthesis as shown vice versa for the BLUF activated PACs [68] and would add a novel membrane localized PAC to the optogenetic toolbox.

A noteworthy subclass of microbial rhodopsins, the xanthorhodopsins [69], employ carotenoid antennas (salinixanthin) to extend the spectral sensitivity to facilitate light-induced proton pumping. This concept is highly interesting for color and sensitivity tuning of potentially any kind of rhodopsin, but has yet to be explored in optogenetics. Another carotenoid employing photoreceptor is the orange carotenoid protein (OCP), which has been found to regulate non-photochemical quenching in cyanobacteria. The carotenoid chromophore (3′-hydroxyechinenone) enables photochromic switching between an orange and a red form [70]. A model involving a large scale translocation of the carotenoid and accompanying structural changes at the C-terminus has been recently proposed [71], which would enable the protein to interact in a regulatory manner with the phycobilisome [72]. OCP belongs to a smaller protein family of about 216 sequences containing a cyanobacteria-specific N-terminal domain (http://pfam.xfam.org/family/PF09150). The remaining protein belongs to the class of Nuclear transport factor 2 (NTF2) domain (~6200 sequences; http://pfam.xfam.org/family/PF02136). So far this photoactivation concept has not been applied in an optogenetic context, but clearly provides interesting novel starting points for the design of optogenetic tools regarding new chromophores and specific light-induced structural changes.

Similarly to the photochemical diversity of retinal binding photoreceptors, also novel photochemical traits of the tetrapyrrole binding receptors have been discovered recently. Although phytochromes already represent near-IR/far-IR bimodal switches, the cyanobacteriochromes can be tuned over a broad range in the visible spectrum and feature well separated spectra of both states. Moreover, these proteins do not require the PHY domain that is needed for efficient photoswitching in phytochromes. These properties are clearly attractive for optogenetics and a widespread use of cyanobacteriochrome in optogenetic application is expected in the future. In addition, novel insights into the monomerization/dimerization prerequisites for phytochromes have been found recently. Inspired by the search for monomeric phytochrome based infrared fluorescent proteins [73], a photochromic monomer/dimer switchable phytochrome module has been described and clearly has a high potential for building tightly regulated bimodal optogenetic mono-/dimerizers [74].

In the first optogenetic experiments, complete visual rhodopsin signaling cascades [75] and modified rhodopsins [76–78] have been employed to stimulate G-protein induced responses. Because of the abundance of GPCRs in higher animals and their relevance for disease and metabolic disorders, GPCRs are naturally attractive targets for optogenetic studies or even therapeutics. A noteworthy completely new type of photoactivatable GPCRs has been found to be responsible for the photoavoidance of *Caenorhabditis elegans* and larvae of *Drosophila melanogaster* [79–82]. Apparently, these gustatory GPCR paralogs (family of 7TM chemosensory receptors; http://pfam.xfam.org/family/PF08395; currently 1900 sequences) facilitate the UV-induced photophobic behavior via different G-protein cascades. LITE-1 found in *C. elegans* mediates photoavoidance via $G_{i/o}$ proteins that stimulate the activity of membrane associated guanylyl cyclases which in turn affect CNG channels [80]. Although action spectra of the LITE-1 mediated photoresponse have been described, the nature of its chromophore is completely unclear. From the crude action spectra recorded by behavioral experiments showing a maximum in the UV-A region at around 350 nm but also a significant response at up to 450 nm, many potential chromophores can be discussed (*see* Fig. 1). Besides using classical prosthetic chromophores LITE-1 may in principle also transiently bind small molecules that have been chemically altered by short wavelength/high-energy light. The discussion whether LITE-1 and the related Gur-3 may also be activated by ROS as a photoinduced by-product is discussed controversially [83]. In *D. melanogaster*, the close homolog Grb28 (isoform B) also employs G-protein mediated signaling, but the classification of the interacting G-protein remains unknown. In contrast to LITE-1, the observed Ca^{2+}photocurrents are not mediated by CNG channels. Instead, the transient receptor potential (TRP) channel TrpA1 was identified to be responsible. Clearly, this class of proteins features promising candidates to stimulate various G-protein related activities in optogenetic approaches.

Another completely new class of soluble photoreceptor that was found in *Myxococcus xanthus* uses a closed ring tetrapyrrole cofactor, Cobalamin (coenzyme B12), and is the first appearance of this chromophore in the photoreceptor world [11]. CarH requires coenzyme B12 for light-dependent gene repressor activity by oligomerization and subsequent DNA binding. Upon illumination the cofactor is photolysed and the oligomer dismantles and dissociates from its cognate operator. The underlying photodynamics and crystal structures in dark and light-activated states have been recently described [84]. Therefore, a solid basis for application of CarH as a novel optogenetic tool has been already established.

4 Concluding Remarks

Up to now nature has provided us with a vast toolbox for optogenetic application and most likely has further so far unknown photo-activation concepts in its repertoire. Moreover, even for well-known photoreceptor concepts novel photochemical traits or signal transduction scenarios are discovered constantly. Most of these proteins provide new opportunities for optogenetic application, either to be harnessed from nature directly or in the sense that we learn about different molecular solutions for photoreception that can be applied in the rational design of new optogenetic tools. Furthermore, combination of synthetic biology with chemical biology is expected to add significantly to the optogenetic toolbox.

Acknowledgments

I would like to thank John Kennis and Peter Hegemann for generous support over the last years.

References

1. Hauser FE, van Hazel I, Chang BSW (2014) Spectral tuning in vertebrate short wavelength-sensitive 1 (SWS1) visual pigments: can wavelength sensitivity be inferred from sequence data? J Exp Zool B 322(7):529–539

2. Thoen HH, How MJ, Chiou TH, Marshall J (2014) A different form of color vision in mantis shrimp. Science 343(6169):411–413

3. Rockwell NC, Duanmu D, Martin SS, Bachy C, Price DC, Bhattacharya D et al (2014) Eukaryotic algal phytochromes span the visible spectrum. Proc Natl Acad Sci U S A 111(10):3871–3876

4. Ishizuka T, Shimada T, Okajima K, Yoshihara S, Ochiai Y, Katayama M et al (2006) Characterization of cyanobacteriochrome TePixJ from a thermophilic cyanobacterium Thermosynechococcus elongatus strain BP-1. Plant Cell Physiol 47(9):1251–1261

5. Losi A, Gärtner W (2012) The evolution of flavin-binding photoreceptors: an ancient chromophore serving trendy blue-light sensors. Annu Rev Plant Biol 63:49–72

6. Bouly JP, Schleicher E, Dionisio-Sese M, Vandenbussche F, Van Der Straeten D, Bakrim N et al (2007) Cryptochrome blue light photoreceptors are activated through interconversion of flavin redox states. J Biol Chem 282(13):9383–9391

7. Carell T, Burgdorf LT, Kundu LM, Cichon M (2001) The mechanism of action of DNA photolyases. Curr Opin Chem Biol 5(5):491–498

8. Beel B, Prager K, Spexard M, Sasso S, Weiss D, Muller N et al (2012) A flavin binding cryptochrome photoreceptor responds to both blue and red light in Chlamydomonas reinhardtii. Plant Cell 24(7):2992–3008

9. Jenkins GI (2014) The UV-B photoreceptor UVR8: from structure to physiology. Plant Cell 26(1):21–37

10. Meyer TE, Kyndt JA, Memmi S, Moser T, Colon-Acevedo B, Devreese B et al (2012) The growing family of photoactive yellow proteins and their presumed functional roles. Photochem Photobiol Sci 11(10):1495–1514

11. Ortiz-Guerrero JM, Polanco MC, Murillo FJ, Padmanabhan S, Elias-Arnanz M (2011) Light-dependent gene regulation by a coenzyme B12-based photoreceptor. Proc Natl Acad Sci U S A 108(18):7565–7570

12. Mathes T, van Stokkum IHM, Kennis JTM (2014) Photoactivation mechanisms of flavin-binding photoreceptors revealed through ultrafast spectroscopy and global analysis methods. Flavins Flavoproteins Methods Protocols 1146:401–442

13. Kennis JTM, Mathes T (2013) Molecular eyes: proteins that transform light into biological information. Interface Focus 3(5):20130005

14. Barlow RB, Birge RR, Kaplan E, Tallent JR (1993) On the molecular-origin of photoreceptor noise. Nature 366(6450):64–66

15. Stujenske JM, Spellman T, Gordon JA (2015) Modeling the spatiotemporal dynamics of light and heat propagation for in vivo optogenetics. Cell Rep 12(3):525–534

16. Mathes T, Heilmann M, Pandit A, Zhu J, Ravensbergen J, Kloz M et al (2015) Proton-coupled electron transfer constitutes the photoactivation mechanism of the plant photoreceptor UVR8. J Am Chem Soc 137(25):8113–8120

17. Heilmann M, Christie JM, Kennis JT, Jenkins GI, Mathes T (2015) Photoinduced transformation of UVR8 monitored by vibrational and fluorescence spectroscopy. Photochem Photobiol Sci 14(2):252–257

18. Ziegler T, Möglich A (2015) Photoreceptor engineering. Front Mol Biosci 2:30

19. Klapoetke NC, Murata Y, Kim SS, Pulver SR, Birdsey-Benson A, Cho YK et al (2014) Independent optical excitation of distinct neural populations. Nat Methods 11(3):338–346

20. Pathak GP, Losi A, Gärtner W (2011) Metagenome-based screening reveals worldwide distribution of LOV-domain proteins. Photochem Photobiol 88(1):107–118

21. Pathak GP, Ehrenreich A, Losi A, Streit WR, Gärtner W (2009) Novel blue light-sensitive proteins from a metagenomic approach. Environ Microbiol 11(9):2388–2399

22. Prigge M, Schneider F, Tsunoda SP, Shilyansky C, Wietek J, Deisseroth K et al (2012) Color-tuned channelrhodopsins for multiwavelength optogenetics. J Biol Chem 287(38):31804–31812

23. Wietek J, Wiegert JS, Adeishvili N, Schneider F, Watanabe H, Tsunoda SP et al (2014) Conversion of channelrhodopsin into a light-gated chloride channel. Science 344(6182):409–412

24. Kleinlogel S, Feldbauer K, Dempski RE, Fotis H, Wood PG, Bamann C et al (2011) Ultra light-sensitive and fast neuronal activation with the Ca(2)+-permeable channelrhodopsin CatCh. Nat Neurosci 14(4):513–518

25. Berndt A, Yizhar O, Gunaydin LA, Hegemann P, Deisseroth K (2009) Bi-stable neural state switches. Nat Neurosci 12(2):229–234

26. Weissenberger S, Schultheis C, Liewald JF, Erbguth K, Nagel G, Gottschalk A (2011) PACalpha--an optogenetic tool for in vivo manipulation of cellular cAMP levels, neurotransmitter release, and behavior in Caenorhabditis elegans. J Neurochem 116(4):616–625

27. Stierl M, Stumpf P, Udwari D, Gueta R, Hagedorn R, Losi A et al (2011) Light modulation of cellular cAMP by a small bacterial photoactivated adenylyl cyclase, bPAC, of the soil bacterium Beggiatoa. J Biol Chem 286(2):1181–1188

28. Schröder-Lang S, Schwarzel M, Seifert R, Strunker T, Kateriya S, Looser J et al (2007) Fast manipulation of cellular cAMP level by light in vivo. Nat Methods 4(1):39–42

29. Chen ZH, Raffelberg S, Losi A, Schaap P, Gärtner W (2014) A cyanobacterial light activated adenylyl cyclase partially restores development of a Dictyostelium discoideum, adenylyl cyclase a null mutant. J Biotechnol 191:246–249

30. Möglich A, Moffat K (2010) Engineered photoreceptors as novel optogenetic tools. Photochem Photobiol Sci 9(10):1286–1300

31. Takahashi F, Yamagata D, Ishikawa M, Fukamatsu Y, Ogura Y, Kasahara M et al (2007) AUREOCHROME, a photoreceptor required for photomorphogenesis in stramenopiles. Proc Natl Acad Sci U S A 104(49):19625–19630

32. Hisatomi O, Nakatani Y, Takeuchi K, Takahashi F, Kataoka H (2014) Blue light-induced dimerization of monomeric aureochrome-1 enhances its affinity for the target sequence. J Biol Chem 289(25):17379–17391

33. Wang X, Chen XJ, Yang Y (2012) Spatiotemporal control of gene expression by a light-switchable transgene system. Nat Methods 9(3):266–269

34. Motta-Mena LB, Reade A, Mallory MJ, Glantz S, Weiner OD, Lynch KW et al (2014) An optogenetic gene expression system with rapid activation and deactivation kinetics. Nat Chem Biol 10(3):196–202

35. Harper SM, Neil LC, Gardner KH (2003) Structural basis of a phototropin light switch. Science 301(5639):1541–1544

36. Wu YI, Frey D, Lungu OI, Jaehrig A, Schlichting I, Kuhlman B et al (2009) A genetically encoded photoactivatable Rac controls the motility of living cells. Nature 461(7260):104–108

37. Schmidt D, Tillberg PW, Chen F, Boyden ES (2014) A fully genetically encoded protein architecture for optical control of peptide ligand concentration. Nat Commun 5:3019

38. Cosentino C, Alberio L, Gazzarrini S, Aquila M, Romano E, Cermenati S et al (2015) Optogenetics. Engineering of a light-gated potassium channel. Science 348(6235):707–710

39. Morgan SA, Al-Abdul-Wahid S, Woolley GA (2010) Structure-based design of a photo-controlled DNA binding protein. J Mol Biol 399(1):94–112

40. Strickland D, Lin Y, Wagner E, Hope CM, Zayner J, Antoniou C et al (2012) TULIPs: tunable, light-controlled interacting protein tags for cell biology. Nat Methods 9(4):379–384

41. Pathak GP, Strickland D, Vrana JD, Tucker CL (2014) Benchmarking of optical dimerizer systems. ACS Synth Biol 3(11):832–838

42. Reis JM, Burns DC, Woolley GA (2014) Optical control of protein-protein interactions via blue light-induced domain swapping. Biochemistry 53(30):5008–5016

43. Kennedy MJ, Hughes RM, Peteya LA, Schwartz JW, Ehlers MD, Tucker CL (2010) Rapid blue-light-mediated induction of protein interactions in living cells. Nat Methods 7(12):973–975, advance online publication

44. Toettcher JE, Gong D, Lim WA, Weiner OD (2011) Light-based feedback for controlling intracellular signaling dynamics. Nat Methods 8(10):837–839

45. Bugaj LJ, Choksi AT, Mesuda CK, Kane RS, Schaffer DV (2013) Optogenetic protein clustering and signaling activation in mammalian cells. Nat Methods 10(3):249–252

46. Taslimi A, Vrana JD, Chen D, Borinskaya S, Mayer BJ, Kennedy MJ et al (2014) An optimized optogenetic clustering tool for probing protein interaction and function. Nat Commun 5:4925

47. Che DL, Duan L, Zhang K, Cui B (2015) The dual characteristics of light-induced cryptochrome 2, homo-oligomerization and heterodimerization, for optogenetic manipulation in mammalian cells. ACS Synth Biol 4(10):1124–1135

48. Kianianmomeni A (2015) UVB-based optogenetic tools. Trends Biotechnol 33(2):59–61

49. Möglich A, Ayers RA, Moffat K (2009) Design and signaling mechanism of light-regulated histidine kinases. J Mol Biol 385(5):1433–1444

50. Gasser C, Taiber S, Yeh C-M, Wittig CH, Hegemann P, Ryu S et al (2014) Engineering of a red-light-activated human cAMP/cGMP-specific phosphodiesterase. Proc Natl Acad Sci U S A 111(24):8803–8808

51. Ryu M-H, Kang I-H, Nelson MD, Jensen TM, Lyuksyutova AI, Siltberg-Liberles J et al (2014) Engineering adenylate cyclases regulated by near-infrared window light. Proc Natl Acad Sci U S A 111(28):10167–10172

52. Han Y, Braatsch S, Osterloh L, Klug G (2004) A eukaryotic BLUF domain mediates light-dependent gene expression in the purple bacterium Rhodobacter sphaeroides 2.4.1. Proc Natl Acad Sci U S A 101(33):12306–12311

53. Strickland D, Moffat K, Sosnick TR (2008) Light-activated DNA binding in a designed allosteric protein. Proc Natl Acad Sci 105(31):10709–10714

54. Ohlendorf R, Vidavski RR, Eldar A, Moffat K, Möglich A (2012) From dusk till dawn: one-plasmid systems for light-regulated gene expression. J Mol Biol 416(4):534–542

55. Qi Y, Rao F, Luo Z, Liang Z-X (2009) A flavin cofactor-binding PAS domain regulates c-di-GMP synthesis in AxDGC2 from Acetobacter xylinum. Biochemistry 48(43):10275–10285

56. Barends TR, Hartmann E, Griese JJ, Beitlich T, Kirienko NV, Ryjenkov DA et al (2009) Structure and mechanism of a bacterial light-regulated cyclic nucleotide phosphodiesterase. Nature 459(7249):1015–1018

57. Kanazawa T, Ren S, Maekawa M, Hasegawa K, Arisaka F, Hyodo M et al (2010) Biochemical and physiological characterization of a BLUF protein–EAL protein complex involved in blue light-dependent degradation of cyclic diguanylate in the purple bacterium Rhodopseudomonas palustris. Biochemistry 49(50):10647–10655

58. Schaap P (2013) Cyclic di-nucleotide signaling enters the eukaryote domain. IUBMB Life 65(11):897–903

59. Broichhagen J, Frank JA, Trauner D (2015) A roadmap to success in photopharmacology. Acc Chem Res 48(7):1947–1960

60. Fehrentz T, Schonberger M, Trauner D (2011) Optochemical genetics. Angew Chem Int Ed Engl 50(51):12156–12182

61. Luck M, Mathes T, Bruun S, Fudim R, Hagedorn R, Nguyen TMT et al (2012) A photochromic histidine kinase rhodopsin (HKR1) that is bimodally switched by ultraviolet and blue light. J Biol Chem 287(47):40083–40090

62. Sexton TJ, Golczak M, Palczewski K, Van Gelder RN (2012) Melanopsin is highly resistant to light and chemical bleaching in vivo. J Biol Chem 287(25):20888–20897

63. van Wyk M, Pielecka-Fortuna J, Lowel S, Kleinlogel S (2015) Restoring the ON switch in blind retinas: opto-mGluR6, a next-generation, cell-tailored optogenetic tool. PLoS Biol 13(5):e1002143

64. Beiert T, Bruegmann T, Sasse P (2014) Optogenetic activation of Gq signalling modulates pacemaker activity of cardiomyocytes. Cardiovasc Res 102(3):507–516

65. Avelar GM, Schumacher RI, Zaini PA, Leonard G, Richards TA, Gomes SL (2014) A rhodopsin-guanylyl cyclase gene fusion functions in visual perception in a fungus. Curr Biol 24(11):1234–1240

66. Scheib U, Stehfest K, Gee CE, Korschen HG, Fudim R, Oertner TG et al (2015) The rhodopsin-guanylyl cyclase of the aquatic fungus Blastocladiella emersonii enables fast optical control of cGMP signaling. Sci Signal 8(389):rs8

67. Gao S, Nagpal J, Schneider MW, Kozjak-Pavlovic V, Nagel G, Gottschalk A (2015) Optogenetic manipulation of cGMP in cells and animals by the tightly light-regulated guanylyl-cyclase opsin CyclOp. Nat Commun 6:8046

68. Ryu MH, Moskvin OV, Siltberg-Liberles J, Gomelsky M (2010) Natural and engineered photoactivated nucleotidyl cyclases for optogenetic applications. J Biol Chem 285(53):41501–41508

69. Balashov SP, Imasheva ES, Boichenko VA, Anton J, Wang JM, Lanyi JK (2005) Xanthorhodopsin: a proton pump with a light-harvesting carotenoid antenna. Science 309(5743):2061–2064

70. Wilson A, Punginelli C, Gall A, Bonetti C, Alexandre M, Routaboul J-M et al (2008) A photoactive carotenoid protein acting as light intensity sensor. Proc Natl Acad Sci U S A 105(33):12075–12080

71. Leverenz RL, Sutter M, Wilson A, Gupta S, Thurotte A, de Carbon CB et al (2015) A 12 angstrom carotenoid translocation in a photoswitch associated with cyanobacterial photoprotection. Science 348(6242):1463–1466

72. Gwizdala M, Wilson A, Kirilovsky D (2011) In vitro reconstitution of the cyanobacterial photoprotective mechanism mediated by the orange carotenoid protein in synechocystis PCC 6803. Plant Cell 23(7):2631–2643

73. Auldridge ME, Satyshur KA, Anstrom DM, Forest KT (2012) Structure-guided engineer-

ing enhances a phytochrome-based infrared fluorescent protein. J Biol Chem 287(10): 7000–7009

74. Takala H, Bjorling A, Linna M, Westenhoff S, Ihalainen JA (2015) Light-induced changes in the dimerization interface of bacteriophytochromes. J Biol Chem 290(26): 16383–16392

75. Zemelman BV, Lee GA, Ng M, Miesenböck G (2002) Selective photostimulation of genetically chARGed neurons. Neuron 33(1):15–22

76. Airan RD, Thompson KR, Fenno LE, Bernstein H, Deisseroth K (2009) Temporally precise in vivo control of intracellular signalling. Nature 458(7241):1025–1029

77. Kim JM, Hwa J, Garriga P, Reeves PJ, RajBhandary UL, Khorana HG (2005) Light-driven activation of beta 2-adrenergic receptor signaling by a chimeric rhodopsin containing the beta 2-adrenergic receptor cytoplasmic loops. Biochemistry 44(7):2284–2292

78. Oh E, Maejima T, Liu C, Deneris E, Herlitze S (2010) Substitution of 5-HT1A receptor signaling by a light-activated G protein-coupled receptor. J Biol Chem 285(40): 30825–30836

79. Xiang Y, Yuan Q, Vogt N, Looger LL, Jan LY, Jan YN (2010) Light-avoidance-mediating photoreceptors tile the Drosophila larval body wall. Nature 468(7326):921–926

80. Liu J, Ward A, Gao J, Dong Y, Nishio N, Inada H et al (2010) C. elegans phototransduction requires a G protein-dependent cGMP pathway and a taste receptor homolog. Nat Neurosci 13(6):715–722

81. Ward A, Liu J, Feng Z, Xu XZ (2008) Light-sensitive neurons and channels mediate phototaxis in C. elegans. Nat Neurosci 11(8):916–922

82. Edwards SL, Charlie NK, Milfort MC, Brown BS, Gravlin CN, Knecht JE et al (2008) A novel molecular solution for ultraviolet light detection in Caenorhabditis elegans. PLoS Biol 6(8):e198

83. Bhatla N, Horvitz HR (2015) Light and hydrogen peroxide inhibit C. elegans Feeding through gustatory receptor orthologs and pharyngeal neurons. Neuron 85(4):804–818

84. Jost M, Fernández-Zapata J, Polanco MC, Ortiz-Guerrero JM, Chen PY-T, Kang G, Padmanabhan S, Elías-Arnanz M & Drennan CL (2015) Structural basis for gene regulation by a B12-dependent photoreceptor. Nature 526, 536–541

85. Muller F, Walker WH, Massey V, Brustlei M, Hemmeric P (1972) Light-absorption studies on neutral flavin radicals. Eur J Biochem 25(3):573–580

86. Gauden M, Yeremenko S, Laan W, van Stokkum IHM, Ihalainen JA, van Grondelle R et al (2005) Photocycle of the flavin-binding photoreceptor AppA, a bacterial transcriptional antirepressor of photosynthesis genes. Biochemistry 44(10): 3653–3662

87. Zirak P, Penzkofer A, Lehmpfuhl C, Mathes T, Hegemann P (2007) Absorption and emission spectroscopic characterization of blue-light receptor Slr1694 from Synechocystis sp. PCC6803. J Photochem Photobiol B 86(1):22–34

88. Zirak P, Penzkofer A, Schiereis T, Hegemann P, Jung A, Schlichting I (2006) Photodynamics of the small BLUF protein BlrB from Rhodobacter sphaeroides. J Photochem Photobiol B 83(3):180–194

89. Kottke T, Heberle J, Hehn D, Dick B, Hegemann P (2003) Phot-LOV1: photocycle of a blue-light receptor domain from the green alga Chlamydomonas reinhardtii. Biophys J 84(2 Pt 1):1192–1201

90. Kennis JTM, Crosson S, Gauden M, van Stokkum IH, Moffat K, van Grondelle R (2003) Primary reactions of the LOV2 domain of phototropin, a plant blue-light photoreceptor. Biochemistry 42(12):3385–3392

91. Müller P, Bouly JP, Hitomi K, Balland V, Getzoff ED, Ritz T et al (2014) ATP binding turns plant cryptochrome into an efficient natural photoswitch. Sci Rep 4

92. Giovani B, Byrdin M, Ahmad M, Brettel K (2003) Light-induced electron transfer in a cryptochrome blue-light photoreceptor. Nat Struct Mol Biol 10(6):489–490

93. Losi A, Gensch T, van der Horst MA, Hellingwerf KJ, Braslavsky SE (2005) Hydrogen-bond network probed by time-resolved optoacoustic spectroscopy: photoactive yellow protein and the effect of E46Q and E46A mutations. Phys Chem Chem Phys 7(10):2229–2236

94. Lincoln CN, Fitzpatrick AE, van Thor JJ (2012) Photoisomerisation quantum yield and non-linear cross-sections with femtosecond excitation of the photoactive yellow protein. Phys Chem Chem Phys 14(45): 15752–15764

95. Fan HY, Morgan SA, Brechun KE, Chen YY, Jaikaran ASI, Woolley GA (2011) Improving a designed photocontrolled

DNA-binding protein. Biochemistry 50(7): 1226–1237

96. Lamparter T, Esteban B, Hughes J (2001) Phytochrome Cph1 from the cyanobacterium Synechocystis PCC6803—purification, assembly, and quaternary structure. Eur J Biochem 268(17):4720–4730

97. Vierstra RD, Quail PH (1983) Photochemistry of 124 kilodalton Avena phytochrome in vitro. Plant Physiol 72(1):264–267

98. Pennacchietti F, Losi A, Xu XL, Zhao KH, Gärtner W, Viappiani C et al (2015) Photochromic conversion in a red/green cyanobacteriochrome from Synechocystis PCC6803: quantum yields in solution and photoswitching dynamics in living E. coli cells. Photochem Photobiol Sci 14(2):229–237

99. Popot JL, Gerchman SE, Engelman DM (1987) Refolding of bacteriorhodopsin in lipid bilayers—a thermodynamically controlled 2-stage process. J Mol Biol 198(4): 655–676

100. Govindjee R, Balashov SP, Ebrey TG (1990) Quantum efficiency of the photochemical cycle of bacteriorhodopsin. Biophys J 58(3): 597–608

101. Pace CN, Vajdos F, Fee L, Grimsley G, Gray T (1995) How to measure and predict the molar absorption-coefficient of a protein. Protein Sci 4(11):2411–2423

102. Chen D, Gibson ES, Kennedy MJ (2013) A light-triggered protein secretion system. J Cell Biol 201(4):631–640

103. Müller K, Engesser R, Schulz S, Steinberg T, Tomakidi P, Weber CC et al (2013) Multichromatic control of mammalian gene expression and signaling. Nucleic Acids Res 41(12):e124

Chapter 3

Algal Photobiology: A Rich Source of Unusual Light Sensitive Proteins for Synthetic Biology and Optogenetics

Arash Kianianmomeni and Armin Hallmann

Abstract

The light absorption system in eukaryotic (micro)algae includes highly sensitive photoreceptors, which change their conformation in response to different light qualities on a subsecond time scale and induce physiological and behavioral responses. Some of the light sensitive modules are already in use to engineer and design photoswitchable tools for control of cellular and physiological activities in living organisms with various degrees of complexity. Thus, identification of new light sensitive modules will not only extend the source material for the generation of optogenetic tools but also foster the development of new light-based strategies in cell signaling research. Apart from searching for new proteins with suitable light-sensitive modules, smaller variants of existing light-sensitive modules would be helpful to simplify the construction of hybrid genes and facilitate the generation of mutated and chimerized modules. Advances in genome and transcriptome sequencing as well as functional analysis of photoreceptors and their interaction partners will help to discover new light sensitive modules.

Key words Algal photobiology, Light perception, Photoreceptors, Synthetic biology, Optogenetics

1 Introduction

Free swimming microalgae use a sophisticated light perception system to measure changes in environmental light conditions and adapt their physiological and developmental processes accordingly. Considering the fact that the ability to sense changes in light quality, quantity, and direction is a fundamental ability to survive in nature, algae possess different classes of photoreceptors that enable them to regulate various light-dependent cellular and physiological processes and to change their swimming behavior in response to environmental light. Detected changes in wavelengths or spectral compositions of light at dusk and dawn serve as an environmental signal cue to swim up or down in the water column. Moreover, changes in wavelengths or spectral compositions regulate developmental and physiological processes, e.g., for day and night adaptation. Thus, efforts towards understanding mechanisms underlying

Arash Kianianmomeni (ed.), *Optogenetics: Methods and Protocols*, Methods in Molecular Biology, vol. 1408,
DOI 10.1007/978-1-4939-3512-3_3, © Springer Science+Business Media New York 2016

phototactic and photophobic responses as well as other light-dependent processes are of particular interest for algal photobiology research. Since more than a century of research, useful insights into physiological, biophysical, and morphological aspects of light responses in algae have been obtained. However, it was not until the early twentieth century that genome and transcriptome sequencing proceeded rapidly and that key proteins involved in light-mediated behavioral responses were discovered. Three groups reported independently from each other that two rhodopsin-like photoreceptors, i.e., channelrhodopsin-1 (ChR1) and channelrhodopsin-2 (ChR2), are involved in light perception and light-induced movements of the unicellular green alga *Chlamydomonas* [1–4]. The ability of these light-gated ion channels to depolarize cells simply by illumination with blue light [1, 2] paved the way to the young field of optogenetics. The field expanded rapidly when new channelrhodopsins were discovered and new modified variants were engineered. These advances enriched the optogenetic toolbox and they provided an excellent basis for precise temporal and spatio temporal control of cell signaling. Concurrently, light sensitive domains were established to control fundamental cellular processes such as protein secretion, nuclear import, chromatin targeting, and gene expression [5, 6]. The fast expansion of the optogenetic field created an increasing need for further light sensitive domains, for instance domains with other absorption spectra and new photocurrent properties. Any new photoreceptor with new properties provides the opportunity for new applications in optogenetics and also in synthetic biology. Current top candidates in this field are photoswitchable enzymes such as light-activated guanylyl or adenylyl cyclases. These enzymes can change the level of the signaling molecules cGMP or cAMP in a light-dependent manner. Another promising class of photoreceptors with enzyme activity is the family of histidine-kinase rhodopsins, which has been identified in the volvocine algae or more specifically in the green microalgae *Volvox* and *Chlamydomonas*. Histidine-kinase rhodopsins might activate a domain with guanylyl (or adenylyl) cyclase activity, which is located on the same polypeptide chain. In this way, external light stimuli could activate histidine-kinase rhodopsins that regulate different light-dependent signaling pathways of physiological and cellular processes through modulation of cGMP (or cAMP) levels. In view of the rapid developments in optogenetics, identification of new photoreceptors, their interaction partners, and signaling pathways will undoubtedly be a key source for new, innovative light sensitive tools for applications in biotechnology and biomedical science. An auspicious group of organisms is predestined to be a search area for light-sensitive proteins involved in the sensing and response to light: the algae.

2 Algae

The term "algae" is not clearly defined. Algae are generally considered to be aquatic oxygen-evolving photosynthetic autotrophs that include different categories of cellular organization: There are unicellular, colonial, and multicellular algae [7, 8]. Some multicellular algae consist only of simple filaments but others even form tissues. In several species, multicellularity was accompanied by the development of differentiated cell types [7–9]. There are both marine and freshwater algae, and algae are found almost everywhere on earth. Some species grow on soil or even on snow; others can even tolerate extreme salinity [10]. Algae also demonstrate big differences in size: the smallest alga and even the smallest eukaryote is the marine alga *Ostreococcus tauri* with a diameter of only 1 μm, whereas the largest alga is the brown alga *Macrocystis pyrifera* with a size of up to 60 m.

More than 70,000 species of algae, including 15 phyla and 54 classes, are listed in Algaebase (http://www.algaebase.org) and this number seems to be only a quite conservative estimation of all existing algae species. Other estimations suggest around one million species of algae including a large fraction of diatoms with at least 200,000 species [9, 10]. The major groups of algae are Phaeophyceae (brown algae), Bacillariophyceae (diatoms), Xanthophyceae (yellow-green algae), Chlorophyta and Charophyta (green algae), Rhodophyta (red algae), Glaucophytes, and Dinoflagellates (Fig. 1). These groups include a handful of algal model systems, namely *Chlamydomonas reinhardtii* (Chlorophyta), *Volvox carteri* (Chlorophyta), *Cyanidioschyzon merolae* (Rhodophyta), *Phaeodactylum tricornutum* (Bacillariophyceae), *Vaucheria frigida* (Xanthophyceae), and *Mougeotia scalaris* (Charophyta), which were the main source for most of the current knowledge about algal photobiology [11]. The extent and diversity of algal species and the colonization of very different habitats promise a rich source of new and unusual light sensitive proteins for synthetic biology and optogenetics.

3 Photoreceptor Proteins

Photoreceptor proteins typically consist of a protein moiety and a nonprotein photosensor. This photosensor is a covalently or noncovalently bound, small organic molecule known as a chromophore. The purpose of any photoreceptor protein is the absorption of photons to directly or indirectly produce a biological effect. Consequently, photoreceptor proteins absorb the wavelengths significantly present in sunlight (from UV-B to far-red). Photoreceptor proteins are either integral membrane proteins (e.g., channelrhodopsins) or cytoplasmic, soluble proteins (e.g., phytochromes).

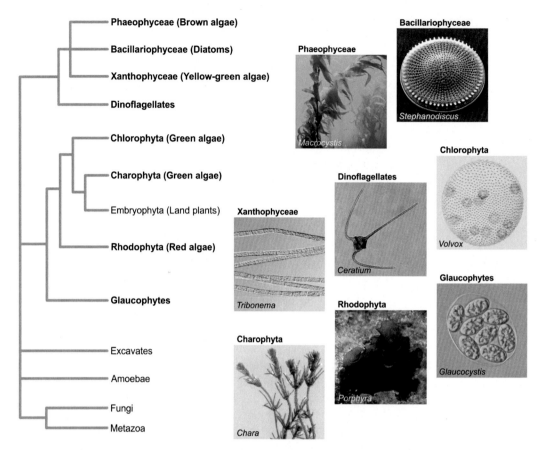

Fig. 1 Eukaryotic tree of life with emphasis on algae showing example pictures of some representatives. The tree was adapted with permission from [11] and is based on results from Kranz et al. [72], Baldauf [73] and Prochnik et al. [74]. Algal groups are in *bold*. Photos by Claire Fackler (*Macrocystis*), Frank E. Round (*Stephanodiscus*), Christian Fischer (*Chara*), Keisotyo (*Ceratium*), and own work

During evolution, in all kingdoms of life, i.e., in Animalia, Plantae, Fungi, Protista, Archaeabacteria, and Eubacteria, species with light-sensitive proteins and the corresponding signaling pathways emerged. Light-sensitive proteins gave those species the benefit to monitor light continuously and to adjust their growth, development, behavior, and reproduction accordingly. For example, photoreceptor proteins are involved in the regulation of adaptive photo responses such as phototaxis and phototropism. They also participate in the regulation of various developmental processes including the control of flowering time, sexual development, and the adaptation of light-dark rhythms. Moreover, light-regulated gene expression mediated by photoreceptors acts as a multifaceted regulator to control the abundance of functional genes at different levels and thereby to regulate many cellular and physiological processes (Fig. 2).

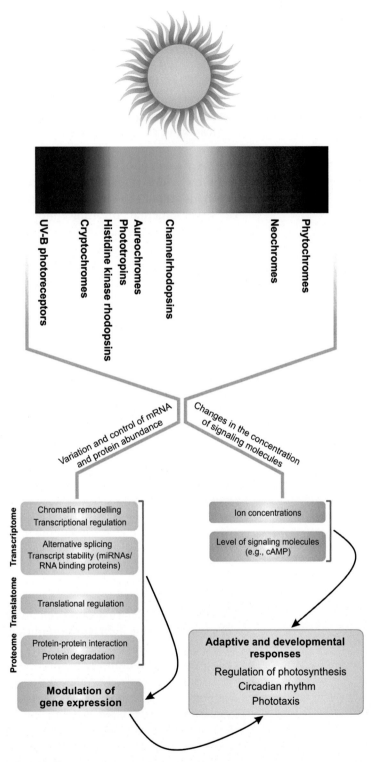

Fig. 2 Multifaceted algal photoreceptor families. Algae possess different photo-receptors related to specific absorption wavelengths (note that the functionality of some photoreceptor types, e.g., cryptochromes or phytochrome, is not strictly limited to the indicated band in the spectrum). Algal photoreceptors are involved in regulation of adaptive and developmental responses during the whole life cycle. Modified after [14]

Photoreceptor proteins may be divided into three major groups: red/far-red light-absorbing (e.g., phytochromes), UV-A/blue light-absorbing (e.g., phototropins and cryptochromes), and UV-B-sensing (e.g., UVR8) photoreceptors. Collectively, these photoreceptors have been shown to orchestrate expression of a significant number of genes during photomorphogenesis in plants (reviewed in [12, 13]). The photoreceptor-mediated control of gene expression occurs at multiple regulatory steps including transcription, posttranscription, translation, and posttranslation. The ultimate objective is the induction of adaptive or developmental photomorphogenic responses (reviewed in [14]). Moreover, recent reports indicated that tissue or developmentally regulated expression of photoreceptors and/or associated signaling components trigger changes in the transcript level of distinct target genes in a tissue-specific manner [15, 16].

4   In Vivo Function of Algal Photoreceptors

Algal photoreceptors include classical photoreceptors, which were originally discovered in higher plants (i.e., phototropins, cryptochromes, phytochromes, and UV-B photoreceptor), rhodopsin-like photoreceptors, which serve as light-gated ion channels, and other photoreceptor families (i.e., aureochromes and neochromes) (reviewed in [11, 17]) (Fig. 2). The existence of different photoreceptor families in algae indicates that light is one of the major environmental signals, which needs to be constantly monitored during their short life cycle. Because algae have to optimize their photosynthetic activities depending on the amount of light available, many cellular processes including sexual development, circadian clock, nitrogen and lipid metabolism, cell cycle, and cellular differentiation are regulated by light [18–20].

Phototropin photoreceptors are blue-light-activated serine/threonine protein kinases. It was shown for *Chlamydomonas* that blue light can induce a delay in cell division [21]. More precisely, in cells grown in blue light, the commitment point is shifted to a later time and, as a consequence, to a larger cell size, when compared to cells grown in red light [22]. This observation is of particular interest because recently, Kreimer and colleagues found out that the eyespots in a *Chlamydomonas* mutant strain with a phototropin knockout (ΔPhot^{G5}) are larger than in wild type strains. Moreover, the eyespot size increased in wild type cells after the cells were incubated in the dark for several days, whereas the eyespot size of the ΔPhot^{G5} mutant remained unchanged under the same condition [23]. In addition, knockdown strains that produce a significantly reduced amount of phototropin were shown to be partially impaired in three steps of the sexual development: in gametogenesis, the maintenance of mating ability, and the germination of zygotes

[19]. These and other observations suggest that the blue light photoreceptor phototropin participates in the mechanisms controlling sexual development and cell size in *Chlamydomonas*. The phototropin photoreceptor is also involved in the blue-light-mediated changes in transcript accumulation of genes encoding enzymes responsible for chlorophyll and carotenoid biosynthesis [24]. In the multicellular alga *Volvox carteri*, which exhibits a division of labor between two completely different cell types (Fig. 3), the phototropin transcript is more expressed in the small somatic cells than in the large, dark green reproductive cells [25]. Interestingly, there is a tendency that blue light leads to accumulation of more transcripts in the somatic cell type, while red light has the opposite effect and leads to accumulation of more transcripts in the reproductive cell type [26]. Blue light also was shown to induce the

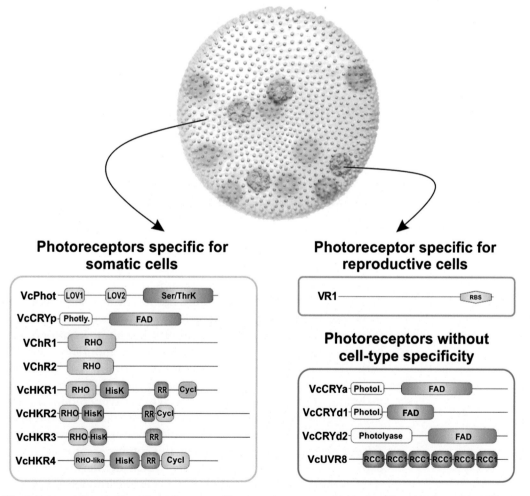

Fig. 3 *Volvox* photoreceptors. Domain compositions are drawn according to the Pfam database. Two different cell types of *Volvox*, i.e., *large dark green* reproductive cells and *small pale*, biflagellate somatic cells are located below and at the surface, respectively. Adapted with permission from [26]

accumulation of two components of the circadian rhythms only in somatic cells, indicating that these clock-relevant components are affected by blue light in a cell-type-specific manner [26]. In *Volvox*, the light-regulated, cell-type-specific gene expression system thus might be triggered by cell-type-specific expression of photoreceptors such as phototropin and could reflect an early development of cell-type-specific signaling mechanisms during evolution to ensure maintenance of differentiation [27].

Members of the algal cryptochrome family typically are blue-light receptors. Among other functions, they have shown to be involved in circadian rhythms and DNA repair. In *Chlamydomonas*, the animal-like cryptochrome regulates the transcript level of genes involved in various pathways not only in response to blue light but also in response to red light [28]. Cryptochromes from the diatoms *Phaeodactylum tricornutum* and *Oestreococcus tauri* show blue-light-regulated gene expression and DNA repair activity [29, 30]. Recently, an unusual cryptochrome of *Phaeodactylum tricornutum*, i.e., CryP, has shown to be involved in regulation of light-harvesting protein expression [31].

Aureochromes have been identified in the stramenopilic alga *Vaucheria frigida*, in the diatoms *Phaeodactylum tricornutum* and *Thalassiosira pseudonana*, the brown algae *Ectocarpus siliculosus* and *Fucus distichus*, and others [32–35]. In *Vaucheria*, aureochromes are required for sex organ formation and blue light-induced branching [36].

Neochromes are hybrid photoreceptors containing domains of both phytochromes and phototropins and have been first identified in the filamentous green alga *Mougeotia scalaris*. Neochromes are likely to be involved in blue-light-regulated chloroplast movement [37].

Already some time ago, rhodopsin-like photoreceptors were shown to be involved in light-induced locomotion of photosynthetic flagellates and they were proposed to be localized in their eyespot apparatus [38, 39]. These rhodopsin-like photoreceptors form the largest family of light-sensitive proteins in the flagellated volvocine algae. At least seven rhodopsin-like photoreceptors could be found in the genomes of *Chlamydomonas* and *Volvox* [17, 25, 26]. This photoreceptor family includes both type I (found in archaea, eubacteria, fungi, and algae) and type II (found in animal including human) rhodopsins. The animal-type rhodopsin was shown to form a 1:1 complex with the chaperone Ycf4 and is likely to be involved in the assembly and biogenesis of photosystem I in *Chlamydomonas* [40]. In *Volvox*, this photoreceptor is highly expressed in the reproductive cells, which demonstrates a high level of photosynthetic activity [25]. Channelrhodopsins, which belong to type I rhodopsins, have shown to be involved in phototaxis or photophobic responses in *Chlamydomonas* [3, 41]. In its multicellular relative *Volvox*, both channelrhodopsins are only expressed in the flagellated somatic cells, most probably due to their involvement in mediating phototactic and/or photophobic responses, which are exclusively restricted to the somatic cells. Because the action spectrum of *Volvox* photocurrent, the action spectra of both negative and positive phototaxis and the absorption spectrum of VChR1, an ion-gated channel, all peak around 520 nm [42–44], it was assumed that VChR1 is the main photoreceptor for phototaxis under vegetative conditions [44]. Another four rhodopsin-like photoreceptors, known as histidine-kinase rhodopsins have been identified in the genomes of both *Volvox* and *Chlamydomonas*. Based on their protein domain structures including a cyclase domain, they have the potential ability to change the cGMP or cAMP concentration in a light dependent manner, particularly after environmental stimuli, what finally could trigger processes in asexual and sexual development [11] (Fig. 4). Cyclic AMP was postulated to be an intracellular second messenger in *Volvox* that is produced in response to the presence of the sex inducer and that triggers a signal cascade leading to sexual development [45]. Signal transduction by cAMP as a second messenger is mediated by intracellular cAMP receptors called cAMP-binding proteins [46]. Although involvement of cAMP in sexual differentiation in *Volvox* was controversially discussed [47], for *Chlamydomonas* the accumulation of cAMP in gametes undergoing agglutination was reported [48, 49]. Pasquale and Goodenough reported that exogenous dibutyryl-cAMP, a membrane-permeable analogue of cAMP, induces all three agglutination-triggered responses: flagellar tip activation, loss of cell walls, and mating

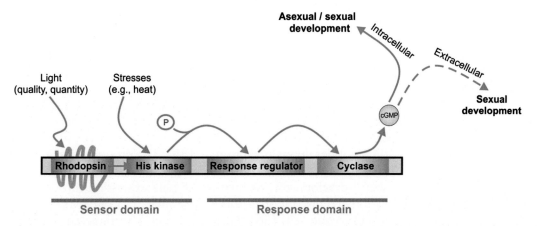

Fig. 4 A model for potential involvement of *Volvox* histidine-kinase rhodopsins in asexual and sexual development. The environmental signals (changes in light and temperature) could be detected by the rhodopsin and the histidine kinase of the sensor domain. The activated kinase transfers a phosphoryl group (from ATP) to the response regulator, which in turn activates the guanylyl (or adenylyl) cyclase to produce cGMP (or cAMP). The produced intracellular cGMP (cAMP) acts as regulator for asexual and sexual development and could be transported to extracellular space as a signal for sexual differentiation in drought periods. Adapted with permission from [11]

structure activation. In addition, the mating of nonagglutinating mutants can be rescued by treatment with dibutyryl-cAMP. These results indicate that cAMP (and/or cGMP) activates not only the known mating responses but all necessary responses [48, 50]. The involvement of cAMP-dependent kinase cascades and/or the gating of cAMP-gated ion channels were assumed in cAMP signal transduction in *Chlamydomonas* [51, 52]. In *Volvox*, disintegration of the sperm packets has been observed after the addition of dibutyryl-cAMP [53]. Therefore, the investigation of histidine-kinase rhodopsins of unknown or unconfirmed function relating to sexual differentiation is of particular interest. These light receptors seem to be able to change the concentration of cAMP and/or cGMP not only in response to light, but also in response to other environmental stimuli such as temperature.

In the natural environment of *Volvox*, a temperature increase in summer can be the environmental stimulus for switching from asexual to sexual development. To obtain information about the day length and the season, the conserved rhodopsin domain of histidine-kinase rhodopsins could be responsible for the detection of changes not only in the relative lengths of light and dark periods, but also changes in light properties like intensity and the ratio between certain wavelengths. Thus, changes in light intensity and wavelengths ratio during different seasons might be another trigger to switch to sexual development, which then is an important interlude that allows for genetic recombination in the diploid zygotes of species that normally reproduce asexually in a haploid life cycle. Even other signals like changes in temperature could be detected by the sensor domain of histidine-kinase rhodopsins. Signal detection could lead to autophosphorylation of the histidine kinase in the sensor domain, which then transfers its high-energy phosphoryl group to the so-called response regulator of the response domain. Finally, the response regulator could activate the guanylyl or adenylyl cyclase in the response domain to produce cGMP/cAMP (Fig. 4). In *Volvox*, the switch to sexual development not only occurs in response to increased temperatures (heat, heat shock) but also in response to oxidative stress [54]. In both cases, light plays a critical role for the success of this switch to a different developmental program [55]. The produced intracellular cAMP/cGMP could induce the required key regulator genes for sexual development. Regulation of gene expression through activation of transcription factors is the primary task of cAMP signaling in eukaryotic cells [56]. Such cAMP/cGMP-triggered transcription factors are likely to be involved in the regulation of cell-type-specific gene expression during development [57]. In *Ostreococcus*, light-dependent changes in the cAMP level have been shown to regulate the synthesis of cyclin A, which interacts with the retinoblastoma protein (RB) to regulate the cell division pathway [58]. Moreover, cAMP-dependent kinase cascades in *Chlamydomonas* are assumed to be involved in cAMP signal transduction and, recently, Boonyareth et al. suggested that an increased level of cAMP might be a result of rhodopsin activation [59].

5 Application of Light-Sensitive Modules in Optogenetics and Synthetic Biology

The application of algal photoreceptors in synthetic biology started with the discovery of channelrhodopsins in 2002 and 2003 [1, 2]. These light-gated ion channels developed into an easy-to-use system for noninvasive control of the membrane potential using external light stimuli (Figs. 5 and 6). Using cell-type-specific promoters, channelrhodopsins can be used to activate specific types of neurons at high spatial and temporal resolution. Although most laboratories currently make use of the wild type channelrhodopsin-2 from *Chlamydomonas*, which absorbs blue light with a peak at about 470 nm, there is a growing body of evidence that channelrhodopsins from other algae, e.g., the ones from *Volvox*, also have a great potential for application in neurobiology. *Volvox* channelrhodopsin-1 is red shifted with a peak at about 520 nm and, therefore, seems to be more suited for applications in animal and human models because red light is less harmful than blue light. During the last years, several other channelrhodopsins have been identified in various algae species due to their sequence similarity to published channelrhodopsins from *Volvox* and *Chlamydomonas*. A red-shifted channelrhodopsin, MChR1, which peaks at 520 nm (at a neutral pH), was found in the flagellated alga *Mesostigma viride* [60]. In the marine alga *Platymonas subcordiformis*, a highly efficient blue-shifted channelrhodopsin was identified, which peaks at 437 nm (at neutral pH). The *Platymonas* channelrhodopsin exhibits an about threefold higher unitary conductance and greater relative permeability for Na^+ ions as compared to the most frequently used *Chlamydomonas* channelrhodopsin-2 [61]. The channelrhodopsins from *Mesostigma* and *Platymonas* belong to the recently found channelrhodopsins with great development potential in optogenetics. Further channelrhodopsins have been identified and characterized base on the gained data from the 1000 plants (1KP) project for application in optogenetics [62–64].

Moreover, a new strategy based on chimeras of *Chlamydomonas* channelrhodopsin-1 and *Volvox* channelrhodopsin-1 was used to generate novel color-tuned channelrhodopsins with a high efficiency. These chimeric channelrhodopsins exhibit larger photocurrents and absorption maxima ranging from 526 to 545 nm [65].

Another type of algal rhodopsin-like photoreceptor, which may be used to control cellular processes in optogenetic applications, comes from the histidine-kinase rhodopsin family. These photoswitchable enzymes can change the level of signaling molecules such as cAMP in a light-dependent manner. Therefore, they

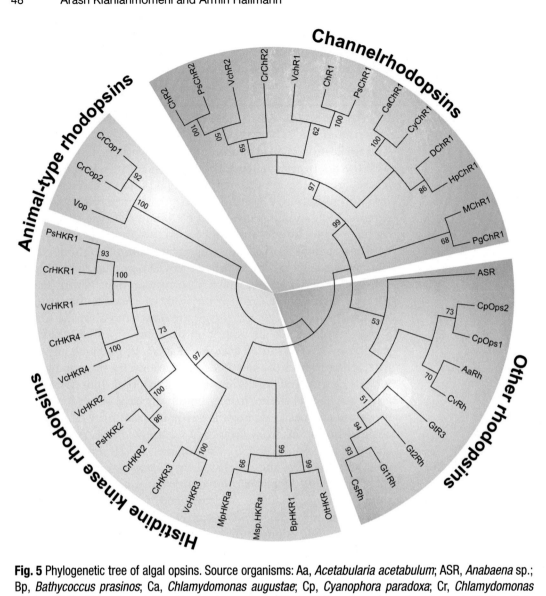

Fig. 5 Phylogenetic tree of algal opsins. Source organisms: Aa, *Acetabularia acetabulum*; ASR, *Anabaena* sp.; Bp, *Bathycoccus prasinos*; Ca, *Chlamydomonas augustae*; Cp, *Cyanophora paradoxa*; Cr, *Chlamydomonas reinhardtii*; Cv, *Chlorella variabilis*; Cs, *Cryptomonas* sp. S2; Cy, *Chlamydomonas yellowstonensis*; D, *Dunaliella salina*; Hp, *Haematococcus pluvialis*; Gt, *Guillardia theta*; M, *Mesostigma viride*; Mp, *Micromonas pusilla*; Msp., *Micromonas* sp. RCC299; Ps, *Pleodorina starrii*; Vc, *Volvox carteri*. Adapted with permission from [11]

are also suited as optogenetic tools for manipulation of animal behavior with external light stimuli. However, a major challenge is the size of these photoreceptors: These large proteins consist of four subunits, which complicates both expression in host cells and, if required, protein purification. Moreover, functional characterization of the histidine-kinase rhodopsins needs to be completed regarding absorption spectra, photocurrent properties, enzyme activity, and biological function.

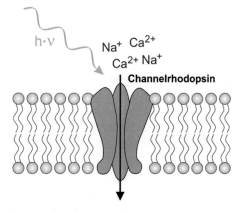

Fig. 6 Application of channelrhodopsins for generation of light-driven membrane depolarization to control neural spiking

Some pioneer works demonstrate that light sensitive proteins can also be used to control protein–protein interactions involved in protein secretion, nuclear import, chromatin targeting, and gene expression with external light stimuli. Plant phytochromes, for example, could be used successfully to control gene expression with red and far-red light [66]. Other experiments were based on the interaction between the photosensitive LOV domain and the amphipathic helix Jα were designed to control protein activity and to produce local or global calcium signals [67, 68].

Another blue light photoreceptor which has been widely used to control gene expression using blue light is cryptochrome together with its interaction partner CIB1. Using these components, an optogenetic two-hybrid system with light-inducible transcriptional effectors (LITEs) has been developed. It is composed of two major parts: an "anchor" (a customizable TALE DNA-binding domain fused to the light-sensitive cryptochrome-2) and an "effector" domain (the interacting partner of cryptochrome-2, CIBI). This LITE system can be used for reversible activation of gene transcription [69].

Discovery of more and more photoreceptors together with their interaction partners can contribute to the design of many optogenetic tools with different absorption properties that allow for multichromatic fine tuning of gene expression in a spatiotemporal fashion. A recent example is the utilization of the UV-B photoreceptor UVR8 and its interaction partner COP1. Based on UVB-induced monomerization of UVR8 which allows interaction with COP1, photoswitches were designed to control protein–protein interactions involved in protein secretion, nuclear import, chromatin targeting, and gene expression using UV light (reviewed in [6]). However, due to the deleterious effects of UVB-irradiation especially for mammalian cells, more efforts should be made to find new UVR8 proteins or to engineer new variants that require smaller doses of UVB for monomerization. In addition, the binding ability and DNA repair activity of DASH cryptochromes could be considered in future for the development of a light-inducible DNA-mismatch-repair tool to induce the repair of (UV induced) mismatches in living cells.

6   Algal Genome and Transcriptome Data as Sources of New Light-Sensitive Modules

During the last decade, many technological advances in genome and transcriptome sequencing, data integration, analysis, and visualization facilitated the access to new sequences of many different species. Concurrently, advances in algal biotechnology brought optimized algae strains for production of biofuel, industrial pigments, vaccines, etc., and, thereby, several new algae species came into focus. The genome and transcriptome data of these species are of particular importance for the biotechnology industry.

These and other algae species are also used as model systems to answer fundamental questions in basic science. For example, the volvocine algae are used as model systems for investigation of multicellularity [47, 70]. Handful of these model algae have been sequenced completely or sequencing is in progress (Table 1). However, although the low sequencing costs and state-of-art technologies allow for sequencing of more organisms in shorter times, the gene annotation process and bioinformatic analyses are still quite time consuming and need well-organized international cooperation. Free and easy access to such sequencing data is also of particular interest for researchers in synthetic biology and optogenetics, who are always on the look-out for new light-sensitive modules with modified properties.

In a recent report, Klapoetke et al. described the sequencing of 127 algal transcriptomes, which led to the identification of 61 new channelrhodopsins. These data belong to the 1000 Plants (1KP) project, which aims at the generation of transcriptome data from over 1000 plants including hundreds of algae species [62, 71]. Among the 61 identified channelrhodopsins are two channelrhodopsins that could be used as powerful tools in optogenetics: the excitation spectrum of the channelrhodopsin Chrimson is red shifted by 45 nm relative to any previous channelrhodopsin, whereas Chronos has faster kinetics than any previous channelrhodopsin. The combination of these two channelrhodopsins allows for crosstalk-free two-color activation of neural spiking [64].

Finally, it should be highlighted that genome and transcriptome data also allow for identification of other new photoreceptors like phototropins, cryptochromes, and phytochromes, or other

Table 1
Genomic resources of some algal model systems

Species	Internet portal
Aureococcus anophagefferens	http://genome.jgi-psf.org/Auran1
Chlamydomonas reinhardtii	http://phytozome.jgi.doe.gov/pz/portal.html#!info?alias=Org_Creinhardtii
Chlorella sp. NC64A	http://genome.jgi-psf.org/ChlNC64A_1/ChlNC64A_1.home.html
Coccomyxa sp. C-169	http://genome.jgi-psf.org/Coc_C169_1
Cyanidioschyzon merolae	http://merolae.biol.s.u-tokyo.ac.jp
Emiliania huxleyi	http://genome.jgi-psf.org/Emihu1
Ectocarpus siliculosus	http://bioinformatics.psb.ugent.be/orcae/overview/Ectsi
Fragilariopsis cylindrus	http://genome.jgi-psf.org/Fracy1/Fracy1.home.html
Galdieria sulphuraria	http://genomics.msu.edu/galdieria/
Guillardia theta	http://genome.jgi.doe.gov/Guith1/Guith1.home.html
Micromonas sp. CCMP1545	http://genome.jgi-psf.org/MicpuC3/MicpuC3.home.html
Micromonas sp. RCC299	http://genome.jgi-psf.org/MicpuN3/MicpuN3.home.html
Nannochloropsis gaditana	http://www.nannochloropsis.org/
Ostreococcus lucimarinus	http://genome.jgi-psf.org/Ost9901_3
Ostreococcus tauri	http://genome.jgi-psf.org/Ostta4/Ostta4.home.html
Ostreococcus sp. RCC809	http://genome.jgi-psf.org/OstRCC809_2/OstRCC809_2.home.html
Phaeodactylum tricornutum	http://genome.jgi-psf.org/Phatr2/Phatr2.home.html
Porphyridium purpureum	http://cyanophora.rutgers.edu/porphyridium/
Thalassiosira pseudonana	http://genome.jgi-psf.org/Thaps3
Volvox carteri	http://phytozome.jgi.doe.gov/pz/portal.html#!info?alias=Org_Vcarteri

proteins with conserved light-sensitive domains such as FAD or LOV. Further sequencing projects, especially with focus on free-swimming microalgae with different light absorption properties and wavelength-dependent photo responses therefore will expand the resources for optogenetics and basic research in light sensitive proteins.

In the future, emphasis should also be placed on the detailed characterization of new photoreceptors, which includes their exact (enzyme) activities, absorption properties, photocurrent properties, and protein structural analysis. The latter helps to reveal the molecular mechanisms and even can be used as a basis for structure-based engineering of light sensitive proteins. There is also a requirement for assays that allow for identification of the interaction partners of the photoreceptors, which is not only important to get new insights into the photoreceptor signaling networks but also to make further progress in synthetic biology and optogenetics.

References

1. Nagel G et al (2002) Channelrhodopsin-1: a light-gated proton channel in green algae. Science 296(5577):2395–2398
2. Nagel G et al (2003) Channelrhodopsin-2, a directly light-gated cation-selective membrane channel. Proc Natl Acad Sci U S A 100(24):13940–13945
3. Sineshchekov OA, Jung KH, Spudich JL (2002) Two rhodopsins mediate phototaxis to low- and high-intensity light in *Chlamydomonas reinhardtii*. Proc Natl Acad Sci U S A 99(13):8689–8694
4. Suzuki T et al (2003) Archaeal-type rhodopsins in *Chlamydomonas*: model structure and intracellular localization. Biochem Biophys Res Commun 301(3):711–717
5. Schmidt D, Cho YK (2015) Natural photoreceptors and their application to synthetic biology. Trends Biotechnol 33(2):80–91
6. Kianianmomeni A (2015) UVB-based optogenetic tools. Trends Biotechnol 33:59–61
7. Barsanti L, Gualtieri P (2014) Algae: anatomy, biochemistry, and biotechnology. CRC, Boca Raton
8. Graham JE, Wilcox LW, Graham LE (2008) Algae, 2nd edn. Benjamin Cummings, San Francisco
9. Guiry MD (2012) How many species of algae are there? J Phycol 48(5):1057–1063
10. Raven JA, Giordano M (2014) Algae. Curr Biol CB 24(13):R590–R595
11. Kianianmomeni A, Hallmann A (2014) Algal photoreceptors: in vivo functions and potential applications. Planta 239(1):1–26
12. Kami C, Lorrain S, Hornitschek P, Fankhauser C (2010) Light-regulated plant growth and development. Curr Top Dev Biol 91:29–66
13. Jenkins GI (2014) The UV-B photoreceptor UVR8: from structure to physiology. Plant Cell 26(1):21–37
14. Kianianmomeni A (2014) More light behind gene expression. Trends Plant Sci 19:488–490
15. Lopez-Juez E et al (2008) Distinct light-initiated gene expression and cell cycle programs in the shoot apex and cotyledons of *Arabidopsis*. Plant Cell 20(4):947–968
16. Ma L et al (2005) Organ-specific expression of *Arabidopsis* genome during development. Plant Physiol 138(1):80–91
17. Hegemann P (2008) Algal sensory photoreceptors. Annu Rev Plant Biol 59:167–189
18. Grossman AR, Lohr M, Im CS (2004) *Chlamydomonas reinhardtii* in the landscape of pigments. Annu Rev Genet 38:119–173
19. Huang KY, Beck CF (2003) Photoropin is the blue-light receptor that controls multiple steps in the sexual life cycle of the green alga *Chlamydomonas reinhardtii*. Proc Natl Acad Sci U S A 100(10):6269–6274
20. Kirk MM, Kirk DL (1985) Translational regulation of protein-synthesis, in response to light, at a critical stage of *Volvox* development. Cell 41(2):419–428
21. Munzner P, Voigt J (1992) Blue light regulation of cell division in *Chlamydomonas reinhardtii*. Plant Physiol 99(4):1370–1375
22. Oldenhof H, Zachleder V, van den Ende H (2004) Blue light delays commitment to cell division in *Chlamydomonas reinhardtii*. Plant Biol (Stuttg) 6(6):689–695
23. Trippens J et al (2012) Phototropin influence on eyespot development and regulation of phototactic behavior in *Chlamydomonas reinhardtii*. Plant Cell 24(11):4687–4702
24. Im CS, Eberhard S, Huang K, Beck CF, Grossman AR (2006) Phototropin involvement in the expression of genes encoding chlorophyll and carotenoid biosynthesis enzymes and LHC apoproteins in *Chlamydomonas reinhardtii*. Plant J 48(1):1–16
25. Kianianmomeni A, Hallmann A (2015) Transcriptional analysis of *Volvox* photoreceptors suggests the existence of different cell-type specific light-signaling pathways. Curr Genet 61(1):3–18
26. Kianianmomeni A (2014) Cell-type specific light-mediated transcript regulation in the multicellular alga *Volvox carteri*. BMC Genomics 15(1):764
27. Kianianmomeni A (2015) Cell-type specific photoreceptors and light signaling pathways in the multicellular green alga *Volvox carteri* and

their potential role in cellular differentiation. Plant Signal Behav 10(4):e1010935

28. Beel B et al (2012) A flavin binding cryptochrome photoreceptor responds to both blue and red light in *Chlamydomonas reinhardtii*. Plant Cell 24(7):2992–3008

29. Heijde M et al (2010) Characterization of two members of the cryptochrome/photolyase family from *Ostreococcus tauri* provides insights into the origin and evolution of cryptochromes. Plant Cell Environ 33(10):1614–1626

30. Coesel S et al (2009) Diatom PtCPF1 is a new cryptochrome/photolyase family member with DNA repair and transcription regulation activity. EMBO Rep 10(6):655–661

31. Juhas M et al (2014) A novel cryptochrome in the diatom *Phaeodactylum tricornutum* influences the regulation of light-harvesting protein levels. FEBS J 281(9):2299–2311

32. Ishikawa M et al (2009) Distribution and phylogeny of the blue light receptors aureochromes in eukaryotes. Planta 230(3):543–552

33. Armbrust EV et al (2004) The genome of the diatom *Thalassiosira pseudonana*: ecology, evolution, and metabolism. Science 306(5693):79–86

34. Bowler C et al (2008) The *Phaeodactylum* genome reveals the evolutionary history of diatom genomes. Nature 456(7219):239–244

35. Cock JM et al (2010) The *Ectocarpus* genome and the independent evolution of multicellularity in brown algae. Nature 465(7298):617–621

36. Takahashi F et al (2007) AUREOCHROME, a photoreceptor required for photomorphogenesis in stramenopiles. Proc Natl Acad Sci U S A 104(49):19625–19630

37. Suetsugu N, Mittmann F, Wagner G, Hughes J, Wada M (2005) A chimeric photoreceptor gene, NEOCHROME, has arisen twice during plant evolution. Proc Natl Acad Sci U S A 102(38):13705–13709

38. Foster KW et al (1984) A rhodopsin is the functional photoreceptor for phototaxis in the unicellular eukaryote *Chlamydomonas*. Nature 311(5988):756–759

39. Foster KW, Smyth RD (1980) Light antennas in phototactic algae. Microbiol Rev 44(4):572–630

40. Ozawa S et al (2009) Biochemical and structural studies of the large Ycf4-photosystem I assembly complex of the green alga *Chlamydomonas reinhardtii*. Plant Cell 21(8):2424–2442

41. Berthold P et al (2008) Channelrhodopsin-1 initiates phototaxis and photophobic responses in *Chlamydomonas* by immediate light-induced depolarization. Plant Cell 20(6):1665–1677

42. Sakaguchi H, Iwasa K (1979) Two photophobic responses in *Volvox carteri*. Plant Cell Physiol 20(5):909–916

43. Schletz K (1976) Phototaxis in *Volvox*—pigments involved in perception of light direction. Z Pflanzenphysiol 77(3):189–211

44. Kianianmomeni A, Stehfest K, Nematollahi G, Hegemann P, Hallmann A (2009) Channelrhodopsins of *Volvox carteri* are photochromic proteins that are specifically expressed in somatic cells under control of light, temperature, and the sex inducer. Plant Physiol 151(1):347–366

45. Kochert G (1981) Sexual pheromones in *Volvox* development. In: O'Day DH, Horgen PA (eds) Sexual interactions in eukaryotic microbes. Academic, New York, pp 73–93

46. Feldwisch O et al (1995) Purification and characterization of a cAMP-binding protein of *Volvox carteri* f. *nagariensis* Iyengar. Eur J Biochem/FEBS 228(2):480–489

47. Kirk D (1998) *Volvox*: molecular-genetic origins of multicellularity and cellular differentiation. Cambridge University Press, Cambridge, UK

48. Goodenough UW (1989) Cyclic AMP enhances the sexual agglutinability of *Chlamydomonas* flagella. J Cell Biol 109(1):247–252

49. Pasquale SM, Goodenough UW (1987) Cyclic Amp functions as a primary sexual signal in gametes of *Chlamydomonas reinhardtii*. J Cell Biol 105(5):2279–2292

50. Goodenough UW, Gebhart B, Mermall V, Mitchell DR, Heuser JE (1987) High-pressure liquid chromatography fractionation of *Chlamydomonas* dynein extracts and characterization of inner-arm dynein subunits. J Mol Biol 194(3):481–494

51. Quarmby LM (1994) Signal transduction in the sexual life of *Chlamydomonas*. Plant Mol Biol 26(5):1271–1287

52. Quarmby LM, Hartzell HC (1994) Dissection of eukaryotic transmembrane signalling using *Chlamydomonas*. Trends Pharmacol Sci 15(9):343–349

53. Waffenschmidt S, Knittler M, Jaenicke L (1990) Characterization of a sperm lysin of *Volvox carteri*. Sex Plant Reprod 3(1):1–6

54. Nedelcu AM, Michod RE (2003) Sex as a response to oxidative stress: the effect of antioxidants on sexual induction in a facultatively sexual lineage. Proc R Soc B 270(Suppl 2):136–139

55. Starr RC, O'Neil RM, Miller CE (1980) L-Glutamic acid as a mediator of sexual morphogenesis in *Volvox capensis*. Proc Natl Acad Sci U S A 77(2):1025–1028

56. McDonough KA, Rodriguez A (2012) The myriad roles of cyclic AMP in microbial pathogens: from signal to sword. Nat Rev Microbiol 10(1):27–38

57. Shaulsky G, Huang E (2005) Components of the *Dictyostelium* gene expression regulatory machinery. In: Loomis WF, Kuspa A (eds) Dictyostelium genomics. Horizon Bioscience, Wymondham, UK, pp 1–22

58. Moulager M, Corellou F, Verge V, Escande ML, Bouget FY (2010) Integration of light signals by the retinoblastoma pathway in the control of S phase entry in the picophytoplanktonic cell *Ostreococcus*. PLoS Genet 6(5):e1000957

59. Boonyareth M, Saranak J, Pinthong D, Sanvarinda Y, Foster KW (2009) Roles of cyclic AMP in regulation of phototaxis in *Chlamydomonas reinhardtii*. Biologia 64(6):1058–1065

60. Govorunova EG, Spudich EN, Lane CE, Sineshchekov OA, Spudich JL (2011) New channelrhodopsin with a red-shifted spectrum and rapid kinetics from *Mesostigma viride*. mBio 2(3):e00115–e00111

61. Govorunova EG, Sineshchekov OA, Li H, Janz R, Spudich JL (2013) Characterization of a highly efficient blue-shifted channelrhodopsin from the marine alga *Platymonas subcordiformis*. J Biol Chem 288(41):29911–29922

62. Wickett NJ et al (2014) Phylotranscriptomic analysis of the origin and early diversification of land plants. Proc Natl Acad Sci U S A 111(45):E4859–E4868

63. Klapoetke NC et al (2014) Addendum: independent optical excitation of distinct neural populations. Nat Methods 11(9):972

64. Klapoetke NC et al (2014) Independent optical excitation of distinct neural populations. Nat Methods 11(3):338–346

65. Prigge M et al (2012) Color-tuned channelrhodopsins for multiwavelength optogenetics. J Biol Chem 287(38):31804–31812

66. Shimizu-Sato S, Huq E, Tepperman JM, Quail PH (2002) A light-switchable gene promoter system. Nat Biotechnol 20(10):1041–1044

67. Pham E, Mills E, Truong K (2011) A synthetic photoactivated protein to generate local or global Ca(2+) signals. Chem Biol 18(7):880–890

68. Strickland D, Moffat K, Sosnick TR (2008) Light-activated DNA binding in a designed allosteric protein. Proc Natl Acad Sci U S A 105(31):10709–10714

69. Konermann S et al (2013) Optical control of mammalian endogenous transcription and epigenetic states. Nature 500(7463):472–476

70. Hallmann A (2011) Evolution of reproductive development in the volvocine algae. Sex Plant Reprod 24(2):97–112

71. Matasci N et al (2014) Data access for the 1,000 plants (1KP) project. GigaScience 3:17

72. Kranz HD et al (1995) The origin of land plants: phylogenetic relationships among charophytes, bryophytes, and vascular plants inferred from complete small-subunit ribosomal RNA gene sequences. J Mol Evol 41(1):74–84

73. Baldauf SL (2003) The deep roots of eukaryotes. Science 300(5626):1703–1706

74. Prochnik SE et al (2010) Genomic analysis of organismal complexity in the multicellular green alga *Volvox carteri*. Science 329(5988):223–226

Chapter 4

Reversible Photoregulation of Gene Expression and Translation

Shinzi Ogasawara

Abstract

Several methods for controlling gene expression by light illumination have been reported. Most of these methods control transcription by regulating the interaction between DNA and transcription factors. The use of a photolabile protecting compound (cage compound) is another promising approach for controlling gene expression, although typically in an irreversible manner. We here describe a new approach for reversibly controlling translation using a photoresponsive 8-styryl cap (8ST-cap) that can be reversibly isomerized by illumination with light of a specific wavelength.

Key words mRNA, Translation, Cap structure, Photoisomerization

1 Introduction

Life processes require the precise control of gene expression, including when, where, and for how long genes are expressed [1, 2]. Light is the most promising external stimulus for the control of gene expression because it allows accurate and easy regulation of the location and time at which gene expression occurs. There have been several attempts to date to develop photoregulation methods for gene expression, most of which are based on controlling the interaction between a transcription factor and its specific promoter sequence on the genomic DNA [3–7]. However, these DNA-based methods have a long lag time between switching the light on or off and the start or stop of protein synthesis. This long lag time is due to mRNA-mediated translation: several hours are required following illumination with light before protein is synthesized, and the protein continues to be synthesized for more than 10 h after the light is turned off because of residual surviving mRNA in the cell. Therefore, the direct regulation of mRNA translation should allow more precise control of rapid cellular functions such as cell fate determination in the early

Arash Kianianmomeni (ed.), *Optogenetics: Methods and Protocols*, Methods in Molecular Biology, vol. 1408,
DOI 10.1007/978-1-4939-3512-3_4, © Springer Science+Business Media New York 2016

development stage. Furthermore, mRNA is not incorporated into the host genome and is completely degraded within a few days; these properties are advantageous for medical applications of mRNA-based translational control. One common strategy for the photoregulation of translation involves the installation of a photolabile protecting compound (cage compound) that can be removed by light illumination [8–11]. Okamoto et al. reported an mRNA caging system that uses 6-bromo-4-diazomethyl-7-hydroxycoumarine (Bhc-diazo) as the caging compound [12]. However, their method could not be used to stop protein expression because the uncaged mRNA liberated upon light illumination could not be recaged: once the Bhc group was removed by light illumination, the uncaged mRNA continued to be translated until it was degraded by nucleases. Here, we describe a method for the reversible photoregulation of translation using *cis–trans* photoisomerization of a photoresponsive-cap. We previously developed several photoresponsive nucleosides, such as "8-styryl-2′-deoxyguanosine," which undergo *cis–trans* photoisomerization upon an external light stimulus and thus reversibly change their photochemical and physical properties [13–15]. Our method focuses on the translation initiation mechanism and comprises the following steps: (1) 7-methylguanosine (cap) at the 5′-end of mRNA binds to eukaryotic initiation factor 4E (eIF4E); (2) several eukaryotic initiation factors act continuously and cooperatively; and (3) the ribosome assembles onto the mRNA complex and translation starts. Since translation does not begin without interaction between the cap and eIF4E, we replaced 7-methylguanosine with the photoresponsive 8-styryl cap (8ST-cap). This allows reversible control of translation by controlling the interaction of 8ST-cap with eIF4E via reversible *cis–trans* photoisomerization of 8ST-cap (*see* Fig. 1).

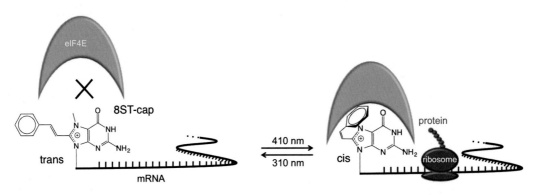

Fig. 1 Schematic illustration for reversible photoregulation of translation by 8-styryl cap (8ST-cap)

2 Materials

2.1 Components for 8ST-Cap Synthesis

1. Guanosine.
2. *N*-bromosuccinimide.
3. Anhydrous acetonerile.
4. *N*-methylpyrrolidone.
5. Tetrakis(triphenylphosphine)palladium.
6. Tributyl(vinyl)tin.
7. Anhydrous dioxane.
8. Palladium(II)acetate.
9. Triethylamine.
10. Bromobenzene.
11. Anhydrous dimethylformamide.
12. 2 M ammonia solution in methanol.
13. Trimethyl phosphate.
14. Phosphorus oxychloride.
15. Anhydrous dimethylsulfoxide.
16. Methyl iodide.
17. 1 M TEAB (pH 7.5): Insufflate CO_2 gas into 1 M triethyl-amine aqueous solution till pH reach 7.5. Store at room temperature.
18. Imidazolide GDP: Prepare imidazolide GDP as reported [16].

2.2 Components for mRNA Preparation

1. DNA polymerase: KOD-plus-DNA polymerase.
2. Venus plasmid: pCS2 Venus plasmid provided by Prof. Atsushi Miyawaki (RIKEN, Japan).
3. Primers: Forward for first step, 5′-GACTCACTATAGGGCGAA TTGGGTACCGGGCCCCCCCTCGAGGTCGA CGGTATCGATAAGCTTGATATGGTGAGCAA GGGCGAGGA-3′; reverse for first step, 5′-AGTTTAGTAG TTGGACTTAGGGAACAAAGGAACCTTTAATAGA AATTGGACAGCAAGAAAGCGAGCTTACTTGTAC AGCTCGTCCAT-3′; forward for second step, ACGTTGT AAAACGACGGCCAGTGAATTCGAGC TCGGTACCCGGGGATCCTCTAGAGATTTTAAT ACGACTCACTATAGGGCGAATT; reverse for second step, GCAATGAAAATAAATGTTTTTTATTAGGCAGAAT CCAGATGCTCAAGGCCCTTCATAA TATCCCCCAGTTTAGTAGTTGGACTTAGG.

2.3 Components for Cell Culture and mRNA Transfection

1. Cell culture medium: Dulbecco's Modified Eagle's Medium, no phenol red containing fetal bovine serum.

2. Transfection reagent: *Trans*IT-mRNA Transfection Kit (Mirus Bio LLC., USA).

3. Glass bottom dish.

3 Methods

3.1 Synthesis of 8-Styryl-Cap (8ST-Cap) (See Fig. 2)

3.1.1 8-Bromoguanosine

1. In a 1-L round-bottom flask equipped with a stir bar, suspend 10.0 g (35.3 mmol) guanosine in 500 mL water.

2. Add 7.54 g (42.4 mmol) N-bromosuccinimide and stir for 2 h at room temperature (*see* **Note 1**).

3. Filter the precipitate and wash the filtered substance with acetone 3× to give 9.07 g (71 %) 8-bromoguanosine (**1**) as a white powder.

3.1.2 2′,3′,5′-Tri-O-Acetyl-8-Bromoguanosine

1. Place 9.03 g (25.0 mmol) 8-bromoguanosine (**1**) in a 500-mL two-neck round-bottom flask equipped with a stir bar and purge flask with nitrogen 3×.

2. Add 100 mL anhydrous acetonerile, 76 mg (0.63 mmol) N-dimethylaminopyridine, 13.9 mL (100 mmol) triethylamine, 9.46 mL (100 mmol) acetic anhydride, and stir for 1.5 h at room temperature under nitrogen.

3. Add 15 mL methanol to reaction mixture to quench the reaction.

4. Evaporate the reaction mixture using rotary evaporator (*see* **Note 2**).

5. Add 100 mL water and filter the resulting precipitate.

6. Wash the filtered substance with water to give 10.3 g (85 %) 2′,3′,5′-tri-O-acetyl-8-bromoguanosine (**2**) as a pinkish white powder.

3.1.3 2′,3′,5′-Tri-O-Acetyl-8-Vinylguanosine

1. In a 100-mL two-neck round-bottom flask equipped with a stir bar, dissolve 4.9 g (10.1 mmol) 2′,3′,5′-tri-O-acetyl-8-bromoguanosine (**2**) in 15 mL N-methylpyrrolidone.

2. Add 1.17 g (1.01 mmol) tetrakis(triphenylphosphine)palladium and purge flask with argon 3×.

3. Add 5.85 mL (20.2 mmol) tributyl(vinyl)tin and stir at 110 °C under argon for 45 min.

4. Cool the reaction mixture to room temperature and add ethyl acetate and water.

5. Filter the reaction mixture and wash the filtrate with 50 mL water, 50 mL saturated sodium bicarbonate aqueous solution and 50 mL brine.

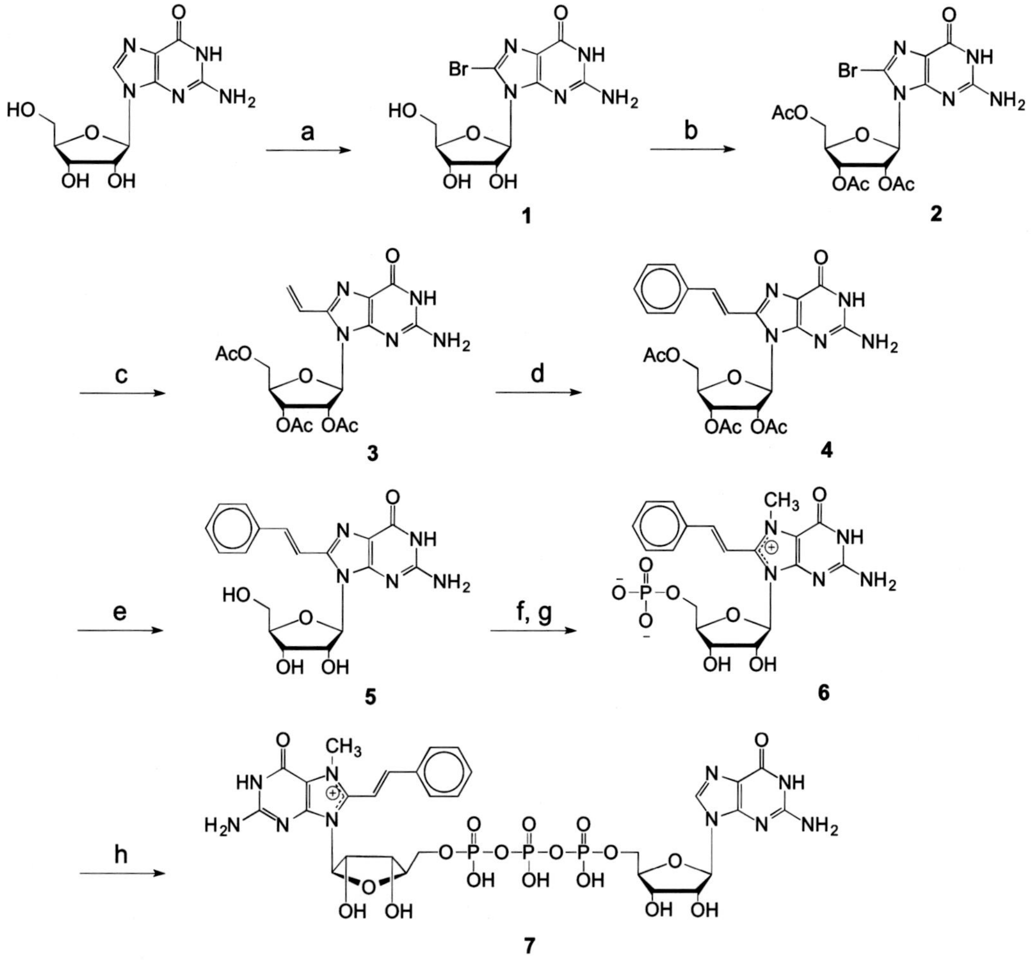

Fig. 2 Synthesis of 8ST-cap. Reagents and conditions: (**a**) N-bromosuccinimide, water, 2 h. (**b**) N-dimethylaminopyridine, triethylamine, acetic anhydride, acetonerile, 1.5 h. (**c**) Tetrakis(triphenylphosphine) palladium, tributyl(vinyl)tin, N-methylpyrrolidone, 110 °C, 45 min. (**d**) Triphenylphosphine, palladium(II)acetate, triethylamine, bromobenzene, dioxane, dimethylformamide, 115 °C, 1 h. (**e**) Ammonia, methanol, 60 °C, 4 h. (**f**) Phosphorus oxychloride, trimethyl phosphate, 2 °C, 20 h. (**g**) Methyl iodide, dimethylsulfoxide, 24 h. (**h**) GDP (imidazolide), zinc chloride, dimethylformamide, 60 h

6. Collect and dry the organic layer over anhydrous sodium sulfate.

7. Evaporate the organic layer using rotary evaporator.

8. Purify the crude product by silica gel column chromatography.

9. Elute the compound with 5–10 % methanol/dichloromethane gradient.

10. Combine and evaporate the appropriate fractions to give 3.43 g (78 %) 2′,3′,5′-tri-O-acetyl-8-vinylguanosine (**3**) as a yellow solid.

3.1.4   2′,3′,5′-Tri-O-
Acetyl-8-Styrylguanosine

1. Place 0.302 g (1.15 mmol) triphenylphosphine in a 100-mL two-neck round-bottom flask equipped with a stir bar and purge flask with nitrogen 3×.

2. Add 15 mL anhydrous dioxane, 0.104 g (0.46 mmol) palladium(II)acetate, 0.956 mL (6.89 mmol) triethylamine and stir at 60 °C for 10 min under nitrogen (*see* **Note 3**).

3. Add 0.721 mL (6.89 mmol) bromobenzene, 2.0 g (4.60 mmol) 2′,3′,5′-tri-*O*-acetyl-8-vinylguanosine (**3**) dissolved in 8 mL anhydrous dimethylformamide and stir at 115 °C for 1 h under nitrogen.

4. Filter the reaction mixture to room temperature and purify the filtrate by silica gel column chromatography.

5. Elute the compound with 3–12 % methanol/chloroform gradient.

6. Combine and evaporate the appropriate fractions to give 858 mg (36 %) 2′,3′,5′-tri-*O*-acetyl-8-styrylguanosine (**4**) as a yellow solid.

3.1.5   8-Styrylguanosine

1. Place 2.70 g (5.28 mmol) 3′,5′-tri-*O*-acetyl-8-styrylguanosine (**4**) in a 300-mL two-neck round-bottom flask equipped with a stir bar and purge flask with nitrogen 3×.

2. Add 45 mL methanol, 45 mL 2 M ammonia solution in methanol and stir at 60 °C for 4 h under nitrogen.

3. Filter the precipitate and wash the filtered substance with methanol 3× to give 1.76 g (87 %) 8-styrylguanosine (**5**) as a white powder (*see* **Note 4**).

3.1.6   7-Methyl-8-
Styrylguanosine
Monophosphate

1. Place 1.76 g (4.54 mmol) 8-styrylguanosine (**5**) in a 100-mL two-neck round-bottom flask equipped with a stir bar and purge flask with nitrogen 3×.

2. Add 25 mL trimethyl phosphate and cool to 4 °C for 4 h.

3. Add 0.832 mL (9.14 mmol) phosphorus oxychloride dropwise over 3 h and stir at 2 °C for 20 h under nitrogen (*see* **Note 5**).

4. Add 20 mL of water and neutralize with saturated sodium bicarbonate aqueous solution.

5. Filter the precipitate and dry the filtered substance in a vacuum desiccators over P_4O_{10} to give 2.24 g monophosphorylated 8-styrylguanosine as a yellow powder.

6. Place 2.24 g (4.82 mmol) crude 8-styrylguanosine monophosphate in a 300-mL two-neck round-bottom flask equipped with a stir bar and purge flask with nitrogen 3×.

7. Add 100 mL anhydrous dimethylsulfoxide, 6.0 mL (96.3 mmol) methyl iodide, and stir for 24 h at room temperature under nitrogen.

8. Pour the reaction mixture into 350 mL cold water slowly.

9. Filter the precipitate and dry the filtered substance in a vacuum desiccators over P_4O_{10} to give 594 mg (27 %) 7-methyl-8-styrylguanosine monophosphate (**6**) as a yellow powder.

3.1.7   8-Styryl Cap

1. Place 30 mg (0.044 mmol) 7-methyl-8-styrylguanosine monophosphate (**6**), 45 mg (0.088) imidazolide GDP in a 50-mL two-neck round-bottom flask equipped with a stir bar, and purge flask with nitrogen 3×.

2. Add 2.0 mL anhydrous dimethylformamide, 36 mg (0.264 mmol) zinc chloride, and stir at room temperature for 60 h under nitrogen.

3. Pour reaction mixture into a flask containing a solution of 172 mg of EDTA in 15 mL cold water and neutralize with 1 M TEAB.

4. Purify the crude product by DEAE Sephadex column.

5. Elute the compound with linear gradient of 0–1 M TEAB.

6. Combine and evaporate the appropriate fractions to give 8-styryl cap TEA salt as a yellow foam.

7. Dissolve the product in 1 mL water and purify by HPLC using the following conditions (*see* **Note 6**):

Column	Inertsil ODS-SP 5 μm (20 × 150 mm)
Buffer A	0.1 M TEAB
Buffer B	Acetonerile
Gradient	95 % buffer A over 5 min, 95–75 % buffer A over 25 min, 0 % buffer A over 5 min, return to 95 % buffer A
Flow rate	10 mL/min
Oven temperature	40 °C

8. Dissolved product in 10 mL water and pass through a Strata-X-AW column.

9. Wash the column with 10 mL water followed by 10 mL methanol and elute the compound with 15 mL of $NH_4OH/MeOH/H_2O$ (2/25/73).

10. Combine and lyophilize the appropriate fractions to give 14 mg (35 %) 8-styryl cap (**7**) as a yellow powder (*see* **Note 7**).

^1H NMR (D_2O, 400 MHz) δ: 7.90 (s, 1*H*), 7.52 (d, *J* = 5.0, 2*H*), 7.34 (s, 3*H*), 7.26 (d, *J* = 16.6, 1*H*), 6.91 (d, *J* = 16.6, 1*H*), 5.88 (d, *J* = 5.8, 1*H*), 5.64 (d, *J* = 5.8, 1*H*), 5.26 (t, *J* = 5.6, 1*H*), 4.61 (t, *J* = 4.1, 1*H*), 4.56 (t, *J* = 5.6, 1*H*), 4.39–4.29 (m, 3*H*), 4.25–4.10 (m, 4*H*), 3.99 (s, 3*H*); ^{31}P NMR

(D_2O, 400 MHz) δ: −11.5 (2P, α, γ), −23.1 (1P, β); FAB MS (M–H)⁻ for $C_{29}H_{35}N_{10}O_{18}P_3$; calculated: 903.13; found: 903.69.

11. Dissolve the 8ST-cap in nuclease-free water as the concentration is 75 mM.

12. Store the 8ST-cap solution at −20 °C.

3.2 Preparation of mRNA Containing 8ST-Cap at 5′-End

1. Prepare the DNA template containing β-globin 3′-UTR sequence by two-step PCR reaction from Venus plasmid. Mix the reaction solution (300 µL total volume) containing 1× reaction buffer, 300 ng plasmid, 200 µM dNTPs, 1 mM $MgSO_4$, 6 U of DNA polymerase and 0.3 µM primers.

2. Amplify the DNA template by following conditions: 96 °C for 2 min (15 s in subsequent cycles), cool to 58 °C for 30 s to anneal the primers, and heat to 68 °C for 60 s for primer extension. Repeat this cycles 35×.

3. Purify the PCR product by Wizard SV Gel and PCR Clean-Up System (Promega).

4. Perform second PCR in a similar fashion.

5. Prepare the mRNA containing 8ST-cap at 5′-end by the MEGAscript™ kit (Ambion). Mix the reaction solution (20 µL total volume) containing 1× reaction buffer, 0.5 µg template DNA, 6 mM each of ATP, CTP, and UTP, 1.2 mM GTP, 4.8 mM 8ST-cap, and 50 U/µL T7 RNA polymerase (*see* **Note 8**).

6. Incubate the reaction mixture at 37 °C for 4 h.

7. Add 1 µL turbo DNase for hydrolysis of the remaining template DNA.

8. Incubate the reaction mixture at 37 °C for 15 min.

9. Perform the poly A tailing at 3′-end by the Poly(A) Tailing Kit (Ambion). Incubate the reaction solution (100 µL total volume) containing 1× reaction buffer, 1 mM ATP, 2.5 mM $MnCl_2$, and 4 µL (2 U/µL) *E*-PAP at 37 °C for 1 h.

10. Purify the transcribed mRNA from the reaction mixture by the MEGAclear™ kit (Ambion).

11. Wash the mRNA pellet twice with 500 µL 70 % ethanol at 4 °C.

12. Dry the mRNA pellet in air and dissolve in nuclease-free water as the concentration of mRNA is 1.0 g/L (*see* **Note 9**).

13. Confirm the incorporation of 8ST-cap into the 5′-end of mRNA by UV/Vis absorptiometer (*see* Fig. 3 and **Note 10**).

14. Store the mRNA solution at −80 °C.

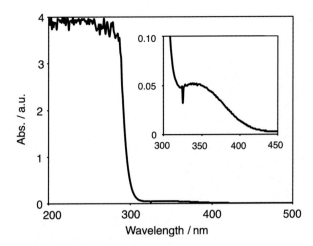

Fig. 3 Absorption spectrum of the mRNA containing 8ST-cap at the 5′-end. The absorption peak at 340 nm comes from 8ST-cap

3.3 Transfection of 8ST-Capped mRNA into Cells

1. Maintain HeLa cells in Dulbecco's modified Eagle's medium with 10 % fetal bovine serum in a 5 % CO_2 at 37 °C.

2. Twenty-four hour before transfection, passage the cells as cell density reach 80 % confluent at the time of transfection.

3. Transfect the 8ST-capped mRNA into the cells with *Trans*IT-mRNA Transfection Kit (Mirus Bio). Two hour before transfection, wash the cells twice with culture medium and refill 3 mL fresh culture medium.

4. Mix 250 μL serum-free medium, 2.5 μg 8ST-capped mRNA, 5 μL transfection reagents, and incubate for 5 min at room temperature.

5. Add transfection mixture dropwise to the culture dish.

6. Incubate the cells for 30 min in a 5 % CO_2 at 37 °C.

7. Replace culture medium with 3 mL of fresh culture medium 3× to wash out reagents and extra mRNA (*see* **Note 11**).

3.4 Photoregulation of the Translation in Living Cells

1. Transfer the culture dish carrying transfected cells to an incubation chamber on the stage of a confocal laser scanning microscope equipped with a 120-W metal halide lamp.

2. Maintain the cells in a 5 % CO_2 at 37 °C.

3. For activation of translation, illuminate the cells with the metal halide lamp (100 % intensity) for 1 min through the 410 nm bandpass filter and the 63× oil-immersion objective lens (*see* Fig. 3 and **Note 12**).

4. For inactivation of activated mRNA, illuminate with 310 nm light for 30 s using a 300-W Xenon lamp equipped with a bandpass filter. Adjust the Xenon lamp to 40 % maximum intensity and place fiber output at 5 cm from the cells (*see* Fig. 4 and **Note 13**).

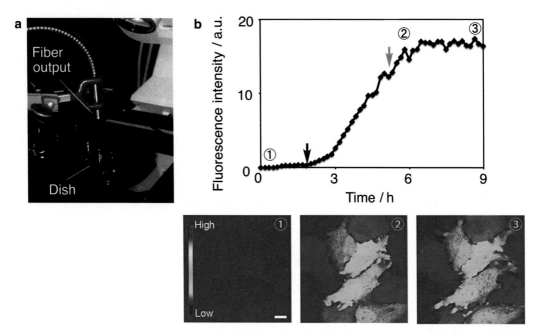

Fig. 4 Photoregulation of fluorescence protein (Venus) expression in living HeLa cells. (**a**) Picture representing 310 nm light illumination system. 310 nm light is illuminated to cells from 5 cm distance. (**b**) Time course and confocal fluorescence images of Venus protein expression. *Black* and *gray arrows* indicate the time of illumination with 410 nm and 310 nm light, respectively. Scale bar, 10 μm

4   Notes

1. The suspension of guanosine in water gradually dissolves after the addition of *N*-bromosuccinimide. 8-Bromoguanosine precipitate as the reaction make progress.

2. Don't dry the product completely.

3. Color of the reaction mixture gradually changes into vine red indicating that palladium(II)acetate convert to bis(triphenylphosphine)palladium(0). This color change is important because only palladium(0) act as catalyst.

4. Wash the 8-styrylguanosine with small amount of methanol. Otherwise, yield of 8-styrylguanosine decreases because 8-styrylguanosine can slightly dissolve in methanol.

5. Add phosphorus oxychloride to the reaction mixture dropwise. Addition of phosphorus oxychloride at once yields guanosine bisphosphate, 3′,5′-bisphosphate or 2′,5′-bisphosphate.

6. Two main peaks of 8ST-cap are observed in some cases. One is the *trans*-form and the other is the *cis*-form. We can easily distinguish those peaks by absorption spectra obtained from diode-array detector. Only the *trans*-form has maximum absorption at 340 nm (*see* Fig. 5). Collect the *trans*-form. For

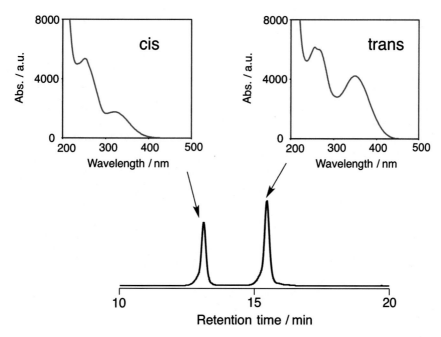

Fig. 5 Chromatogram and absorption spectrum of 8ST-cap obtained by HPLC equipped with diode-array detector. The peak at 13 min is the *cis*-form. The other is the *trans*-form

the *cis*-form, collect it after conversion to the *trans*-form by illumination with 310 nm light.

7. Dissolve the 8ST-cap into nuclease-free water and store at −20 °C.

8. When the 8ST-cap precipitate, warm the 8ST-cap solution at 37 °C for 10 min before addition. Add the 8ST-cap to the reaction solution at the end of mixing to avoid agglomeration.

9. Don't dry mRNA pellets completely because overdried pellets cannot be dissolved in water.

10. We can observe the absorption around at 340 nm from 8ST-cap. Measure the absorbance spectrum of mRNA solution in high concentration (more than 500 ng/μL) because absorption at 340 nm of 8ST-cap is very weak.

11. According to manufacturer's protocol, it is not necessary to replace the transfection medium with fresh medium. However, long time treatment with transfection mixture causes serious cell death.

12. For another way of photoactivation, illuminate the cells with 405 nm light from a semiconductor laser diode (LP405-SF30, Thorlabos) mounted on the side port of the microscope. Adjust laser power to 0.04 mW by controlling the current and using an ND filer.

13. In my system, illumination with 310 nm light for a long time (over 30 s), at a high power (over 40 %) or from a short distance (less than 5 cm form the cells) caused cell death. Find appropriate conditions for photoinactivation in your system.

Acknowledgments

The present work was supported by Precursory Research for Embryonic Science and Technology (PRESTO), Japan Science and Technology Agency (JST) and JSPS KAKENHI Grant Number 25104526.

References

1. Hirata H, Bessho Y, Kokubu H, Masamizu Y, Yamada S, Lewis J, Kageyama R (2004) Instability of Hes7 protein is crucial for the somite segmentation clock. Nat Genet 36:750–754

2. Driever W, Volhard CN (1988) A gradient of bicoid protein in Drosophila embryos. Cell 54:83–93

3. Polstein LR, Gersbach CA (2012) Light-inducible spatiotemporal control of gene activation by customizable zinc finger transcription factors. J Am Chem Soc 134:16480–16483

4. Wang X, Chen X, Yang Y (2012) Spatiotemporal control of gene expression by a light-switchable transgene system. Nat Methods 9:266–269

5. Kennedy MJ, Hughes RM, Peteya LA, Schwartz JW, Ehlers MD, Tucker CL (2010) Rapid blue-light-mediated induction of protein interactions in living cells. Nat Methods 7:973–975

6. Yazawa M, Sadaghiani AM, Hsueh B, Dolmetsch RE (2009) Induction of protein-protein interactions in live cells using light. Nat Biotechnol 27:941–945

7. Levskaya A, Chevalier AA, Tabor JJ, Simpson ZB, Lavery LA, Levy M, Davidson EA, Scouras A, Ellington AD, Marcotte EM, Voigt CA (2005) Synthetic biology: engineering Escherichia coli to see light. Nature 438:441–442

8. Hemphill J, Liu Q, Uprety R, Samanta S, Tsang M, Juliano RL, Deiters A (2015) Conditional control of alternative splicing through light-triggered splice-switching oligonucleotides. J Am Chem Soc 137:3656–3662

9. Yamazoe S, Liu Q, McQuande LE, Deiters A, Chen JK (2014) Sequential gene silencing using wavelength-selective caged morpholino oligonucleotides. Angew Chem Int Ed 53:10114–10118

10. Wu L, Wang Y, Wu J, Lv C, Wang J, Tang X (2013) Caged circular antisense oligonucleotides for photomodulation of RNA digestion and gene expression in cells. Nucleic Acids Res 41:677–686

11. Shestopalov IA, Sinha S, Chen J (2007) Light-controlled gene silencing in zebrafish embryos. Nat Chem Biol 3:650–651

12. Ando H, Furuta T, Tsien RY, Okamoto H (2001) Photo-mediated gene activation using caged RNA/DNA in zebrafish embryos. Nat Genet 28:317–325

13. Ogasawara S, Maeda M (2009) Reversible photoswitching of a G-quadruplex. Angew Chem Int Ed 48:6671–6674

14. Ogasawara S, Maeda M (2008) Straightforward and reversible photoregulation of hybridization using photochromic nucleoside. Angew Chem Int Ed 47:8839–8842

15. Ogasawara S, Saito I, Maeda M (2008) Synthesis and reversible photoisomerization of photoswitchable nucleoside, 8-styryl-2′-deoxyguanosine. Tetrahedron Lett 49:2479–2482

16. Jemielity J, Fowler T, Zuberek J, Stepinski J, Lewdorowicz M, Niedzwiecka A, Stolarski R, Darzynkiewicz E, Rhoads RE (2003) Novel "anti-reverse" cap analogs with superior translation properties. RNA 9:1108–1122

Chapter 5

Controlling Protein Activity and Degradation Using Blue Light

Anne P. Lutz, Christian Renicke, and Christof Taxis

Abstract

Regulation of protein stability is a fundamental process in eukaryotic cells and pivotal to, e.g., cell cycle progression, faithful chromosome segregation, or protein quality control. Synthetic regulation of protein stability requires conditional degradation sequences (degrons) that induce a stability switch upon a specific signal. Fusion to a selected target protein permits to influence virtually every process in a cell. Light as signal is advantageous due to its precise applicability in time, space, quality, and quantity. Light control of protein stability was achieved by fusing the LOV2 photoreceptor domain of *Arabidopsis thaliana* phototropin1 with a synthetic degron (cODC1) derived from the carboxy-terminal degron of ornithine decarboxylase to obtain the photosensitive degron (psd) module. The psd module can be attached to the carboxy terminus of target proteins that are localized to the cytosol or nucleus to obtain light control over their stability. Blue light induces structural changes in the LOV2 domain, which in turn lead to activation of the degron and thus proteasomal degradation of the whole fusion protein. Variants of the psd module with diverse characteristics are useful to fine-tune the stability of a selected target at permissive (darkness) and restrictive conditions (blue light).

Key words Optogenetics, Protein degradation, Proteasome, Ubiquitin-independent degradation, Protein stability, Synthetic biology, LOV2 domain, Blue light, Degron

---

1 Introduction

Regulated proteolysis by the proteasome is fundamental in eukaryotic cells among other things for protein quality control, cell cycle progression, signal transduction, and transcriptional regulation. In summary, the ubiquitin–proteasome system is one of the key players involved in virtually every regulatory step [1]. Thus, selective proteolysis of a specific target protein is a versatile tool to achieve synthetic regulation of cell functions [2]. Several methods have been developed to switch a protein from a stable to an unstable form relying on external signals like heat, nutrients, or other chemical compounds [3]. Common to all of the methods is the activation of a degradation-inducing sequence (degron). In most cases, the degron is recognized by an ubiquitin–protein ligase, resulting

Arash Kianianmomeni (ed.), *Optogenetics: Methods and Protocols*, Methods in Molecular Biology, vol. 1408, DOI 10.1007/978-1-4939-3512-3_5, © Springer Science+Business Media New York 2016

in polyubiquitylation of the substrate protein. The latter process requires an ubiquitin-conjugating enzyme loaded with activated ubiquitin. Ultimately, the polyubiquitylated substrate is recognized and degraded by the proteasome [4]. This mechanism is contrasted by ubiquitin-independent degradation by the proteasome; one example for this class of substrates is ornithine decarboxylase (ODC). In ODC, the carboxy-terminal 37 amino acids (cODC) function as a degron that is directly recognized by the proteasome [5]. A cODC-derived sequence called cODC1 was used to engineer a synthetic conditional degron to regulate protein abundance in vivo [6].

Light has been recognized as an almost ideal signal to regulate cellular functions in microorganisms and cell culture due to its unmatched preciseness, in which quality and quantity can be controlled spatially and temporally. In recent years, many different constructs have been developed to control protein synthesis, localization, activity, or stability [7, 8]. Regulation of protein stability by light has been achieved with the photosensitive degron (psd) module and derivatives thereof [9–11]. The psd module contains two domains, a photoreceptor of the LOV (light oxygen voltage) family and the cODC1 degron (Fig. 1a). The light-receptor domain LOV2 binds flavin mononucleotide (FMN) as cofactor, which is the primary light-activated compound. Excitation of the FMN cofactor by blue light leads to formation of a bond between FMN and a cysteine of the LOV2 domain. This induces a structural rearrangement of the LOV2 domain that leads to the unfolding of a carboxy-terminal-located α-helix, which makes the cODC1 degron accessible for recognition by the proteasome and ubiquitin-independent degradation of the whole construct [9]. By fusion of the psd module to the carboxy-terminus of key regulators, cellular functions can be controlled by blue light irradiation (Fig. 1b, c). In yeast, growth, secretion, cell cycle events, and enzymatic activity have been regulated with this technique [9]. The method has been further improved by the generation of psd module variants with increased degradation rates [11]. For light control of protein stability in higher eukaryotes, a similar construct has been generated [10]. Here, we describe in detail the steps necessary to obtain a protein fused to the photosensitive degron and how to adapt experimental conditions for successful application of this conditional degron in yeast.

2   Materials

2.1   Yeast Transformation

1. SORB buffer: 100 mM lithium acetate, 10 mM Tris–HCl pH 8.0, 1 mM EDTA/NaOH pH 8.0, 1 M sorbitol, adjusted with acetic acid to pH 8.0 and filter sterilized.

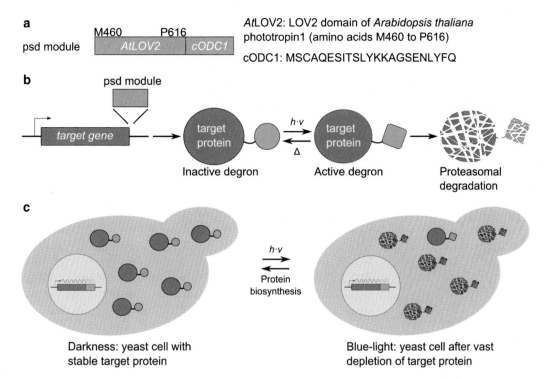

a

psd module

M460   P616

| AtLOV2 | cODC1 |

*At*LOV2: LOV2 domain of *Arabidopsis thaliana* phototropin1 (amino acids M460 to P616)

cODC1: MSCAQESITSLYKKAGSENLYFQ

b

psd module

target gene → target protein (Inactive degron) $\overset{h \cdot v}{\underset{\Delta}{\rightleftharpoons}}$ target protein (Active degron) → Proteasomal degradation

c

$\overset{h \cdot v}{\underset{\text{Protein biosynthesis}}{\rightleftharpoons}}$

Darkness: yeast cell with stable target protein

Blue-light: yeast cell after vast depletion of target protein

Fig. 1 Control of protein stability by light. (**a**) The photosensitive degron consists of the LOV2 domain (amino acids M460 to P616) of *Arabidopsis thaliana* phototropin1 fused to 23 amino acids of a synthetic degron derived from the carboxy terminus of murine ornithine decarboxylase (cODC1). The psd module variants (not depicted) contain point mutations in the LOV2 domain as indicated in Table 1. (**b**) To generate a light-sensitive mutant, the photosensitive degron (psd) module is attached to the 3′-end of the target gene. This leads to the generation of a target protein modified at the carboxy terminus with an inactive degron. Blue light (465 nm, 30 µmol/m²/s) will excite the LOV2 domain, which leads to exposure of the ornithine decarboxylase-like degron cODC1 and proteasomal degradation of the whole fusion protein. (**c**) Light control of protein abundance with the psd module. In darkness, the target protein is stable and functional. Its abundance depends on its biosynthesis rate and its turnover. Blue light exposure of the cells leads to an increase in target protein degradation due to activation of the photosensitive degron. After establishment of equilibrium, target protein abundance will be much lower compared to the situation in darkness, due to the increased target protein turnover. Under ideal circumstances, the target protein will be depleted almost completely. Again, residual abundance of the target protein depends on its synthesis rate versus its degradation rate

2. Carrier DNA: 10 mg/ml herring sperm DNA, denatured at 100 °C for 10 min, and immediately cooled on ice.

3. Polyethylene glycol (PEG) buffer: 100 mM lithium acetate, 10 mM Tris–HCl pH 8.0, 1 mM EDTA/NaOH pH 8.0, 40 % PEG 4000; filter sterilized.

2.2 Cell Lysis and Immunoblotting

1. Alkaline lysis buffer: 1.85 M NaOH, 7.5 % β-mercaptoethanol.

2. High urea buffer: 5 % sodium dodecylsulfate (SDS), 8 M urea, 200 mM Na₂HPO₄/NaH₂PO₄ pH 6.8, 0.1 mM EDTA, 0.01 % bromophenol blue.

3. Antibodies against tagRFP can be obtained from evrogen (www.evrogen.com), antibodies against the myc epitope are available from many suppliers.

2.3 Preparation of Yeast Chromosomal DNA

1. Breaking buffer: 2 % (v/v) Triton X-100, 1 % (w/v) SDS, 100 mM NaCl, 10 mM Tris–HCl pH 8.0, 1 mM EDTA/NaOH pH 8.0.

2.4 Yeast Media

1. Yeast complex complete medium (YPD): 1 % yeast extract, 2 % peptone, 2 % glucose, and 200 mg/l geneticin G418 in case selection for *kanMX* is necessary.

2. Low-fluorescence medium [11]: 5 g/l $(NH_4)_2SO_4$, 1 g/l KH_2PO_4, 0.5 g/l $MgSO_4$, 0.1 g/l NaCl, 0.1 g/l Ca_2Cl, 0.5 mg/l H_3BO_4, 0.04 mg/l $CuSO_4$, 0.1 mg/l KI, 0.2 mg/l $FeCl_3$, 0.4 mg/l $MnSO_4$, 0.2 mg/l Na_2MoO_4, 0.4 mg/l $ZnSO_4$, 2 mg/l biotin, 0.4 mg/l calcium pantothenate, 2 mg/l inositol, 0.4 mg/l niacin, 0.2 mg/l 4-aminobenzoic acid (PABA), 0.4 mg/l pyridoxine HCl, 0.4 mg/l thiamine, 2 % glucose, supplemented with amino acids, and/or other nutrients as required by the yeast strain.

2.5 Yeast Strains

No specific yeast strain is required, the psd module can be introduced in a wild-type strain (e.g., ESM356-1 [12], S288C background), or any existing mutant strain.

2.6 Cassette Plasmids, Oligos, and Polymerase

1. An overview of the psd module plasmids to generate carboxy-terminal fusions to a protein of interest can be found in Table 1.

2. A pair of primers (S2 and S3 primers) that contain sequences of homology to the target gene within their 5′ regions, which are used to amplify the psd module cassette of choice; target gene-specific and psd module-specific primers (C1–C4) to check for correct chromosomal integration.

3. A polymerase chain reaction (PCR) kit with a high-fidelity polymerase (e.g., Phusion, KOD, or Herculase) to generate a psd module cassette for chromosomal integration.

2.7 Illumination of Yeast with Light

1. Light emitting diode (LED) stripes or clusters (output wavelength 465 nm, e.g., 3 stripes of 6 high power LEDs or 6 clusters of 42 StrawHat LEDs) for illumination of yeast cells grown on plate or in liquid medium. The LED setup should include a dimmer to achieve a photon flux of 30 μmol/m²/s at the level of the yeast cells.

2. An optometer (e.g., P2000, equipped with light detector D-9306-2, Gigahertz-Optik, Türkenfeld, Germany) to measure the photon flux.

Table 1
Plasmids for carboxy-terminal tagging of target proteins and characteristics of psd module variants

Name	Construct	Half-life in darkness (min)[a]	Half-life in 30 μmol/m^2/s blue light (min)[a]	Reference
pCT337	*tagRFP–AtLOV2–cODC1::kanMX*	123 ± 21	20 ± 1	[9]
pDS96	*3myc–AtLOV2–cODC1::kanMX*	123 ± 21	20 ± 1	[9]
pDS135	*3myc–AtLOV2$^{K121M\ N128Y}$– cODC1::kanMX*	44 ± 8	8.5 ± 0.3	This work
pSU22	*3myc–AtLOV2$^{K92R\ E132A\ N148D\ E155G}$– cODC1::kanMX*	103 ± 26	10.5 ± 0.3	This work
pSU24	*3myc–AtLOV2$^{K92R\ E132A\ E139N\ N148D\ E155G}$– cODC1::kanMX*	66 ± 10	9.8 ± 0.4	This work
pDS121	*3myc–AtLOV2^{E139N}–cODC1::kanMX*	89 ± 22	11 ± 0.5	This work
pDS170	*3myc–AtLOV2$^{K92R\ E132A\ E155G}$– cODC1::kanMX*	102 ± 41	12 ± 0.4	This work
pDS157	*3myc–AtLOV2$^{K121M\ N128Y\ G138A}$– cODC1::kanMX*	92 ± 28	13 ± 1	This work
pDS1	*tagRFP::kanMX*	–	–	This work
pYM4	*3myc::kanMX*	–	–	[21]

[a]Measured using following construct: *tagRFP–LOV2–cODC1* and derivatives thereof under control of the constitutive *ADH1* promoter [11]

3. Clear plastic cell culture flasks for suspension cultures equipped with a ventilated cap to grow yeast cells in liquid medium during illumination with light.

4. Light-tight boxes to protect yeast cells from ambient light during growth.

5. A box with LEDs attached to the lid, spaced evenly to ensure uniform illumination. The bottom and the walls of the box may be covered with reflective material to optimize light usage. The box should be large enough to host several Petri dishes or cell culture flasks.

6. Heat shrinkable tubing (light-tight) adjusted to a diameter which will hold a test tube inside to shield light-sensitive yeast cells from ambient light during growth. To create such tubing, test tubes were covered by aluminum foil, put inside the heat shrinkable tubing, and exposed to heat to adjust the diameter of the tubing. After cooling, the aluminum foil was removed with forceps.

3 Methods

3.1 Generation of Conditional Mutants with the psd Module

The strategy to obtain conditional mutants with the psd module is outlined in Fig. 2; below are detailed protocols for each step.

1. Chromosomal modification of the target gene with the psd module is achieved by transformation of specific PCR products into yeast cells (protocol see below). The PCR products with homologous sequences are obtained with a psd module plasmid (*see* Table 1 for the properties of the psd module variants) and target gene-specific S2/S3 primers (Fig. 2). After transformation of the PCR products into yeast, the gene-specific sequences direct the integration of the psd module at the stop codon of the target gene (Fig. 2).

2. Modification of the target gene with a psd module can be verified by analytical PCR using gene-specific oligos (C1, C4) in combination with cassette-specific oligos (C2, C3) and chromosomal DNA as template (Fig. 2). Also, immunoblotting can be used to control the modification of the target protein with the psd module. The cells should be kept in darkness prior to generation of cell extracts.

3.2 Design of Target Gene-Specific Primers

Target gene-specific primers are designed according to the directions given for S2 and S3 primers [13]. Sequences homologous to the target gene locus are added to the 5′ ends of sequences specific for psd module plasmids as follows:

1. S2 primer, the reverse complement of 45–55 bases downstream of the stop codon (including stop) of the target gene, followed by 5′-ATCGATGAATTCGAGCTCG-3′.

2. S3 primer, 45–55 bases upstream of the stop codon (excluding stop) of the target gene, followed by 5′-CGTACGCTGC AGGTCGAC-3′ (Fig. 2).

3.3 PCR

Standard PCR conditions can be used to generate the PCR products used for yeast transformation [13]. A kit containing a high-fidelity polymerase is advisable to reduce the introduction of errors during the reaction. Before transformation, PCR products are concentrated tenfold by ethanol precipitation or usage of a PCR cleanup kit.

3.4 Yeast Transformation

The protocol for yeast transformation is based on the lithium acetate method [14] and has been described previously [13, 15, 16].

1. Yeast cells are inoculated from an overnight preculture (approx. 1:50 dilution) and grown to an optical density (A_{600}) of 0.8–1.0 at 30 °C in 50 ml of YPD medium.

Workflow for the Generation of Conditional Mutants with the psd Module

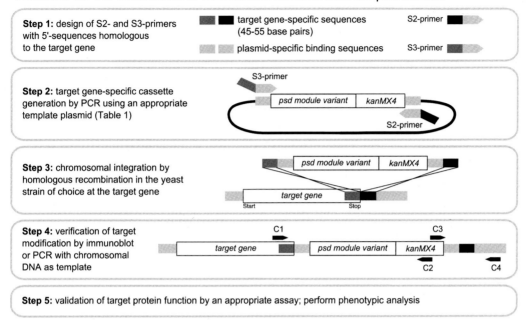

Step 1: design of S2- and S3-primers with 5'-sequences homologous to the target gene

■■ ■ target gene-specific sequences (45–55 base pairs)

S2-primer ■■ ▶

░░ ░░ plasmid-specific binding sequences

S3-primer ■▶

S3-primer

Step 2: target gene-specific cassette generation by PCR using an appropriate template plasmid (Table 1)

psd module variant | kanMX4

S2-primer

Step 3: chromosomal integration by homologous recombination in the yeast strain of choice at the target gene

psd module variant | kanMX4

target gene
Start   Stop

Step 4: verification of target modification by immunoblot or PCR with chromosomal DNA as template

C1   C3
target gene | psd module variant | kanMX4
C2   C4

Step 5: validation of target protein function by an appropriate assay; perform phenotypic analysis

Fig. 2 Workflow scheme illustrating the generation of conditional mutants with the psd module. Step 1: Design target gene-specific primers necessary to amplify the psd module by polymerase chain reaction (PCR). Primer design: S2 primer, the reverse complement of 45–55 bases downstream of the stop codon (including stop) of the target gene, followed by 5′-ATCGATGAATTCGAGCTCG-3′; S3 primer, 45–55 bases upstream of the stop codon (excluding stop) of the target gene, followed by 5′-CGTACGCTGCAGGTCGAC-3′. Step 2: Perform the PCR with the template plasmid containing the psd module variant of choice (for characteristics of the variants, *see* Table 1) and the S2 and S3 primers to obtain target gene-specific PCR products. Several templates can be tested in parallel without a considerable increase of workload. Step 3: Transformation of the selected yeast strain with the PCR products. The target gene-specific sequences flanking the PCR product result in insertion of the cassette upstream of the stop codon by homologous recombination. Step 4: Verify the correct integration at the target gene locus. Use immunoblotting or PCR with appropriate primers (C1–C4). The appearance of specific PCR products (C1–C2, C3–C4, and C1–C4) confirms the successful integration. Step 5: Make a test for the functionality of the target protein at permissive conditions (darkness) and at restrictive conditions (blue light illumination) together with an appropriate control strain

2. Yeast cells are harvested by centrifugation (3 min, $500 \times g$), washed once with sterile water (0.1–0.5 volumes) followed by a washing step with sterile SORB buffer (0.1–0.2 volumes).

3. Cells are suspended in 450 μl SORB buffer and 50 μl of carrier DNA is added. Cells are divided into appropriate aliquots and placed at −80 °C (no shock freezing).

4. Usually, 50 μl of competent cells are used for the transformation with a PCR product. Thawed competent cells are mixed with PCR product (5–15 μl of DNA for 50 μl of cells) and six volumes of PEG buffer are added.

5. Cells are incubated at room temperature for approximately 30 min.

6. Cells are incubated at 42 °C for 5–20 min (15 min works well with most strains).

7. Cells are sedimented (3 min, $500 \times g$), washed once with YPD, and resuspended in 3 ml of YPD.

8. Cells are incubated on a shaker for 3–16 h at 30 °C and spread on a plate containing 200 mg/l Geneticin (G418). After transformation, the plates should be kept in darkness during growth of the yeast cells. Selection of drug-resistant yeast clones on plates often requires replication onto the same selective medium after 2 days at 30 °C due to the high background of transiently transformed cells.

3.5 Verification of Chromosomal Integrations

A modified standard protocol for isolation of yeast chromosomal DNA is used [17].

1. An amount of yeast cells corresponding to the size of a match head is scraped off a plate and dissolved in 500 μl of breaking buffer.

2. 200 μl of phenol/chloroform/isoamyl alcohol mixture (25/24/1, buffered with TE, pH 7.5–8) and 0.3 g of glass beads (~200 μl) are added.

3. Cells are disrupted by vortexing (5 min, highest speed) and phases are separated in a microcentrifuge (10 min, $16,000 \times g$).

4. 10 μl of the aqueous layer are taken and diluted with water (1:10).

5. 1 μl of the diluted solution is used for the analytical PCR.

3.6 Detection of psd Module Constructs by Immunoblotting

The target protein can be detected by immunoblotting using antibodies directed against the target protein, myc, or tagRFP. For immunoblotting, crude cell extracts can be prepared by alkaline lysis [18].

1. 1 ml of logarithmically growing cells ($A_{600} = 1$) is treated with 150 μl of alkaline lysis buffer and kept on ice for 10 min.

2. Proteins are precipitated by addition of 150 μl 55 % (w/v) trichloroacetic acid (TCA) followed by incubation on ice for 10 min.

3. The pellet obtained by centrifugation (10 min, $16,000 \times g$) is dissolved in 60 μl of high urea buffer by heating the sample (65 °C) and mixing it frequently by vortexing.

4. For SDS PAGE, the extracts are cleared from cell debris by centrifugation (10 min, $16,000 \times g$). 10–20 μl of sample is loaded per lane. Standard procedures can be used for SDS PAGE and blotting [19, 20].

3.7 Illumination of Yeast Cells with Blue Light

Blue light (465 nm, 30 µmol/m²/s) is used to activate the photosensitive degron (*see* **Note 1**). Typically, 4–6 h are sufficient to inactivate a target protein (*see* **Note 2**). For a description how to expose cells growing on solid medium, please *see* **Note 3**.

1. The yeast strains are grown in darkness until mid-log phase is reached in low fluorescence medium supplemented with 2 % glucose (*see* **Note 4**).

2. Activation of the psd module is induced by exposure to blue light (465 nm, 30 µmol/m²/s). Depletion of the target protein is typically observed after 4 h, but the actual time might vary (*see* **Note 2**). An immunoblot or a functional assay are possible tests to control inactivation of a target protein (*see* **Note 5**). Efficiency of target protein depletion by light can be assessed by comparing the abundance of the target protein in cells grown at permissive conditions (darkness) to the abundance in cells kept at restrictive conditions (blue light) for an appropriate time. The time span necessary to inactivate a target protein cannot be predicted in advance; it depends on the synthesis rate of the target protein and the minimal level of target protein that is necessary to perform its function. The ongoing target protein synthesis will lead to minimal amounts of target protein even after prolonged exposure to blue light (*see* **Note 6**). A control strain, in which the target protein is modified with the 3myc tag or tagRFP, is helpful to visualize the overall change in target protein abundance.

3. In case an essential gene is modified with the psd module, a serial dilution assay at restrictive and permissive conditions tests successful inactivation of the target protein (*see* **Note 7**). Appropriate control strains are necessary to ensure that the illumination conditions as such are not harmful for the cells (*see* **Note 8**).

4. For experiments, the strain containing the target gene modified with the psd module at permissive conditions might not always be suitable as control. The modification of the target gene with a psd module might change the abundance or the functionality of the target protein in darkness (permissive conditions). This should be tested for each target protein with an appropriate assay. An isogenic yeast strain without modification of the target gene is a good choice as control for *in vivo* functionality assays. Another possibility is to fuse the target gene to the red fluorescent protein or to the 3myc tag (plasmids pDS1 and pYM4, Table 1), to modify the target protein at the carboxy-terminal end, but without the actual degron. Such a strain serves as control during measurement of the target protein abundance and in a functionality assay.

4   Notes

1. The quality and the quantity of the blue light used for illumination can be varied. The LOV2 domain, which is part of the psd module, can be excited by blue light in the range of 400–500 nm, the extinction maximum is at 450 nm. If efficient activation of the psd module is ensured, the wavelength of the illumination source can be varied without losing the protein destabilization activity of the psd module. In case of the psd module variants K92R E132A E155G (pDS170), K121M N128Y (pDS135), K121M N128Y G138A (pDS157), no difference in half-lives has been observed for illumination with two different light fluxes (465 nm, 5 and 30 $\mu mol/m^2/s$) using tagRFP as target protein. In case of the other constructs, illumination with light of low intensity prolonged the half-lives compared to exposure to the higher light dose [11]. Thus, full depletion of a target protein fused to the variants K92R E132A E155G (pDS170), K121M N128Y (pDS135), or K121M N128Y G138A (pDS157) might be achieved even with low intensities of blue light. However, this should be tested individually for each target by immunoblotting.

2. The time window necessary to deplete a target protein from a cell can vary considerably. It is dependent on the cellular target, its synthesis rate and the selected psd module variant. The half-lives of the psd module variants range from 8.5 to 20 min under blue light illumination [11]. Considerable loss of target protein abundance can be expected to happen during 1–2 h, but this should be verified by immunoblotting. It might be necessary to add some illumination time to allow the development of an observable phenotype.

3. The medium composition can be chosen freely in case yeast cells are grown on solid medium. The Petri dishes should be oriented in a way that the cells are directly exposed to the light source to avoid unnecessary shading.

4. In case cells are grown in liquid medium, low fluorescence medium is recommended. The carbon source can be selected freely in this case. If a complex medium is necessary for growth, light absorption by the medium should be taken into account and the light flux should be increased accordingly.

5. Control of target protein depletion: Target protein degradation can be monitored by immunoblotting using target protein-specific antibodies. Commercially available antibodies directed against tagRFP can be used in case pCT337 was used as template for PCR, antibodies directed against myc in case one of the other plasmids was used. If a target protein modi-

fied with tagRFP-psd is detectable by fluorescence microscopy at permissive conditions, target protein depletion might be followed by live-cell imaging. This works very well for targets that show a specific cellular localization like Cdc14 (our unpublished observation).

6. Inactivation of a target protein with blue light is compatible with many molecular biology and cell biology assays. Fluorescence-based methods that rely on the detection of green fluorescence protein (GFP), which is excited by blue light as well, have been performed successfully (our unpublished observation). However, conditions should be used to ensure that the blue light used to inactivate the target protein does not interfere with the detection of the GFP due to premature bleaching of the fluorophore.

7. Complete inactivation of target proteins failed in some cases using the wild-type psd module (pCT337, pDS96). The variants with lower half-life at restrictive conditions (pSU22: K92R E132A N148D E155G; pSU24: K92R E132A E139N N148D E155G; pDS121: E139N; pDS170: K92R E132A E155G; pDS135: K121M N128Y; pDS157: K121M N128Y G138A) circumvented this problem for some cases (our unpublished observation). The characteristics of the psd module variants are given in Table 1. To be accessible for degradation, the carboxy terminus of the target protein has to be localized to the cytosol or the nucleus. A problem that we encountered sometimes was the appearance of spontaneous suppressor colonies, either in strains that did not cease to grow completely after exposure to blue light or in strains with a slow-growth phenotype in darkness. To minimize the formation of suppressors, we used yeast cells freshly grown from a permanent culture (kept at –80 °C in 15 % glycerol) before starting an experiment.

8. Blue light in high doses is toxic to yeast cells. Illumination of cells with higher light intensities (above 50 $\mu mol/m^2/s$, 465 nm) resulted in slower growth or even loss of viability, whereas the illumination regimen described here (465 nm, 30 $\mu mol/m^2/s$) did not result in growth defects [9, 11]. Even a very light-sensitive yeast strain ($yap1\Delta$) was able to grow under these conditions, although at reduced rate [11].

Acknowledgements

We thank D. Störmer for her excellent technical assistance. This work was supported by the DFG grant TA320/3-1 and the DFG-funded graduate school GRK1216.

References

1. Hershko A, Ciechanover A (1998) The ubiquitin system. Annu Rev Biochem 67:425–479

2. Rakhit R, Navarro R, Wandless TJ (2014) Chemical biology strategies for posttranslational control of protein function. Chem Biol 21(9):1238–1252

3. Kanemaki MT (2013) Frontiers of protein expression control with conditional degrons. Pflugers Arch 465(3):419–425

4. Ravid T, Hochstrasser M (2008) Diversity of degradation signals in the ubiquitin-proteasome system. Nat Rev Mol Cell Biol 9(9):679–690

5. Jariel-Encontre I, Bossis G, Piechaczyk M (2008) Ubiquitin-independent degradation of proteins by the proteasome. Biochim Biophys Acta 1786(2):153–177

6. Jungbluth M, Renicke C, Taxis C (2010) Targeted protein depletion in Saccharomyces cerevisiae by activation of a bidirectional degron. BMC Syst Biol 4:176

7. Gautier A, Gauron C, Volovitch M, Bensimon D, Jullien L, Vriz S (2014) How to control proteins with light in living systems. Nat Chem Biol 10(7):533–541

8. Zhang K, Cui B (2015) Optogenetic control of intracellular signaling pathways. Trends Biotechnol 33(2):92–100

9. Renicke C, Schuster D, Usherenko S, Essen LO, Taxis C (2013) A LOV2 domain-based optogenetic tool to control protein degradation and cellular function. Chem Biol 20(4):619–626

10. Bonger KM, Rakhit R, Payumo AY, Chen JK, Wandless TJ (2014) General method for regulating protein stability with light. ACS Chem Biol 9(1):111–115

11. Usherenko S, Stibbe H, Musco M, Essen LO, Kostina EA, Taxis C (2014) Photo-sensitive degron variants for tuning protein stability by light. BMC Syst Biol 8:128

12. Pereira G, Tanaka TU, Nasmyth K, Schiebel E (2001) Modes of spindle pole body inheritance and segregation of the Bfa1p-Bub2p checkpoint protein complex. EMBO J 20(22):6359–6370

13. Janke C, Magiera MM, Rathfelder N, Taxis C, Reber S, Maekawa H, Moreno-Borchart A, Doenges G, Schwob E, Schiebel E, Knop M (2004) A versatile toolbox for PCR-based tagging of yeast genes: new fluorescent proteins, more markers and promoter substitution cassettes. Yeast 21(11):947–962

14. Schiestl RH, Gietz RD (1989) High efficiency transformation of intact yeast cells using single stranded nucleic acids as a carrier. Curr Genet 16(5-6):339–346

15. Taxis C, Knop M (2006) System of centromeric, episomal, and integrative vectors based on drug resistance markers for Saccharomyces cerevisiae. Biotechniques 40(1):73–78

16. Taxis C, Knop M (2012) TIPI: TEV protease-mediated induction of protein instability. Methods Mol Biol 832:611–626

17. Ausubel FM, Kingston RE, Seidman FG, Struhl K, Moore DD, Brent R, Smith FA (eds) (1995) Current protocols in molecular biology. Wiley, New York

18. Yaffe MP, Schatz G (1984) Two nuclear mutations that block mitochondrial protein import in yeast. Proc Natl Acad Sci U S A 81(15):4819–4823

19. Laemmli UK (1970) Cleavage of structural proteins during the assembly of the head of bacteriophage T4. Nature 227(5259):680–685

20. Towbin H, Staehelin T, Gordon J (1979) Electrophoretic transfer of proteins from polyacrylamide gels to nitrocellulose sheets: procedure and some applications. Proc Natl Acad Sci U S A 76(9):4350–4354

21. Knop M, Siegers K, Pereira G, Zachariae W, Winsor B, Nasmyth K, Schiebel E (1999) Epitope tagging of yeast genes using a PCR-based strategy: more tags and improved practical routines. Yeast 15(10B):963–972

Chapter 6

Photo Control of Protein Function Using Photoactive Yellow Protein

Jakeb M. Reis and G. Andrew Woolley

Abstract

Photoswitchable proteins are becoming increasingly common tools for manipulating cellular processes with high spatial and temporal precision. Photoactive yellow protein (PYP) is a small, water-soluble protein that undergoes a blue light induced change in conformation. It can serve as a scaffold for designing new tools to manipulate biological processes, but with respect to other protein scaffolds it presents some technical challenges. Here, we present practical information on how to overcome these, including how to synthesize the PYP chromophore, how to express and purify PYP, and how to screen for desired activity.

Key words Photoactive yellow protein, PYP, Optogenetic, Protein design, Protein engineering

1 Introduction

Naturally occurring photoswitchable proteins exhibit changes in structure and activity in response to light. Through advances in protein engineering and design, these proteins can now be used as tools to manipulate cellular processes with high spatial and temporal precision [1]. Photoactive yellow protein (PYP) is a small, water-soluble protein with a cysteine linked *p*-coumaric acid chromophore. Blue light triggers *trans*-to-*cis* isomerization of the chromophore and production of a light state with altered conformational dynamics [2]. In wild-type PYP, this light state relaxes back to the dark-adapted state thermally with a half-life of a few seconds. This light induced conformational change can be transmitted to a target protein by creating an appropriate linkage to PYP. Protein–protein interactions can be controlled by fusing a binding motif to PYP in such a way its accessibility is altered upon irradiation. For example, our laboratory has designed tools to photo control homodimerization and activation of the GCN4 transcription factor [3, 4], a dominant negative inhibitor for the CREB transcription factor [5] and binding to a designed coiled-coil motif [6]. The success of a particular design depends on the

Arash Kianianmomeni (ed.), *Optogenetics: Methods and Protocols*, Methods in Molecular Biology, vol. 1408,
DOI 10.1007/978-1-4939-3512-3_6, © Springer Science+Business Media New York 2016

details of the linkage between PYP and the target protein and is a major challenge that is specific to each target protein. Here, we provide a general protocol for producing a candidate PYP-fusion protein. In addition, we provide experimental strategies for assessing the structure, dynamics, and activity of a candidate PYP-fusion protein.

2   Materials

2.1   Chromophore Synthesis

All solvents used in the reaction should be anhydrous.

2.1.1   Reagents and Reaction Solvents

Tetrahydrofuran (anhydrous).

Dichloromethane (anhydrous).

Dimethylformamide (anhydrous).

Oxalyl chloride.

p-Coumaric acid (Sigma–Aldrich).

n-Butyllithium 1.6 M in hexanes (Aldrich).

Thiophenol.

Pyridine.

Other materials:

Silica 60 gel TLC plate F_{254}.

UV light for visualization.

Methanol.

Balloons.

10 mL Luer-Lock Syringe.

Dry ice.

Nitrogen gas.

Septa for 44/40 size flasks.

20 Gauge needle.

Stainless steel 316 syringe (Aldrich).

1× 250 mL Round bottom flask (44/40).

2× 150 mL Round bottom flask (44/40).

Thermometer capable of reading –30 °C.

2.1.2   Work Up/ Purification

Diethyl ether.

Dichloromethane.

$MgSO_4$ (anhydrous).

10 % Sodium bicarbonate.

pH paper.

Silica.

Rotary evaporator.

Separatory flask.

Conical flask.

P8 Coarse Filter Paper (Fisher Scientific).

Funnel.

2.2   Protein Expression

Chemically competent BL21(DE3) *E. coli* cells.

pET24b PYP plasmid.

Activated chromophore (*see* Procedure below).

IPTG (Sigma–Aldrich).

1 L Erlenmeyer flask.

Lysogeny Broth.

Kanamycin.

2.3   Protein Purification

Ni^{2+}-NTA Agarose resin.

Lysis buffer: 50 mM sodium phosphate pH 8.0, 300 mM NaCl, 5 mM $MgCl_2$.

Salt buffer: Lysis buffer + 2 M NaCl.

Low imidazole buffer: Lysis buffer + 5 mM imidazole.

Elution buffer: Lysis buffer + 200 mM imidazole.

Dialysis bag (6000–8000 MWCO).

Protein buffer: 40 mM tris acetate pH 7.5, 1 mM EDTA, 100 mM NaCl.

Size exclusion chromatography system.

2.4   In Vitro Characterization

10 mm Cuvette.

UV–Visible spectrometer.

2.4.1   UV–Visible Spectroscopy

Luxeon III Star Royal Blue Lambertian LED (455 nm) or similar high power royal blue LED.

2.4.2   NMR

M9 Media: 1× M9 salts, 0.3 % d-glucose, 10 mg/L biotin, 10 mg/L thiamine, 1 mM $MgSO_4$, 1 mM $CaCl_2$, 30 µg/mL kanamycin, 0.1 % NH_4Cl. For 1 L media use 0.1 % ^{15}N NH_4Cl (Cambridge Isotope Laboratories) and supplement with 1 mL BioExpress ^{15}N Concentrate (Cambridge Isotopes Laboratories).

10× M9 Salts: 128 g $Na_2HPO_4 \cdot 7H_2$, 30 g KH_2PO_4, 5.0 g NaCl, water to 1 L.

3 mm NMR tube.

Fiber optic cable (Thorlabs).

3   Method

3.1   Protein Design

An effective target-PYP-fusion protein couples the blue light driven conformational change of PYP to a change in the target protein. This may lead to differential exposure of the target protein in the dark versus the light state, or a change in the stability of a binding competent conformer. Specific sequence optimization may be required for a given target protein, but there are general strategies that can be followed. A target protein can be fused to the N-terminus of PYP, which undergoes disengagement from the protein core upon exposure to light. This strategy is applicable to targets that share some sequence homology with the N-terminus of PYP, such as coiled-coil transcription factors [3–5]. Target protein sequences can also be inserted into surface loops of PYP. This should be done in a way such that the target sequence structure is incompatible with the dark state of PYP. A circularly permuted version of PYP that places the target of interest between the wild-type N and C termini can be used [6, 7]. Using one of these strategies, one may obtain a protein with a small change in photoswitchable activity (i.e., binding in light versus dark) that can then serve as a starting point for further sequence refinement through structurally guided or random library methods.

3.2   Chromophore Synthesis

E. coli does not produce the PYP chromophore and thus it must be introduced exogenously. It is possible to do this through coexpression of two biosynthetic genes that produce the chromophore from l-tyrosine, though this requires specific optimization to achieve full incorporation in each PYP variant designed [4]. A thiophenyl activated chromophore (4-hydroxycinnamic acid *S*-thiophenyl ester) can be obtained through a simple synthetic procedure [8] and is readily incorporated into PYP variants expressed in *E. coli*.

The reaction is carried out in three flasks. They must be dried to remove moisture and filled with inert atmosphere prior to the reaction. A method to create an insert atmosphere in the reaction flasks with readily available equipment is described below.

3.2.1   Producing an Inert Atmosphere for the Reaction Flasks

1. Remove the plunger from a 10 mL Luer-lock syringe and cut off the flared end with a pair of scissors, being sure to remove any sharp edges.

2. Wrap the cut end with two layers of Parafilm to make a smooth edge.

3. Cover the end with a balloon and wrap it securely with Parafilm.

4. Fill the syringe and balloon with nitrogen gas.

5. Place a stir bar into a 250 mL 44–40 size round bottom flask and cap it with a septum.

6. Screw in a 20 gauge needle into the syringe and pierce the septum to transfer the nitrogen gas. Another needle can be placed in the septum to allow the gas to exit the round bottom flask. Purge with 3× the volume of the flask.

3.2.2 Synthesis of p-Coumaric Acid S-Thiophenyl Ester

Reaction flask 1

1. Place a stir bar into a 250 mL round bottom flask and purge with nitrogen as described above.

2. Place the round bottom flask in ice so the reaction proceeds at 0 °C.

3. Add 18 mL dichloromethane to the flask using a syringe and needle.

4. Transfer 2 mL dimethylformamide into the flask using a fresh syringe and needle.

5. Add 3 mL oxalyl chloride dropwise to the solution. During this time, gas will evolve and may build enough pressure to push off the septum (*see* **Note 1**).

6. After the evolution of gases, allow the reaction mixture to stir for 1 h at 0 °C, during this time prepare flask #2.

Reaction flask 2

1. Weigh 1.7 g *p*-coumaric acid and transfer into a 150 mL round bottom flask with a magnetic stir bar.

2. Cap with a septum and purge as described above.

3. Transfer 18 mL anhydrous tetrahydrofuran and 1 mL pyridine to the flask (*see* **Note 2**).

4. Stir until the *p*-coumaric acid dissolves and then move the round bottom flask onto ice to continue stirring at 0 °C

Reaction flask 1

1. After 1 h of stirring, use a rotary evaporator to dry the reaction mixture. Recap the round bottom flask with a septum and purge with nitrogen.

2. Place a Dewar in the fumehood and place the thermometer into the Dewar, secured by a clamp.

3. Pour methanol into the Dewar and add dry ice until the temperature reaches −30 °C (*see* **Note 3**).

4. Place both round bottom flask #1 and #2 into the Dewar.

5. Transfer the entire contents of flask #2 into flask #1.

6. Allow the reaction mixture to stir for 1 h at −30 °C.

7. During this time prepare flask 3.

Reaction flask 3

1. Prepare a 150 mL round bottom flask with a magnetic stir bar and septum.

2. Secure the flask and place in ice.

3. Purge with 3 volumes of nitrogen gas using the method described above.

4. Transfer 20 mL tetrahydrofuran into the flask.

5. Transfer 0.8 mL thiophenol to the flask.

6. Transfer 5 mL *n*-butyllithium to the flask with a stainless steel syringe.

7. Transfer the entire contents of flask 3 into flask 1 using a needle and syringe.

8. Allow the mixture to stir at −15 °C for 2 h.

3.2.3 Reaction Workup

1. Quench the reaction by adding 150 mL water.

2. Add 200 mL diethyl ether and transfer to a separatory flask.

3. Transfer the organic phase into a 500 mL Erlenmeyer flask and wash with 10 % sodium bicarbonate until the pH is 6.

4. Add anhydrous $MgSO_4$ to the flask until the diethyl ether is dried. Remove $MgSO_4$ by filtration with P8 coarse filter paper.

5. Remove the solvent by rotary evaporation and store the dried product at −20 °C until purification.

3.2.4 Product Purification by Column Chromatography

1. Fill a 500 mL beaker up to 250 mL with 230–400 mesh silica gel and add dichloromethane until a slurry is formed.

2. Transfer to a glass column (approximately 70 cm height 15 cm circumference) being sure that no bubbles are formed as the slurry sets. Leave a layer of ~5 cm dichloromethane so that the column does not run dry.

3. Dissolve the dried product in a minimal volume of dichloromethane. Note that some white solid may not dissolve in dichloromethane. This is not the product.

4. Analyze the crude product dissolved in dichloromethane using thin layer chromatography (100 % dichloromethane mobile phase). Visualize under UV light, the desired product should be the third spot from the solvent front, but this should be confirmed after purification.

5. Allow the dichloromethane to fully enter the column and immediately add the crude product. Allow it to enter the column and add a layer of 3 cm of sand.

6. Fill the column with dichloromethane and begin collecting the flow through in a 500 mL flask. The first two visible bands do not contain product, this can be confirmed by thin layer chromatography during purification.

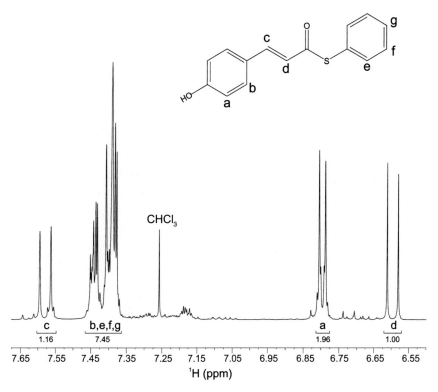

Fig. 1 ¹H NMR spectrum of activated PYP chromophore after purification. Each unique ¹H signal is labeled, integrations are listed underneath

7. After the first two compounds elute from the column, begin collecting the eluent in 10 mL test tubes, check the fractions for product by TLC. Combine the all fractions containing purified product and remove the solvent by rotary evaporation. The product can be further dried by vacuum overnight.

8. Verify the identity of the compound by ESI–MS (expected mass = 256.1 Da) and ¹H NMR (δ = 7.67 ppm ^3J = 15.7 Hz, 1H) (δ = 7.30–7.55 ppm, m, 7H), (δ = 6.87 ppm, d, ^3J = 8.6 Hz, 2H), (δ = 6.70 ppm d, ^3J = 15.7 Hz, 1H) (Fig. 1).

3.3 Protein Expression

PYP-fusion proteins can be produced in *E. coli* BL21 (DE3) using an IPTG inducible T7 based pET24b system. This can be used to produce a PYP-fusion protein with a C-terminal His Tag that can be used for purification. TEV cleavable His tags can also be used. Intein mediated tag cleavage should be avoided due to chromophore loss under cleavage conditions. Export tags have also been reported, but these require purification from large volumes of solution [9]. Coexpression with biosynthetic genes for the PYP chromophore is possible, but requires optimization for each PYP construct expressed [10]. Thus, we recommend addition of exogenous synthetically produced chromophore to the growth media.

1. Transform chemically competent BL21 (DE3) *E. coli* with 2 ng pET24b PYP plasmid.

2. Plate on LB-Agar media containing 30 µg/mL kanamycin.

3. The next day select a single colony from the plate and grow in 25 mL LB containing 30 µg/mL kanamycin.

4. Use 1 mL from the overnight culture to inoculate a 1 L culture.

5. Grow this culture at 37 °C until $OD_{600} = 0.6$.

6. Lower the temperature to 25 °C and add ITPG to a final concentration of 1 mM.

7. Grow for 1.5 h and add 25 mg activated chromophore dissolved in 1 mL absolute ethanol.

8. Grow for an additional 5 h and harvest the cells by centrifugation $3300 \times g$ for 30 min.

9. Resuspend the pelleted cells in 50 mL lysis buffer. Store at −20 °C.

3.4 Protein Purification

1. Thaw the resuspended pellet at room temperature and then place on ice.

2. Sonicate on ice in 15 s pulses for 5 min.

3. Centrifuge at $17,000 \times g$ for 1 h to remove debris.

4. Apply the supernatant to a Ni^{2+}-NTA column preequilibrated with lysis buffer (*see* **Note 4**).

5. Allow the entire supernatant to flow through the Ni^{2+}-NTA column.

6. Wash the column with 10 CV lysis buffer (50 mM sodium phosphate, 300 mM NaCl, 5 mM $MgCl_2$).

7. Wash with 5 CV salt buffer remove weakly bound proteins.

8. Wash with 5 CV Low imidazole buffer.

9. Elute in 10 CV elution buffer.

10. Dialyze in a 6000–8000 MWCO dialysis bag against protein buffer.

Store the protein at 4 °C protected from light in aluminum foil wrapped vials.

3.4.1 Size Exclusion Chromatography

Some PYP-fusion proteins self-associate to form higher order oligomers, and these may have different biological activities. In particular, some PYP-target fusions may relieve strain caused by target insertion by domain swapping to form slowly interconverting oligomers [6]. Therefore, it is important to separate Ni^{2+}-NTA

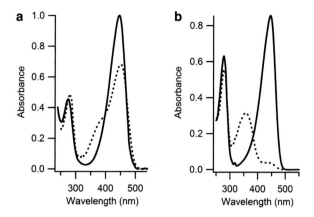

Fig. 2 UV–visible spectra of (**a**) fully dark adapted, *trans*, deprotonated PYP (*dark line*), a spectrum showing some fraction of *trans*-protonated chromophore (*dashed line*). (**b**) Opto-DN-CREB in the fully dark-adapted state (*solid line*) and blue light irradiated state (*dashed line*)

purified proteins by a further size exclusion step (*see* **Note 5**). If multiple oligomeric species are formed, their rate of interconversion should be determined and they should be characterized separately. If multiple oligomeric species are suspected, special notice should be given to the sample history, particularly to light exposure and temperature that may accelerate interconversion.

3.5 In Vitro Characterization

3.5.1 UV–Visible Spectroscopy

PYP contains a covalently linked chromophore that can act as a reporter on the protein structure. In the dark, the chromophore shows a maximal absorbance at 450 nm and is in the trans, deprotonated state (Fig. 2). Destabilizing mutations, or insertion of some target sequences may distort the chromophore binding pocket or alter H-bonding patterns involving the chromophore hydroxyl group. In these cases, a shoulder at ~380 nm (Fig. 2) is observed in the UV–vis spectrum [11]. The spectrum can be recorded with respect to temperature, denaturant, salt, and ligand concentration to uncover information on protein stability. The UV–vis spectrum can also be used to detect apoprotein by comparing the ratio of the 446–280 nm absorption bands. Blue light irradiation causes *trans*-to-*cis* isomerization and a change in maximal absorption from 446 to 350 nm (Fig. 2b). The light state spectrum may be technically difficult to obtain if the thermal relaxation rate of a given PYP construct is too fast. We use a diode array spectrophotometer to simultaneously irradiate the sample with high intensity blue light and record the full UV–vis spectrum.

1. Wrap the protein samples in aluminum foil and leave at 4 °C overnight to ensure the protein is fully dark adapted. If using a diode array instrument, be careful of changes to the spectrum due to irradiation from the measuring beam.

2. Dilute the sample in protein buffer. Several dilutions may need to be made until the absorbance is less than 1.

3. Transfer the sample to a quartz cuvette and record the spectrum from 550 to 260 nm.

3.5.2 Thermal Relaxation Kinetics

Blue light causes *trans*-to-*cis* isomerization of the *p*-coumaric acid chromophore and production of a "light state." This state relaxes back to the dark-adapted state thermally on the order of seconds to hours depending on the specific PYP construct. The thermal relaxation rate is sensitive to mutations in PYP and also to binding interactions in the context of designed PYP constructs. Thus, it can be used as a preliminary screen for binding to a target protein (i.e., there may be a change in thermal relaxation rate upon binding to a target). The rate can be measured by irradiating with blue light and following the recovery of *trans* chromophore (446 nm) or loss of *cis* chromophore (350 nm) over time. As an example, we show the thermal relaxation of a PYP-fusion protein with increasing concentrations of its target, CREB. This protein shows a biexponential thermal relaxation rate with a fast (1.4 min) and slow (20 min) component. Addition of CREB reduces the fraction of the slow relaxation component. The thermal relaxation rate of a particular PYP construct may be too fast for a given application, and can be changed by a mutation of M100E or M100A (wild-type PYP numbering) [12].

1. Prepare a 10 μM protein solution in a quartz cuvette and place in a temperature controlled holder.

2. Shine blue light (e.g., Luxeon III Star Royal Blue Lambertian) on the sample for 1 min. If using a diode array instrument you can watch for the appearance of at 350 nm absorbance band during irradiation. If you cannot read the absorbance in real time, watch for the solution to turn from yellow to clear.

3. Turn off the light source and start recording.

4. Record consecutive scans from 550 to 260 nm with a set time interval between each scan. This may need to be adjusted for each new construct. For some constructs, the thermal relaxation rate may be too fast to record entire spectra at each time point. In these cases, monitor a single wavelength (446 or 350 nm) over time (*see* **Note 6**).

5. Plot the absorbance at 446 and 350 nm versus time. Fit these data to an exponential function to obtain a rate constant for thermal relaxation.

6. Repeat with increasing concentrations of target ligand. If thermal relaxation rates are too fast, a lower temperature can be used.

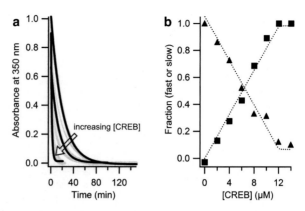

Fig. 3 Thermal relaxation kinetics of a PYP designed to bind CREB (opto-DN-CREB). (**a**) Decrease in *cis* chromophore (350 nm absorbance) over time with increasing concentrations of CREB. No CREB added, 0.5 equiv. and 1 equiv. are *bolded*. The thermal relaxation of opto-DN-CREB is a biexponential process. (**b**) The fraction of slow (*triangles*, $\tau_{1/2} = 20$ min) and fast (*squares*, $\tau_{1/2} = 1.4$ min) plotted as a function of CREB concentration

3.5.3 NMR

Dark-adapted PYP is a well folded protein in a single conformational state, and the chemical shift assignments are available. An NH HSQC experiment of a particular designed construct can yield information about the structural similarity with wild-type PYP. If the target insert does not perturb the structure of the PYP core, a well dispersed NH HSQC spectrum (e.g., Fig. 3) should be observed. Since assignments of PYP are known, one can assess the similarity of a designed construct to a normal PYP-like fold. For example, insertion of a coiled-coil forming eHelix sequence into a circular permutant of PYP (c-eHelix-PYP) causes little change in the NH HSQC TROSY spectrum (Fig. 4b). Additionally, if assignments are known for the target sequence, its structure can be probed. The light state HSQC spectrum shows fewer resonances due to conformational exchange on the millisecond time scale [2]. A designed PYP-fusion protein should show a similar characteristic loss of signals if it undergoes a similar blue light induced conformational change, as shown in Fig. 4a. Light state NMR spectra can be obtained by placing a fiber optic cable into a 3 mm NMR tube, but careful attention must be given to the sample concentration, sample height, and light intensity so that full photoswitching takes place. In addition to loss of PYP resonances upon irradiation, one can observe structural changes in the insert domain.

3.5.4 Isotopically Labeled Protein Expression

1. Transform *E. coli* with a pET24b plasmid containing a PYP construct, plate on LB-Agar containing 30 μg/mL kanamycin.

2. Select a single colony from this plate and grow overnight in 25 mL LB containing 30 μg/mL kanamycin (37 °C, 250 rpm).

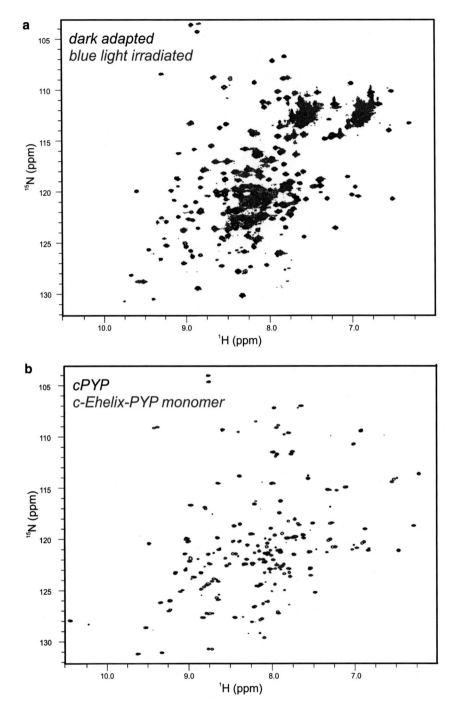

Fig. 4 (**a**) NH HSQC spectra of zENG-cPYP in the fully dark-adapted state (*black*) and under blue light irradiation (*red*). (**b**) NH HSQC TROSY Spectra of cPYP containing a flexible insert linker (*black*) and with the eHelix insert (*red*), showing minimal perturbation to the PYP core with eHelix insertion

3. Pellet 1 mL LB by centrifuging at $3300 \times g$ for 10 min, discard the supernatant, and resuspend the pellet in 50 mL nonlabeled M9 media.

4. Grow the 50 mL M9 Culture at 37 °C 250 rpm until it $OD_{600} = 0.9$ is reached.

5. Pellet the entire 50 mL culture by centrifuging at $3300 \times g$ and resuspend in a 1 L volume of ^{15}N labeled M9 media.

6. Grow the 1 L culture until it reaches $OD_{600} = 0.6$ and add IPTG to a final concentration of 1 mM. Continue growing at 25 °C for 1.5 h and add 25 mg activated chromophore.

7. Lower the temperature to 18 °C and grow for 12 h shaking at 250 rpm.

8. Harvest the cells by centrifugation at $3300 \times g$ for 30 min. Purify by the method described above.

4   Notes

1. This reaction step should be carried out slowly because it evolves gas. A separate empty balloon attached to a syringe (as described in Subheading 3.2.1) can be used during this step. If oxalyl chloride is added too quickly, the septum may be pushed off of the reaction flask. If this occurs, replace the septum and purge the reaction flask with nitrogen gas before continuing.

2. Be sure to allow the p-coumaric acid to dissolve fully before cooling the reaction, otherwise it will be difficult to dissolve.

3. Add small pieces of dry ice slowly as this step causes vigorous boiling. Also be careful that placing the reaction flask in the cooled solution does not cause it to overflow.

4. PYP associated to the column may cause the color to turn yellow in this step. A lack of color does not indicate that no protein is associated with the column, because the chromophore may be photo bleached by ambient light during purification.

5. Higher oligomers may be formed if the protein is concentrated prior to size exclusion chromatography. These species may not form at concentrations relevant to the protein's activity, and thus may not need to be characterized.

6. It is better to determine rate constants by following absorbance at 350 nm. This avoids light scattering from the source used to isomerize PYP and isomerization by the measuring beam of the spectrometer.

Acknowledgments

This work has been supported by the Natural Sciences and Engineering Research Council of Canada and by NIH (R01MH086379).

References

1. Moglich A, Moffat K (2011) Engineered photoreceptors as novel optogenetic tools. Photochem Photobiol Sci 9(10):1286–1300
2. Ramachandran PL et al (2011) The short-lived signaling state of the photoactive yellow protein photoreceptor revealed by combined structural probes. J Am Chem Soc 133(24): 9395–9404
3. Fan HY et al (2011) Improving a designed photocontrolled DNA-binding protein. Biochemistry 50(7):1226–1237
4. Morgan SA, Woolley GA (2010) A photoswitchable DNA-binding protein based on a truncated GCN4-photoactive yellow protein chimera. Photochem Photobiol Sci 9(10): 1320–1326
5. Ali AM et al (2015) Optogenetic control of CREB signaling. Chem Biol. 22(11): 1531–9
6. Reis JM, Burns DC, Woolley GA (2014) Optical control of protein-protein interactions via blue light induced domain swapping. Biochemistry 53(30):5008–5016
7. Kumar AB, Burns DC, Al-Abdul-Wahid MS, Woolley GA (2013) A circularly permuted photoactive yellow protein as a scaffold for photoswitch design. Biochemistry 52(19): 3320–3331
8. Changenet-Barret P et al (2002) Excited-state relaxation dynamics of a PYP chromophore model in solution: influence of the thioester group. Chem Phys Lett 365(3–4):285–291
9. Genick UK et al (1997) Active site mutants implicate key residues for control of color and light cycle kinetics of photoactive yellow protein. Biochemistry 36(1):8–14
10. Kyndt JA et al (2003) Heterologous production of Halorhodospira halophila holo-photoactive yellow protein through tandem expression of the postulated biosynthetic genes. Biochemistry 42(4):965–970
11. Kumar A, Woolley GA (2015) Origins of the intermediate spectral form in M100 mutants of photoactive yellow protein. Photochem Photobiol 91(4):985–991
12. Sasaki J, Kumauchi M, Hamada N, Oka T, Tokunaga F (2002) Light-induced unfolding of photoactive yellow protein mutant M100L. Biochemistry 41(6):1915–1922

Chapter 7

A Fluorometric Activity Assay for Light-Regulated Cyclic-Nucleotide-Monophosphate Actuators

Charlotte Helene Schumacher, Heinz G. Körschen, Christopher Nicol, Carlos Gasser, Reinhard Seifert, Martin Schwärzel, and Andreas Möglich

Abstract

As a transformative approach in neuroscience and cell biology, optogenetics grants control over manifold cellular events with unprecedented spatiotemporal definition, reversibility, and noninvasiveness. Sensory photoreceptors serve as genetically encoded, light-regulated actuators and hence embody the cornerstone of optogenetics. To expand the scope of optogenetics, ever more naturally occurring photoreceptors are being characterized, and synthetic photoreceptors with customized, light-regulated function are being engineered. Perturbational control over intracellular cyclic-nucleotide-monophosphate (cNMP) levels is achieved via sensory photoreceptors that catalyze the making and breaking of these second messengers in response to light. To facilitate discovery, engineering and quantitative characterization of such light-regulated cNMP actuators, we have developed an efficient fluorometric assay. Both the formation and the hydrolysis of cNMPs are accompanied by proton release which can be quantified with the fluorescent pH indicator 2′,7′-bis-(2-carboxyethyl)-5-(and-6)-carboxyfluorescein (BCECF). This assay equally applies to nucleotide cyclases, e.g., blue-light-activated bPAC, and to cNMP phosphodiesterases, e.g., red-light-activated LAPD. Key benefits include potential for parallelization and automation, as well as suitability for both purified enzymes and crude cell lysates. The BCECF assay hence stands to accelerate discovery and characterization of light-regulated actuators of cNMP metabolism.

Key words BCECF, bPAC, Cyclic nucleotide monophosphate, cNMP phosphodiesterase, LAPD, Microtiter plate, Nucleotide cyclase, Optogenetics, Sensory photoreceptor

1 Introduction

Optogenetics encompasses manifold transformative applications of sensory photoreceptors to control by light cellular physiology and behavior in spatiotemporally precise, reversible, and noninvasive fashion [1]. Whereas the first optogenetics experiments were implemented exclusively using light-gated ion channels and light-driven ion pumps [2–6], nowadays optogenetics can resort to a broad repertoire of different light-regulated actuators (i.e. sensory photoreceptors). In particular, the optogenetics toolbox now extends to soluble proteins and enzymes, thus enabling applications

Arash Kianianmomeni (ed.), *Optogenetics: Methods and Protocols*, Methods in Molecular Biology, vol. 1408,
DOI 10.1007/978-1-4939-3512-3_7, © Springer Science+Business Media New York 2016

transcending neurobiology. As a case in point, the so-called photo-activated cyclases (PACs) are stimulated by light to catalyze the formation of the universal second messenger 3′,5′-cyclic adenosine monophosphate (cAMP) and have proven particularly powerful candidates for optogenetics [7–11]. The most commonly applied PAC, dubbed bPAC, derives from the bacterium *Beggiatoa* sp. and shows low cyclase activity in darkness that can be enhanced by about two orders of magnitude via blue-light irradiation [8, 9].

In a bid to further expand the scope of optogenetics, researchers have engineered novel sensory photoreceptors with customized light-regulated function [12]. Although specific details widely differ, the engineering of novel photoreceptors generally entails the recombination of photosensor modules possessing sensitivity to the desired light quality with (originally light-insensitive) effector modules possessing the desired biological activity. Two pertinent photoreceptors for optogenetic intervention in cyclic-nucleotide-monophosphate (cNMP) metabolism have recently been engineered [13, 14]. We connected the N-terminal red-light sensor from *Deinococcus radiodurans* bacteriophytochrome [15], encompassing PAS, GAF and PHY domains [16], to the effector of the *Homo sapiens* phosphodiesterase 2A (*Hs*PDE2A) [17]. The resultant engineered photoreceptor LAPD hydrolyzed the cyclic nucleotides cAMP and 3′,5′-cyclic adenosine monophosphate (cGMP) with catalytic efficiency and substrate affinity en par with wild-type *Hs*PDE2A; whereas red-light exposure enhanced LAPD activity by up to sevenfold, far-red-light exposure attenuated LAPD activity. Allosteric regulation by cGMP in the parental *Hs*PDE2A enzyme was hence reprogrammed to regulation by light in LAPD [13]. Employing a closely similar rationale, a related bacteriophytochrome photosensor was fused to the catalytic domain of a bacterial adenylate cyclase to yield the PAC IlaC. Red light effected an approximately sixfold activation of IlaC that could be reversed by exposure to far-red light, but specific activity in IlaC was decreased by two orders of magnitude relative to the parental adenylate cyclase [14]. Full functionality of both LAPD and IlaC has been demonstrated in eukaryotic cells and organisms, and the optogenetic arsenal is hence complemented by hitherto unavailable, orthogonal light-regulated actuators.

Given that LAPD and IlaC represent the first iterations of general design strategies, we expect that in future versions key parameters, e.g., overall activity, substrate specificity, dynamic range of light regulation, response kinetics, and spectral sensitivity, can be improved/altered, e.g., by mutagenesis or replacement of photosensor and effector modules. Formation and hydrolysis of cyclic nucleotides is routinely quantified by high-performance liquid chromatography (HPLC) and enzyme-linked immunosorbent assays (ELISA) [8, 9]. To facilitate the discovery and quantitative characterization of light-regulated cNMP actuators, we sought to establish alternative functional assays for nucleotide cyclase and cNMP phosphodiesterase (PDE) activities. Ideally, such assays should be easy to conduct, should be inexpensive, should provide a continuous readout to facilitate kinetic measurements, should be compatible with illumination during measurements, should be suitable for automation and parallelization, and should apply to both initial screening of candidate constructs and quantitative characterization. To this end, we have established a fluorometric assay that meets these criteria and that applies to the quantitative characterization of both nucleotide cyclase and cNMP phosphodiesterase activities. We demonstrate that the assay can be used for measurements of purified light-activated nucleotide cyclases (*see* Subheading 3.1) and cNMP phosphodiesterases (*see* Subheading 3.2), as well as for measurements of PDE activities in *Escherichia coli* (*see* Subheading 3.3) and *Drosophila melanogaster* (*see* Subheading 3.4) crude lysates.

The fluorometric assay capitalizes on the release of protons that is concomitant with both the formation and the hydrolysis of cyclic nucleotides (Fig. 1a). Provided the buffering capacity is sufficiently low, proton release leads to acidification of assay solutions which can be detected via the pH-sensitive fluorescence of $2',7'$-bis-(2-carboxyethyl)-5-(and-6)-carboxyfluorescein (BCECF, Fig. 1b). Notably, BCECF possesses an aryl hydroxyl group with a pK_a of around 7 and hence undergoes (de)protonation reactions at pH values near physiological (i.e. around neutral pH) [18, 19]. When excited with a wavelength λ_{Ex} of about 500 nm, the BCECF fluorescence emission at $\lambda_{Em} = 535$ nm is strongly enhanced in the deprotonated state relative to the protonated state (Fig. 1c). By contrast, for excitation at the isosbestic wavelength $\lambda_{Ex} = 440$ nm, the fluorescence intensity at 535 nm is independent of protonation state. To compensate for concentration differences and

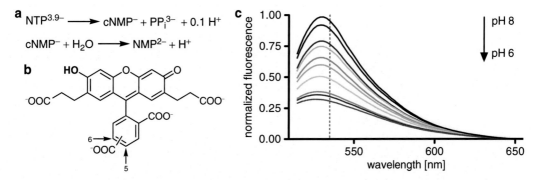

Fig. 1 BCECF fluorometric assay for nucleotide cyclase and cNMP phosphodiesterase activities. (**a**) Both the production and hydrolysis of cNMP lead to acidification of the reaction buffer. The average charge of ATP at pH 7.6 was calculated assuming a relevant pK_a value of 6.6. (**b**) BCECF is a fluorescein derivative with a pK_a of ~7 for its hydroxyl group. (**c**) In acidified buffers the hydroxyl group of BCECF is predominantly protonated, and the fluorescence excited at $\lambda_{Ex} = 500$ nm decreases with decreasing pH values between 8.0 (*black*) and 6.0 (*magenta*). The *vertical line* denotes a wavelength of 535 nm at which BCECF fluorescence emission is routinely measured

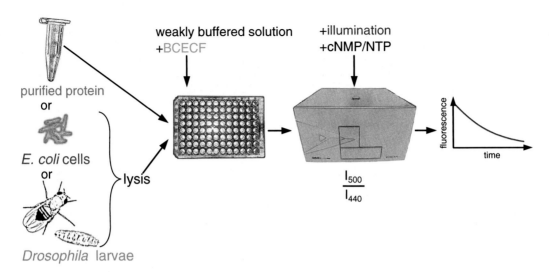

Fig. 2 BCECF assay work-flow. The assay can either be performed with purified enzyme or with crude cell lysates, which are produced by hypotonic lysis in case of *E. coli* or by mechanic homogenization using a pestle in case of *D. melanogaster* larvae. Purified protein or crude lysate is mixed with a weakly buffered solution containing BCECF and is transferred to a microtiter plate. The samples are illuminated as needed, and the reaction is started by addition of the substrate. Fluorescence emission at 535 nm is measured over time for excitation wavelengths of either 440 or 500 nm

instrument fluctuations, BCECF measurements are hence routinely conducted in ratiometric fashion, and fluorescence intensities I_{500} acquired for excitation at 500 nm are normalized by those acquired for excitation at 440 nm (I_{440}). As pointed out above, the BCECF fluorometric assay is applicable to both measurements on purified enzymes (*see* Subheadings 3.1 and 3.2) and measurements in crude lysates (*see* Subheadings 3.3 and 3.4) (Fig. 2). The assay can be conducted in 96-well microtiter format using a conventional microplate reader with suitable fluorescence optics. If the microplate reader possesses injection capabilities, successive measurements can be programmed via automation of substrate injection. Furthermore, a recently developed, custom-built upgrade allows programmable illumination inside of the plate reader and thus permits convenient measurements of light-regulated cNMP actuators under both dark and light conditions [20]. Finally, the BCECF assay can be calibrated against HPLC (*see* Subheading 4) or ELISA measurements, so as to convert relative fluorescence changes into absolute changes in cNMP concentration.

2 Materials

2.1 Buffers and Stock Solutions

1. Assay Buffer I (AB I): 2 mM potassium phosphate pH 7.6, 5 mM magnesium chloride, 100 mM potassium chloride, 10 mM dithiothreitol, 0.04 % (w/v) bovine serum albumin (*see* **Note 1**).

2. Assay Buffer II (AB II): 0.5 µM BCECF in AB I (Life Technologies, Frankfurt, Germany).

3. Lysis Buffer: 5 mM potassium phosphate pH 7.6, 10 mM dithiothreitol, 0.1 mg/mL lysozyme, protease inhibitor (*see* **Note 2**).

4. 4.8 mM ATP, GTP, cAMP, or cGMP in AB I.

2.2 Lab Equipment

1. Microplate reader with fluorescence optics and temperature control, e.g., as used here, Tecan Infinite M200 pro or F200 pro instruments.

2. *Optional*: If desired and required hardware capabilities are available, illumination can be programmed and conducted inside of the microplate reader [20].

3. Light-emitting diodes (or other light sources) for illumination of light-regulated nucleotide cyclases and cNMP phosphodiesterases.

4. 96-well plates, black wall, clear bottom (µClear plates, Greiner Bio-One, Frickenhausen, Germany).

3   Methods

3.1   Determining the Activity of Purified Light-Activated Cyclases

1. Preheat plate reader to desired temperature, e.g., 29 °C, for 30 min (*see* **Note 3**). All following steps should be carried out in the dark or under non-activating light conditions, e.g., dim red light for bPAC.

2. Prepare desired dilution of purified light-activated cyclase in AB I (*see* **Note 4**).

3. For n samples, mix $(n+1) \times 10$ µL of the protein dilution with $(n+1) \times 60$ µL AB II and pipet 70 µL of this mixture into each well to be measured (*see* **Note 5**).

4. Fill an additional well with a mixture of 10 µL AB I and 60 µL AB II as a reference sample (buffer only) (*see* **Note 6**).

5. Transfer the plate to the plate reader and equilibrate temperature for 5–10 min. Place ATP or GTP solution at desired temperature.

6. Eject the plate. Add 50 µL of ATP or GTP solution using a multichannel pipette. Mix fast and thoroughly while avoiding bubbles (*see* **Notes 7** and **8**).

Optional: To automate substrate addition, 50 µL of ATP or GTP solution can instead be delivered via the injection port of the microplate reader.

Optional: If measurements are to be conducted on cyclases in their light-adapted state, individual wells in the plate reader can be illuminated using an automated customized setup [20].

If necessary, the illumination can be repeated during the measurement.

7. Measure the fluorescence intensity at intervals of 10 s to 1 min for a total duration of 10 min to 2 h, depending on the activity of the protein of interest, e.g., every 15 s for a total time of 15 min for 1 μM bPAC. Use settings of $\lambda_{Ex} = 500$ nm/$\lambda_{Em} = 535$ nm for measurements of I_{500}, and of $\lambda_{Ex} = 440$ nm/$\lambda_{Em} = 535$ nm for measurements of I_{440}.

8. Normalize by taking the ratio of I_{500} over I_{440}. Normalize to initial value at $t = 0$, correct for reference sample (*cf.* **step 4**) and plot normalized fluorescence against time (Fig. 3a).

9. Fit the linear range of the measurement by least-squares optimization to determine the initial velocity.

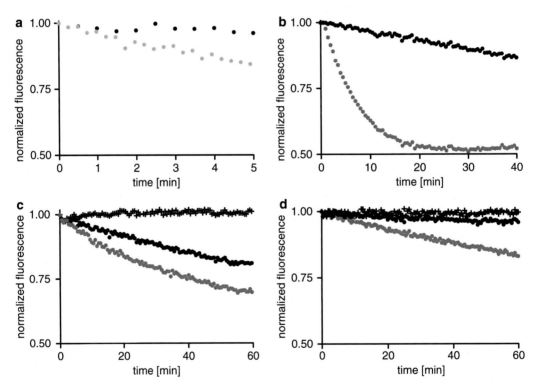

Fig. 3 Measurement of light-regulated adenylate cyclase and cNMP phosphodiesterase activities using the fluorometric BCECF assay. (**a**) cAMP production either catalyzed by 1 μM purified, 450-nm illuminated bPAC (*blue*), or by 5 μM dark-adapted purified bPAC (*black*) leads to a decrease in fluorescence intensity. Since bPAC recovery is too fast ($\tau \sim 25$ s at 20 °C) to maintain full activation throughout the measurement, the apparent dynamic range for light activation of around 25-fold does not necessarily reflect the actual stimulation factor. (**b**) cGMP hydrolysis catalyzed by 20 nM purified LAPD leads to a decrease in fluorescence intensity. This decrease is approximately 14-fold faster after illumination for 30 s with a 670 nm LED (*red*) relative to the dark-adapted sample (*black*). (**c**) When incubated with the BCECF assay components, lysate of *E. coli* cells containing LAPD displays a light-dependent fluorescence decrease indicative of PDE activity (*red*: 30 s illumination with 670 nm light, *black*: dark-adapted sample), which is completely inhibited by addition of 30 μM BAY-60-7550 (*crosses*). (**d**) Similarly, the light-dependent fluorescence decrease observed for *D. melanogaster* larvae homogenate is completely inhibited by BAY-60-7550 (colors: *see* **c**)

3.2 Determining the Activity of Purified Light-Activated PDEs

1. Preheat plate reader to desired temperature, e.g., 29 °C, for 30 min (*see* **Note 3**). All following steps should be carried out in the dark or under non-activating light conditions, e.g., dim green light for LAPD.

2. Prepare desired dilution of purified light-activated PDE in AB I (*see* **Note 4**).

3. For n samples, mix $(n+1) \times 10$ μL of the protein dilution with $(n+1) \times 60$ μL AB II and pipet 70 μL of this mixture into each well to be measured (*see* **Note 5**).

4. Fill an additional well with a mixture of 10 μL AB I and 60 μL AB II as a reference sample (buffer only) (*see* **Note 6**).

5. Transfer the plate to the plate reader and equilibrate temperature for 5–10 min. Place cAMP or cGMP solution at desired temperature.

6. Eject the plate. For determination of light-activated PDE activity, illuminate selected wells with light of the desired wavelength for 5 s to 1 min, depending on the light sensitivity of the protein of interest, e.g., for 30 s with 670-nm light for LAPD. Protect all other wells from illumination by covering them to be able to determine the activity of dark-adapted protein (*see* **Note 9**).

7. Add 50 μL of cAMP or cGMP solution using a multichannel pipette. Mix fast and thoroughly while avoiding bubbles (*see* **Note 7**).

   Optional: To automate substrate addition, 50 μL of cAMP or cGMP solution can instead be delivered via the injection port of the microplate reader.

8. Measure the fluorescence intensity at intervals of 10 s to 1 min for a total duration of 10 min to 2 h, depending on the activity of the protein of interest, e.g., every 30 s for a total time of 40 min for 20 nM LAPD. Use settings of $\lambda_{Ex} = 500$ nm/$\lambda_{Em} = 535$ nm for measurements of I_{500}, and of $\lambda_{Ex} = 440$ nm/$\lambda_{Em} = 535$ nm for measurements of I_{440}.

9. Normalize by taking the ratio of I_{500} over I_{440}. Normalize to initial value at $t = 0$, correct for reference sample (*cf.* **step 4**) and plot normalized fluorescence against time (Fig. 3b).

10. Fit the linear range of the measurement by least-squares optimization to determine the initial velocity (*see* **Note 10**).

3.3 Determining PDE Activity in E. coli Lysates

1. All following steps should be carried out in the dark or under non-activating light conditions, e.g., dim green light for LAPD. Harvest 2 mL of cell culture after expression of the phosphodiesterase of interest [13] by centrifugation at $11,000 \times g$ for 3 min, discard the supernatant, and resuspend the cells in 150 μL of lysis buffer. Slightly shake the suspension at 22 °C for 1–2 h.

2. Preheat plate reader to desired temperature, e.g., 29 °C, for 30 min (*see* **Note 3**).

3. Prepare the desired dilution of the lysate in AB I or directly proceed with undiluted lysate.

4. For n samples, mix $(n+1) \times 10$ µL of the lysate dilution with $(n+1) \times 60$ µL AB II and pipet 70 µL of this mixture into each well to be measured (*see* **Note 5**).

5. Fill an additional well with a mixture of 10 µL AB I and 60 µL AB II as a reference sample (buffer only) (*see* **Note 6**).

6. Transfer the plate to the plate reader and equilibrate temperature for 5–10 min. Place cAMP or cGMP solution at desired temperature.

7. Eject the plate. For determination of light-regulated PDE activity, illuminate selected wells with light of the desired wavelength for 5 s to 1 min, depending on the light sensitivity of the protein of interest, e.g., for 30 s with 670-nm light for LAPD. Protect all other wells from illumination by covering them to be able to determine the activity of dark-adapted protein (*see* **Note 9**).

8. Add 50 µL of cAMP or cGMP solution using a multichannel pipette. Mix fast and thoroughly while avoiding bubbles (*see* **Note 7**).

   Optional: To automate substrate addition, 50 µL of cAMP or cGMP solution can instead be delivered via the injection port of the microplate reader.

9. Measure the fluorescence intensity at intervals of 10 s to 1 min for a total duration of 10 min to 2 h, depending on the activity of the protein of interest, e.g., every 30 s for a total time of 40 min for LAPD. Use settings of $\lambda_{Ex} = 500$ nm/$\lambda_{Em} = 535$ nm for measurements of I_{500}, and of $\lambda_{Ex} = 440$ nm/$\lambda_{Em} = 535$ nm for measurements of I_{440}.

10. Normalize by taking the ratio of I_{500} over I_{440}. Normalize to initial value at $t = 0$, correct for reference sample (*cf.* **step 5**) and plot normalized fluorescence against time (Fig. 3c).

11. Fit the linear range of the measurement by least-squares optimization to determine the initial velocity (*see* **Note 10**).

12. To discriminate genuine PDE activity from other processes in crude lysate that could cause pH changes, a PDE-specific inhibitor can be added in control experiments. For example, at a concentration of 30 µM the inhibitor BAY-60-7550 [21] abolishes LAPD activity, and BCECF fluorescence should hence stay constant over time (crosses in Fig. 3c).

3.4 Determining PDE Activity in Homogenate Derived from D. melanogaster Tissue

1. Preheat plate reader to desired temperature, e.g., 29 °C, for 30 min (*see* **Note 3**). All following steps should be carried out in the dark or under non-activating light conditions, e.g., dim green light for LAPD.

2. Collect ten larvae expressing the phosphodiesterase (*see* **Note 11**) in a 1.5 mL reaction tube, wash with ddH$_2$O twice, and homogenize on ice with a homogenization pestle in 250 μL AB I for 1 min. Centrifuge homogenate at 12,000×g for 4 min. Use undiluted supernatant or desired dilution in AB I for the assay (*see* **Note 4**).

3. For n samples, mix $(n+1)\times10$ μL of the homogenate dilution with $(n+1)\times60$ μL AB II and pipet 70 μL of this mixture into each well to be measured (*see* **Note 5**).

4. Fill an additional well with a mixture of 10 μL AB I and 60 μL AB II as a reference sample (buffer only) (*see* **Note 6**).

5. Transfer the plate to the plate reader and equilibrate temperature for 5–10 min. Place cAMP or cGMP solution at desired temperature.

6. Eject the plate. For determination of light-activated PDE activity, illuminate selected wells with light of the desired wavelength for 5 s to 1 min, depending on the light sensitivity of the protein of interest, e.g., for 30 s with 670-nm light for LAPD. Protect all other wells from illumination by covering them to be able to determine the activity of dark-adapted protein (*see* **Note 9**).

7. Add 50 μL of cAMP or cGMP solution using a multichannel pipette. Mix fast and thoroughly while avoiding bubbles (*see* **Note 7**).

   Optional: To automate substrate addition, 50 μL of cAMP or cGMP solution can instead be delivered via the injection port of the microplate reader.

8. Measure the fluorescence intensity at intervals of 10 s to 1 min for a total duration of 10 min to 2 h, depending on the activity of the protein of interest, e.g., every 30 s for a total time of 40 min for LAPD. Use settings of $\lambda_{Ex}=500$ nm/$\lambda_{Em}=535$ nm for measurements of I_{500}, and of $\lambda_{Ex}=440$ nm/$\lambda_{Em}=535$ nm for measurements of I_{440}.

9. Normalize by taking the ratio of I_{500} over I_{440}. Normalize to initial value at $t=0$, correct for reference sample (*cf.* **step 4**) and plot normalized fluorescence against time (Fig. 3d).

10. Fit the linear range of the measurement by least-squares optimization to determine the initial velocity (*see* **Note 10**).

11. To discriminate genuine PDE activity from other processes in crude lysate that could cause pH changes, a PDE-specific inhibitor can be added in control experiments. For example, at a concentration of 30 µM the inhibitor BAY-60-7550 [21] abolishes LAPD activity, and BCECF fluorescence should hence stay constant over time (crosses in Fig. 3d).

3.5 Conclusion

In summary, we have presented the BCECF fluorometric assay that provides a convenient and inexpensive means of measuring light-dependent nucleotide cyclase and cNMP phosphodiesterase activities and that thus supplements existing assays for cNMP determination, such as HPLC and ELISA analysis. Of key advantage, the BCECF assay provides a continuous readout (as opposed to discrete time points) and can be automated. Moreover, the assay can be conducted for both purified proteins and crude cell lysates, as we demonstrate for the specific cases *E. coli* and *D. melanogaster* lysates. Based on these aspects, we expect the BCECF assay to facilitate the search for and quantitative characterization of new light-regulated cyclases and cNMP phosphodiesterases. Arguably, the assay will also prove of utility for light-insensitive nucleotide cyclases and cNMP phosphodiesterases. More generally, the BCECF assay could also apply to enzymes catalyzing formation and breakdown of other cyclic nucleotides, e.g., of cyclic diguanosine monophosphate [22].

4 Notes

1. All buffers should be prepared freshly. The necessarily low buffering capacity promotes fast acidification of the buffers due to CO_2 uptake from the environment. Additionally, dithiothreitol hydrolyzes over time and should therefore not be stored in solution for longer periods of time. The buffer composition can be varied, e.g., regarding the presence of divalent cations or reducing agents. Further, the buffering agent may be exchanged, ideally for compounds that are relatively temperature-insensitive.

2. Add one small spatula tip of a powdered complete protease inhibitor cocktail tablet, EDTA-free (Roche, Mannheim, Germany) per mL lysis buffer to avoid protein degradation by proteases during prolonged incubation at ambient assay temperatures.

3. The temperature for the reaction should be chosen according to the conditions best suited for the protein of interest. To ensure constant temperature conditions, the temperature should be at least 5 °C above room temperature for plate readers lacking active cooling.

4. The total reaction volume in the well is 120 μL. As 10 μL of the diluted protein will be added, the additional dilution factor of 1:12 should be taken into account when calculating the required dilution.

   The concentration of the protein should be chosen dependent on the specific activity of the enzyme. On the one hand, for proteins with fast recovery kinetics it is favorable that the reaction be completed before a significant portion of the protein has recovered to its dark-adapted state after illumination. On the other hand, the initial velocity of the reaction should be fitted based on as many data points as possible. Therefore at least five data points should be in the linear range of the reaction, thus precluding overly fast reactions. For light-activated LAPD or bPAC, 20 nM or 1 μM of purified enzyme were used, respectively.

5. When measuring illuminated and dark-adapted samples in parallel, choose wells with sufficient spacing so as to protect the dark-adapted samples from unintended illumination.

6. The reference sample is used for correction of fluctuations, e.g., in temperature, during the measurement. When doing measurements with concentrated protein solutions, we recommend adding 10 μL of protein storage buffer instead of AB I to the reference sample to compensate for solutes contained in the protein storage buffer.

7. Before first using the substrate solution, make sure that the pH is still at 7.6 after addition of ATP, GTP, cAMP, or cGMP and neutralize with base if necessary. Also add the substrate to the reference well which only contains buffers and no enzyme. Substrate solutions should be kept on ice except for temperature equilibration directly prior to the experiment.

8. When examining a protein with slow relaxation kinetics, it is also possible to illuminate the protein before addition of ATP or GTP solution as described in Subheading 3.2, **step 6**.

9. When examining a protein with fast relaxation kinetics, we recommend using an automated setup for repeated illumination during the measurement [20] as described in Subheading 3.1, **step 7**.

10. The assay can be calibrated by a combined BCECF-HPLC measurement (Fig. 4). To this end, it is necessary to prepare separate wells for each time point. Start the measurement by adding substrate but eject the plate from the plate reader after each measurement at desired time points. After the first measurement, transfer 100 μL of the solution in the first well to a 1.5 mL reaction tube and immediately stop the reaction at 95 °C for 10 min. Record the actual duration of the reaction (starting with addition of cNMP, ending with denaturation at 95 °C).

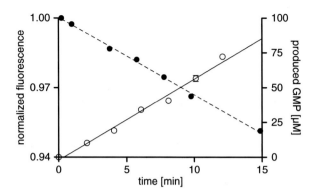

Fig. 4 Calibration of the BCECF assay for LAPD activity measurements by HPLC. Over reaction time, the normalized fluorescence intensity (*filled dots*) decreases and the amount of produced GMP, as determined by HPLC analysis (*open dots*), increases in linear fashion during at least the first 15 min of the measurement. Based on three independent measurements, a decrease in normalized fluorescence intensity of 0.1 corresponds to hydrolysis of 198 ± 25 µM cGMP at these specific assay conditions

Repeat this procedure for all other wells and time points. Dilute the denatured samples 50-fold in HPLC running buffer, centrifuge at $10,000 \times g$ and 4 °C for 10 min and filtrate the supernatant (0.2 µm pore size, Chromafil, Macherey-Nagel, Düren, Germany). Analyze on HPLC using published conditions [13].

11. We used wandering third instar *Drosophila* larvae. Genetically modified animals expressed LAPD in muscle cells as marked by the MHC-Gal4 trans activator; LAPD was ectopically expressed under UAS control.

Acknowledgements

Financial support by the Cluster of Excellence in Catalysis 'Unicat' (A.M.) of the Deutsche Forschungsgemeinschaft (DFG), through a Sofja-Kovalevskaya Award by the Alexander-von-Humboldt Foundation (A.M.), by the Caesar institute (H.G.K. and R.S.) as well as by a Heisenberg Fellowship by the DFG (M.S.) is gratefully acknowledged.

References

1. Deisseroth K, Feng G, Majewska AK, Miesenböck G, Ting A, Schnitzer MJ (2006) Next-generation optical technologies for illuminating genetically targeted brain circuits. J Neurosci 26:10380–10386

2. Nagel G, Ollig D, Fuhrmann M, Kateriya S, Musti AM, Bamberg E, Hegemann P (2002) Channelrhodopsin-1: a light-gated proton channel in green algae. Science 296: 2395–2398

3. Nagel G, Szellas T, Huhn W, Kateriya S, Adeishvili N, Berthold P, Ollig D, Hegemann P, Bamberg E (2003) Channelrhodopsin-2, a directly light-gated cation-selective membrane channel. Proc Natl Acad Sci U S A 100:13940–13945

4. Boyden ES, Zhang F, Bamberg E, Nagel G, Deisseroth K (2005) Millisecond-timescale, genetically targeted optical control of neural activity. Nat Neurosci 8:1263–1268

5. Li X, Gutierrez DV, Hanson MG, Han J, Mark MD, Chiel H, Hegemann P, Landmesser LT, Herlitze S (2005) Fast noninvasive activation and inhibition of neural and network activity by vertebrate rhodopsin and green algae channelrhodopsin. Proc Natl Acad Sci U S A 102:17816–17821

6. Zhang F, Wang L-P, Brauner M et al (2007) Multimodal fast optical interrogation of neural circuitry. Nature 446:633–639

7. Schröder-Lang S, Schwärzel M, Seifert R, Strünker T, Kateriya S, Looser J, Watanabe M, Kaupp UB, Hegemann P, Nagel G (2007) Fast manipulation of cellular cAMP level by light in vivo. Nat Methods 4:39–42

8. Ryu M-H, Moskvin OV, Siltberg-Liberles J, Gomelsky M (2010) Natural and engineered photoactivated nucleotidyl cyclases for optogenetic applications. J Biol Chem 285: 41501–41508

9. Stierl M, Stumpf P, Udwari D et al (2011) Light-modulation of cellular cAMP by a small bacterial photoactivated adenylyl cyclase, bPAC, of the soil bacterium beggiatoa. J Biol Chem 286:1181–1188

10. Raffelberg S, Wang L, Gao S, Losi A, Gärtner W, Nagel G (2013) A LOV-domain-mediated blue-light-activated adenylate (adenylyl) cyclase from the cyanobacterium Microcoleus chthonoplastes PCC 7420. Biochem J 455:359–365

11. Avelar GM, Schumacher RI, Zaini PA, Leonard G, Richards TA, Gomes SL (2014) A rhodopsin-guanylyl cyclase gene fusion functions in visual perception in a fungus. Curr Biol 24:1234–1240

12. Möglich A, Moffat K (2010) Engineered photoreceptors as novel optogenetic tools. Photochem Photobiol Sci 9:1286–1300

13. Gasser CF, Taiber S, Yeh C-M, Wittig CH, Hegemann P, Ryu S, Wunder F, Möglich A (2014) Engineering a red-light-activated human cAMP/cGMP-specific phosphodiesterase. Proc Natl Acad Sci U S A 111:8803–8808

14. Ryu M-H, Kang I-H, Nelson MD, Jensen TM, Lyuksyutova AI, Siltberg-Liberles J, Raizen DM, Gomelsky M (2014) Engineering adenylate cyclases regulated by near-infrared window light. Proc Natl Acad Sci U S A 111:10167–10172

15. Hughes J, Lamparter T, Mittmann F, Hartmann E, Gärtner W, Wilde A, Börner T (1997) A prokaryotic phytochrome. Nature 386:663

16. Rockwell NC, Su YS, Lagarias JC (2006) Phytochrome structure and signaling mechanisms. Annu Rev Plant Biol 57:837–858

17. Pandit J, Forman MD, Fennell KF, Dillman KS, Menniti FS (2009) Mechanism for the allosteric regulation of phosphodiesterase 2A deduced from the X-ray structure of a near full-length construct. Proc Natl Acad Sci U S A 106:18225–18230

18. Rink TJ, Tsien RY, Pozzan T (1982) Cytoplasmic pH and free Mg2+ in lymphocytes. J Cell Biol 95:189–196

19. Boens N, Qin W, Basarić N, Orte A, Talavera EM, Alvarez-Pez JM (2006) Photophysics of the fluorescent pH indicator BCECF. J Phys Chem A 110:9334–9343

20. Richter F, Scheib US, Mehlhorn J, Schubert R, Wietek J, Gernetzki O, Hegemann P, Mathes T, Möglich A (2015) Upgrading a microplate reader for photobiology and all-optical experiments. Photochem Photobiol Sci 14:270–279

21. Boess FG, Hendrix M, van der Staay F-J, Erb C, Schreiber R, van Staveren W, de Vente J, Prickaerts J, Blokland A, Koenig G (2004) Inhibition of phosphodiesterase 2 increases neuronal cGMP, synaptic plasticity and memory performance. Neuropharmacology 47: 1081–1092

22. Gomelsky M (2011) cAMP, c-di-GMP, c-di-AMP and now cGMP: bacteria use them all! Mol Microbiol 79:562–565

Chapter 8

Optogenetic Control of Pancreatic Islets

Thomas M. Reinbothe and Inês G. Mollet

Abstract

In light of the emerging diabetes epidemic, new experimental approaches in islet research are needed to elucidate the mechanisms behind pancreatic islet dysfunction and to facilitate the development of more effective therapies. Optogenetics has created numerous new experimental tools enabling us to gain insights into processes little was known about before. The spatial and temporal precision that it can achieve is also attractive for studying the cells of the pancreatic islet and we set out to explore the possibilities of this technology for our purposes. We here describe how to use the islets of an "optogenetic beta-cell" mouse line in islet batch incubations and Ca^{2+} imaging experiments. This protocol enables light-induced insulin release and provides an all-optical solution to control and measure intracellular Ca^{2+} levels in pancreatic beta-cells. The technique is easy to set up and provides a useful tool for controlling the activity of distinct islet cell populations.

Key words Pancreas, Islets, Beta-cell, Diabetes, Insulin, Channelrhodopsin-2, Calcium, Optogenetics, All-optical

---

1 Introduction

Optogenetics has lead to groundbreaking discoveries in medical science [1] and the number of related articles being published per year has grown exponentially over the last 5 years. However, its main advantage, the cell-specific and spatiotemporally precise control of distinct cell subpopulations, has to date been mainly exploited in the field of neuroscience [2]. Although the brain is central to the regulation of energy homeostasis and metabolism [3], and optogenetics has also contributed to advances in this area of research [4], the sole tissue that controls the release of the key blood glucose-regulating hormones are the islets of Langerhans. Their two main cell-types, the beta- and alpha-cells are responsible for maintaining glucose homeostasis by releasing insulin and glucagon, respectively. Under normal circumstances, insulin is released from beta-cells upon increased blood glucose levels while glucagon is secreted when the glucose concentration falls below a critical point during fasting. In patients suffering from type-2 diabetes, the

Arash Kianianmomeni (ed.), *Optogenetics: Methods and Protocols*, Methods in Molecular Biology, vol. 1408,
DOI 10.1007/978-1-4939-3512-3_8, © Springer Science+Business Media New York 2016

counter-regulatory pattern of the two hormones is greatly perturbed, with a lack of insulin and an excess of glucagon both leading to hyperglycemia, the main hallmark of the disease [5]. The pathways leading to the opposing effects of these two hormones, under the same pathophysiological stimuli, are incompletely understood [6]. To experimentally isolate externally induced glucose-dependent effects from their intrinsic (basal) activity, islets are often routinely isolated from rodents and tested for their in vitro secretory capacity in batch incubations. This approach however does neither allow for the precise stimulation of one distinct cell population within the islet, let alone one cell, nor to swiftly manipulate the stimulus, i.e. both the spatial and temporal precision is greatly limited. To overcome this, electrophysiological ("patch-clamping") measurements are traditionally performed. However, this requires the islets to be dissociated into single cells with the consequent loss of physiologically important interactions inherent to their architecture. Alternatively, the (electrical) activity of individual cells can also be assessed in intact islets [7], but this allows measurement of only one (peripheral) cell at the time and is both a very difficult and time-consuming undertaking.

The prospect to control both the activity of a complete islet cell subpopulation and to adjust the stimulatory pattern on a millisecond time scale encouraged us to explore the suitability of optogenetics for islet studies, that we recently reported [8]. In the following, we describe this approach in detail to allow other researchers in the field to take advantage of this possibility.

2   Materials

2.1   Buffers and Solutions

Buffers should be prepared with double-deionized water (ddH$_2$O) and can be stored at room temperature unless otherwise specified.

1. 1 M Tris-HCl: Weigh 60.6 g Tris Base into ddH$_2$O, adjust pH to 8 with HCl and the volume to 500 ml, autoclave.

2. 1 M D-Glucose: Add 9 g of D-glucose to 25 ml ddH$_2$O and adjust to a final volume of 50 ml while stirring. Sterile filter by passing the solution through a 0.2 μm filter. Store at 4 °C.

3. 1 M NaOH: Add 2 g of NaOH to 30 ml ddH$_2$O, adjust to 50 ml and sterile filter.

   Caution: *Corrosive—may damage skin and eyes. Wear safety equipment!*

4. 50 mM NaOH: Prepare 100 ml by dilution with ddH$_2$O from 1 M NaOH (*see* Subheading 3).

   Caution: *Corrosive—may damage skin and eyes. Wear safety equipment!*

5. Hank's buffer salt solution (HBSS): Dissolve a vial of Hank's buffer salt in ddH$_2$O, add 0.35 g NaHCO$_3$, adjust to 1 l, if necessary adjust pH to 7.4 with 1 M NaOH, keep at 4 °C.

6. 13.3 U/ml collagenase type V stock (Sigma): Add 10.5 ml HBSS to 100 mg collagenase. Mix well. Store at –20 °C as 1 ml aliquots.

7. 0.1 M NaHCO$_3$: Prepare a solution of 0.1 M NaHCO$_3$ by adding 0.42 g of NaHCO$_3$ to a final volume of 50 ml ddH$_2$O and adjust pH to 8.

8. Islet medium: RPMI 1640 supplemented with 4 mM L-Glutamine, 10 mM glucose, 5 % FBS, 100 μg/ml streptomycin and 100 units/ml penicillin.

9. Krebs-Ringer's bicarbonate buffer (KRBB): 120 mM NaCl (7.01 g), 4.7 mM KCl (0.35 g), 1.2 mM KH$_2$PO$_4$ (0.16 g), 1.2 mM MgSO$_4$ (0.14 g), 25 mM NaHCO$_3$ (2.1 g), 2.5 mM CaCl$_2$ (0.28 g), 10 mM HEPES (2.38 g) in ddH$_2$O, adjust pH to 7.4 with NaOH and volume to 1 l. Store at 4 °C for a maximum of 3 days. Add 0.1 % [w/v] BSA (1 g) and 1 M D-glucose to the desired concentration on the day of use. Waterbath and carbogen gas (95 % O$_2$ and 5 % CO$_2$) are needed for final preparation. *See* **Note 1**.

10. Calcium imaging buffer 1: 140 mM NaCl (8.18 g), 3.6 mM KCl (0.35 g), 0.5 mM NaH$_2$PO$_4$ (0.06 g), 0.5 mM MgSO$_4$ (0.06 g), 2 mM NaHCO$_3$ (0.17 g), 2.5 mM CaCl$_2$ (0.28 g), 5 mM HEPES (1.19 g), adjust pH to 7.4 with NaOH, adjust volume to 1 l, and add D-glucose to desired concentration freshly.

11. Calcium imaging buffer 2 (High potassium buffer): Same as above but with 50 mM KCl (3.73 g for 1 l) and 93.6 mM NaCl (5.47 g) instead of 140 mM NaCl.

2.2 Animals and Genotyping

Important: For animal breeding and experiments, national animal welfare regulations and guidelines must be followed and ethical permission be granted prior to commencement.

1. Mouse strain #012569 (The Jackson Laboratory, Bar Harbor, ME, USA).

2. Mouse strain #003573 (The Jackson Laboratory).

3. Mouse ear puncher.

4. Table-top centrifuge.

5. RipCre PCR: Forward primer 5′TGC CAC GAC CAA GTG3′ and reverse primer 5′CAA GGT TAC GGA TAT3′ in 10 mM mix (each).

6. ChR2-YFP PCR: Wild-type allele primer forward 5′AAG GGA GCT GCA GTG GAG TA3′ and reverse 5′CCG AAA ATC

TGT GGG AAG TC3'; for the transgenic allele forward/ reverse primers 5'ACA TGG TCC TGC TGG AGT TC3' and 5'GGC ATT AAA GCA GCG TAT CC3' in 10 mM mix.

7. Standard Taq polymerase.

8. Hot Start polymerase.

2.3 Islet Preparation and Culture

1. Noncoated plastic petri dishes 35×10 mm.

2. Autoclavable glass or nonstick plastic petri dishes 60×10 mm.

3. Centrifugation tubes PS 11 ml for washing islets.

4. Waterbath, instruments for mouse surgery, dissection microscope.

5. Picking microscope and dimmable light source.

6. Petri dishes 35 mm with No. 1 thickness (0.13–16 mm) glass bottom.

7. Cell-Tak (BD Biosciences, Franklin Lakes, NJ, USA).

8. Bovine serum albumin (BSA).

9. Humidified incubator 37 °C, 5 % CO_2.

2.4 Batch Incubation LED Illuminations

1. Ellerman plastic tubes 55×11 mm and 11 mm plug caps.

2. Pointy forceps e.g. Dumont #3C or needle.

3. Light-emitting diodes (LED) 470.

4. Laboratory power supply such as Easy Power PS2000B (Elektro-Automatik GmbH, Germany).

5. Plastic housing or an empty pipette tip box.

6. Resistors 160 Ω.

7. Highly flexible twin cable 0.75 mm².

8. Banana plugs 4 mm.

9. Silicone and superglue.

10. Soldering equipment, fine file, small hand saw with fine teeth.

11. No. 1 cover glasses 18×18 mm.

12. Optical power meter such as PM100D with S120C Photodiode (Thorlabs Sweden AB, Göteborg, Sweden).

13. BNC Female to RCA Male Adapter.

2.5 Ca^{2+} Sensitive Dyes

1. Leak-resistant Ca^{2+} sensitive Fura-2 (Teflabs, Blackdog Technical Services Inc., Jackson Springs, NC, USA) or other suitable dye. *See* **Note 2**.

To make a stock dissolve a 50 μg vial of leak-resistant Fura-2 in 50 μl DMSO (1μg/μl, 0.8 mM stock). Store in 10 μl aliquots at –20 °C. Each aliquot is sufficient for loading islets in four glass-bottom dishes. Once thawed, do not refreeze.

2.6 Ca²⁺ Imaging Setup

1. Inverted epifluorescence microscope and illumination/detection system with trigger out (TTL signal) port.

2. Fiber-coupled blue LED (WT&T, Pierrefonds, QC, Canada) with an attached miniature collimator (WT&T).

3. YFP and Fura-2 excitation and emission filterset. *See* **Note 2**.

4. Micromanipulator system with head stage holder as used for patch clamp (e.g. MHW-3, Narishige, London, UK) to hold the fiber-coupled LED. *See* **Note 3**.

5. Perfusion pump allowing bi-directional (in/out) fine-tuning of the flow.

6. Heated perfusion tube and temperature control system (Alascience, NY, USA) with temperature control sensor mounted on the imaging chamber perfusion inlet to sense the buffer temperature in real-time.

3 Methods

3.1 Mouse Breeding and Genotyping

To leverage cellular activation by blue light-induced depolarization, the ChR-2(H134R) variant is readily available as part of a (Cre-dependent) CAG promoter driven loxP-STOP-loxP-ChR2(H134R)YFP transgene in a mouse line [9] distributed by The Jackson Laboratory (Strain no. 012569). To achieve expression specifically in beta-cells, breed mice with Ins2Cre ("RipCre") mice or an alternative cell-specific Cre-driver line. We used an in-house colony, originally created in the Herrera lab [10] but The Jackson Laboratory holds strain numbers 018960 and 003573 for purchase that may be used instead. The presence of Cre recombinase in beta-cells causes the excision of the STOP cassette from the CAG-loxP-STOP-loxP-ChR2(H134R)YFP transgene, leading to cell-specific expression of the transgene ("ChR2-YFP" in the following). All offspring must be genotyped. *See* **Note 4**.

1. At weaning take mouse ear biopsies.

2. Add 300 μl 50 mM NaOH to each biopsy.

3. Incubate for 1 h at 98 °C, vortex and return to heat for 30 min.

4. Vortex again and neutralize by adding 30 μl 1 M Tris-HCl to each biopsy, vortex.

5. Centrifuge for 10 min at >10,000×g to precipitate remaining tissue.

6. Transfer the supernatant to a new tube.

7. Use 1 μl of the supernatant in the PCR reaction and 10 mM primers.

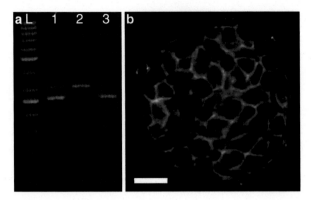

Fig. 1 (**a**) Genotyping result for ChR2-YFP PCR: 10 μl PCR products run on 1.5 % gel. L: 50 bp DNA ladder. Lane 1: Hemizygous transgenic, lane 2: WT, lane 3: Homozygous transgenic. (**b**) Fluorescent signal in islets of transgenic animals (ChR2-YFP$^{+/-}$RipCre$^{+/-}$). Scale bar 20 μm

8. For RipCre PCR, with Taq polymerase, use an annealing temperature of 55 °C and 30 cycles following the polymerase manufacturer's instructions.

9. For ChR2-YFP PCR use Hot Start polymerase with an annealing temperature of 62 °C and 35 cycles following the polymerase manufacturer's instructions.

10. Run 10 μl of the products on a 1.5–2 % agarose gel and expect bands of 250 bp for the ChR2-YFP transgene, 300 bp for the WT allele, and a band of 750 bp for RipCre (*see* Fig. 1).

3.2 Islet Preparations

1. Prepare collagenase working solution after thawing an aliquot on ice and add 9 ml HBSS, which is sufficient for three mice.

2. Prepare mouse islets by collagenase injection into the common bile duct according to established protocols [11, 12] with the following amendments:

   - After incubation of the pancreas at 37 °C, add 30 ml ice-cold HBSS to stop the reaction, shake vigorously for >30 s to dissociate the tissue, and immediately pour 10 ml of the homogenized organ into 11 ml PS centrifugation tubes placed on ice.

   - Instead of centrifugation to remove exocrine tissue, let the islets sink down in the tubes for 4 min and carefully remove the top 9 ml of buffer containing mostly exocrine tissue by pipetting or vacuum. Add 9 ml HBSS, invert the tube several times and incubate again. Repeat 3–5 times until the top buffer appears clear at the end of the 4 min incubation. Handpick several times to remove all exocrine cells and collect islets in a petri dish with HBSS on ice. To limit direct light exposure of the transgenic islets, dim the picking microscope illumination system to the minimum while still enabling comfortable vision. *See* **Notes 5** and **6**.

3.3 Preparing Dishes for Whole Islet Imaging

During the collagenase incubation (*see* Subheading 3.2) coat glass-bottom dishes under a biological work cabinet as follows:

1. To prepare one dish, add 57 µl of 0.1 mM $NaHCO_3$ to the dish, add 2 µl of Cell-Tak, mix by pipetting, add 1 µl of 1 M NaOH (*see* Subheading 3.2, **step 1**) and mix once again. Scale up as necessary. *Critical*: Work swiftly; after adding NaOH, Cell-Tak may start binding to the reaction tube.

2. Incubate for a minimum of 20 min.

3. Remove the solution by vacuum or pipetting and wash 2× with 500 µl ddH_2O.

4. Let the dishes dry and use directly or wrap with Parafilm and store at 4 °C for max. 1 week until use.

5. Add 2.5 ml RPMI 1640 medium to each dish and continue work under the picking microscope. Collect >20 islets from the collection petri dish on ice in a volume not exceeding 20 µl and place them carefully close together in the center of the glass-bottom dish. *Critical*: To avoid that islets stick to the inside of the pipette tip, rinse it in medium by pipetting prior to use.

6. When placing the islets, lift the lid of the dish containing the medium only minimally to avoid contamination. Ideally use a microscope mounted under a biological work cabinet.

7. Place all dishes into a larger 15 cm dish or tray to even out eventual horizontal shaking during transport.

8. Incubate overnight in a humidified incubator at 37 °C, 5 % CO_2 to let islets attach properly. Open and close incubator doors carefully to avoid vibrations.

3.4 Preparing a LED Illumination System for Islet Batch Incubations

Caution: *If you do not have experience with building and soldering on electrical systems, please consult an electrician and/or your local lab workshop. Exposed electrical connections and mistakes during soldering could lead to fire and serious injury. All soldered unprotected leads must be insulated using electrical tape or by heat shrink tubing.*

1. Cut or file small openings into the sides of the plastic housing to accommodate the cables.

2. Cut twin cable into 50 cm long pieces, separate the leads slightly on each end and remove about 1 cm of insulation from both ends.

3. Put drops of silicone into the housing, 3–4 cm apart, two for each resistor.

4. Trim the resistors to length leaving about 2 cm on each end.

5. Place each resistor into a pair of silicone drops and place the resistors parallel to each other. Leave about 1.5 cm space on each side and leave about 1 cm space between each resistor. Let the silicone solidify.

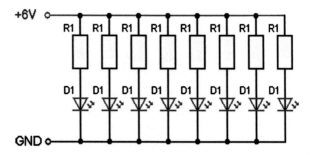

Fig. 2 Circuit diagram for soldering eight LEDs in parallel. R1 = resistors 160 Ω, GND = Ground, D1 = Light-emitting diode (LED) 470 nm

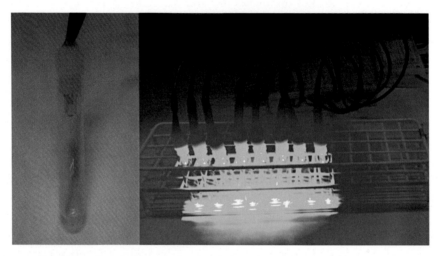

Fig. 3 LED mounted to cap of incubation tube (*left*) and as set of 8 ready for incubation with blue light switched on (*right*). Illumination is set to maximum for demonstration purposes

6. Solder one lead of each twin cable to one end of the resistor (–), trim the other end in length and solder to a common pole connecting all secondary poles of the resistor to (+) (*see* Fig. 2). Connect (+) and (–) leads to each one of the banana plugs.

7. Close the housing properly with screws.

8. Pinch two holes (2.5 mm apart) into the tube caps and push the LEDs through the holes. *See* Fig. 3.

9. To hold them in place, put tubes into a rack, attach the caps, and fill the open side of the caps with silicone and let solidify overnight. *Critical*: The LEDs should face straight downwards to deliver the light on to the islets.

10. Put heat shrink tubing sleeves around the unconnected twin cables originating from the plastic housing before they are soldered to the LEDs. The longer leg of the LED marks (+), the shorter one (–). After soldering, heat up the heat shrink tube to properly isolate the exposed electrical contacts.

11. Testing the illumination system: Cut off the round bottom of an Ellerman plastic tube using a small saw and glue the even (noncut) end of the tube to a No. 1 cover glass.

12. Place a cap with LED on the tube and press it in while holding the tube. Placing it on the table while pressing may break the glass.

13. Switch on the power meter and set it to the appropriate wavelength. Switch on the power source driving the LEDs and set it to 6 V. At this setting the voltage at the LEDs will measure 3.2 V. *See* **Note 7**.

14. Place the tube with attached coverslip onto the power meter, note the intensity, and check that it is identical for all the LEDs.

15. Mark the light spot outer diameter with a pen while illuminating on a piece of paper. Switch off the LED and use a scale to measure the diameter and to calculate intensity per area.

3.5 In Vitro Batch Islet Stimulations with LED Light

Caution: *Do not look directly into the beam of power LEDs as the bright light may damage the eyes.*

1. Freshly prepare KRBB with 2.8 and 20.2 mM D-glucose, bubble for 2 min with carbogen gas (95 % O_2, 5 % CO_2) using a glass Pasteur pipette and add BSA to 0.1 % [w/v]. Let the BSA sink in for 5 min and mix by turning the tube upside down several times. Keep at RT for the duration of the experiment and discard any leftover.

2. Place 1 ml KRBB with basal glucose concentration (2.8 mM) into Ellerman plastic tubes (*see* Subheading 2.3) in appropriate racks and place 12 islets into each tube.

3. Pick islets under a picking microscope using a 10 µl pipette and place them centrally at the bottom of the tubes.

4. Gas each tube with carbogen to place a 95 % O_2, 5 % CO_2 atmosphere above the buffer and quickly close the tube with the cap (without LED). Do not gas the solution.

5. Place the rack into a pre-warmed 37 °C waterbath and pre-incubate for 30 min. *See* **Note 8**.

6. Remove the rack from the waterbath, remove the caps, slowly aspirate 0.8 ml of the solution while observing the islets under the picking microscope.

7. To half of the tubes add 0.8 ml KRBB with 2.8 mM glucose to serve as reference, to the rest add 0.8 ml KRBB containing 20.2 mM glucose (giving a final concentration of 16.7 mM glucose). *Critical*: Remove and add the buffer slowly (add towards the wall of the tubes) to minimize turbulence.

8. Overlay the solution in each tube with carbogen (*see* Subheading 4) and immediately close with caps containing the LEDs (or without LED as control) to maintain the 95 % O_2, 5 % CO_2 atmosphere.

9. Put the rack into the waterbath and incubate for 1 h while stimulating with blue light. *Critical*: Place control nonilluminated tubes in a separate rack protected by aluminum foil to prevent any blue light reaching the control islets.

10. Switch on the power supply and set to 6 V. This will deliver the maximum of 3.2 V to the LEDs. *See* **Note 9**.

11. Disconnect the power source, remove the rack from the waterbath, remove the caps, and while looking against the light slowly sample 400 μl buffer from each tube by using the lower marked line on the tubes as reference. *Critical*: The islets should not move.

12. Analyze for hormones directly using ELISA or RIA or freeze the samples.

3.6 Ca²⁺ Indicator Loading of Islets

Important: *Work in low light.*

1. Thaw a vial of leak resistant Fura-2 stock and bring to room temperature.

2. Adjust Calcium imaging buffer 1 freshly to 2.8 mM glucose. The volume of buffer needed depends on the duration of the experiment and perfusion speed (e.g. 1 ml/min). Also prepare Calcium imaging buffer 1 with 16.7 mM glucose and Calcium imaging buffer 2 with 2.8 mM glucose, both needed as positive controls.

3. Fura-2 working solution: Dissolve 2.5 μl of 0.8 mM Fura-2 stock in 1 ml of Calcium imaging buffer 1 supplemented with 2.8 mM glucose (final concentration 2 μM Fura-2). Vortex to mix thoroughly.

4. Remove the medium from the side of the plate without emptying the center glass bottom. Wash once with 2 ml of 2.8 mM glucose Calcium imaging buffer 1 by carefully pipetting the buffer to the side of the dish and removing it slowly. Add the 1 ml of prepared Fura-2 to the side of the dish and incubate for 40 min at 37 °C protected from light.

5. Remove the Fura-2 incubation buffer from the side of the dish without emptying the center glass bottom. Wash once with 2 ml of 2.8 mM glucose calcium imaging buffer 1. Add 2 ml of fresh buffer without Fura-2 to de-esterify. Incubate for at least 30 min at 37 °C prior to imaging.

3.7 Preparing for the Light Stimulation and Ca²⁺ Imaging

Caution: *Do not look directly at or through the microscope into the beam of power LEDs as the bright light may damage the eyes.*

Important: *Do not bend the fiber.*

An inexpensive way to stimulate the islets directly on an imaging microscope without the need to purchase an additional mirror-based illumination system operating through the side ports of the microscope is to use a fiber-coupled LED.

Before starting the experiment, the illuminating LED has to be calibrated to deliver the desired power to the biological specimen. This is achieved by detecting the emitted light intensity using a (handheld or mounted) power meter at the imaging plane and by measuring the size of the light spot. Since you must not align the LED while looking into the microscope or while the islets are mounted, as this will already stimulate them, this is done in advance. The instructions given below describe the procedure for an inverted microscope.

1. Fasten the fiber-coupled LED to a mounting post (6 mm diameter and 7 cm length) using tape. Attach the post directly to the micromanipulator holder of the headstage. You may need to thicken the post with tape to make it tightly fit into the holder. *See* Fig. 4 and **Note 3**.

2. Mark a glass bottom dish with a dot in the center using a waterproof marker and place it on the microscope stage.

3. Check under the microscope that the marked dot is directly in the visual center corresponding to the central point of where

Fig. 4 Fiber-coupled LED mounted on post with fitted collimator and filter. Custom-made plastic insert with grooves for the perfusion inlets and outlets and temperature sensor. Note the distinct illumination in the center of the dish

images are acquired. You may also check in live mode using the camera.

4. Remove the center plate of the microscope stage, maximally wind down the objective and place the power meter probe in the theoretical specimen mounting position/focal plane. Switch on the LED, measure the light intensity, and adjust the power of the LED to desired value. Mark the power knob position for later reference. Switch off the LED.

5. Switch the mirror to the acquisition (camera) position or otherwise block the beampath to the eyepiece.

6. Switch on the fiber-coupled LED and direct it onto the marked center of the glass bottom dish while observing from a lateral position. Lower the intensity and do not look directly into the beam. Fix the LED in this position. Mark the outer diameter of the beam with a fine waterproof marker on the dish. Switch the LED power source to remote control mode and now change neither the LED's position nor the position of the microscope stage. Connect the LED remote control port to the trigger box (TTL out) of the microscope acquisition system. You will eventually need a BNC Female to RCA Male Adapter.

3.8 Islet Perfusion and Temperature Control

1. Carefully place the plastic perfusion insert (*see* Fig. 4) into the glass-bottom dish containing the dye-loaded islets.

2. Place the dish on the microscope stage, switch mirror to the visual (eyepiece) position and position the islets in the visual center by moving the dish. Use low transmitted light to visualize. Try to get several islets into the field of view. *Critical*: Do not move the microscope stage, as this will also change the position of the LED meaning that you will have to re-calibrate the LED's position.

3. Place the perfusion in- and outlet nozzles into the grooves of the insert. Begin perfusion with 2.8 mM Calcium imaging buffer 1 at a rate ~1 ml/min.

4. Switch on the perfusion pump and temperature control system for the tube heating element. Make sure the temperature sensor is in contact with the buffer.

5. Verify that the temperature is stable at a value below 37 °C to avoid accidental overheating of the islets due to temperature fluctuations. *Critical*: Whenever switching off the perfusion, turn off the tube-heating element first.

6. Perfuse for 5–10 min until temperature and buffer levels have stabilized.

   Critical: The temperature and buffer level must be stable during the experiment.

3.9 Image Acquisition and Light Stimulation

1. Verify a positive YFP signal in your transgenic mouse islets and absence in the control islets. You may actually place and image them simultaneously in the same dish.

2. Take a test image of the islets using the Fura-2 filter set with illumination at 340 and 380 nm settings. Take note of the ideal exposure times and keep them constant for the whole set of experiments.

3. Within the acquisition software create a light trigger protocol [3.5–5 V TTL signal out] in order to remote control the fiber-coupled LED for the desired duration and frequency. Start with between 1 and 10 ms. Take into account the exposure time in order not to stimulate the islets simultaneously with image acquisition. This prevents recording the light emitted by the blue LED light. *See* **Notes 2** and **9**.

4. Mark regions for analysis: Select a background region and regions of interest (ROI) covering one or several islets or parts thereof. Save the region selection.

5. Towards the end of the light stimulation protocol perfuse with Calcium imaging buffer 1 (high glucose) or 2 (high potassium) serving as positive controls. *See* **Note 10**.

3.10 Calibration

Absolute Ca^{2+} concentrations can be calculated after calibration with cells (islets) and ionophores such as ionomycin or A23187, cell-free using a dilution series of buffers with known Ca^{2+} concentration or both in combination. Commercially available (Fura-2) Ca^{2+} imaging calibration kits may be a convenient alternative. The absolute Ca^{2+} concentration can then be inferred by using the calculations as in [13].

3.11 Data Analysis

The data analysis features of imaging acquisition software differ widely and this is only a short example for processing the data.

Open the data file produced during export from the acquisition software using spreadsheet software. Subtract the background values and calculate the 340/380 nm ratio for each ROI using the following formula: $\text{Ratio}_{340/380}$ $(R_{340/380}) = (\text{Mean intensity}_{340} - \text{Mean intensity Background}_{340})/(\text{Mean intensity}_{380} - \text{Mean intensity background}_{380})$. Plot against time. To determine the position and intensity of the peak of each calcium oscillation, calculate the % difference in ratio $(R_{340/380})$ for each successive pair of points: $\text{Percent increase} = (Ri_{+1} - Ri)/(100 \times Ri)$.

4 Notes

1. As an alternative buffer to KRBB (*see* Subheading 2.1, **item 9**), Krebs Ringer's HEPES buffer (KRBH) with 2 mM $NaHCO_3$ and 25 mM HEPES [14] may be used, not requiring gassing with carbogen. Results may differ.

2. Using a red-shifted calcium indicator instead of Fura-2 (*see* Subheading 3.6, **step 1**) and an additional narrowing bandpass emission filter (e.g. FB460-10-12.5-SP and FC/PC collimation package F230FC-aligned to 460 nm, both Thorlabs) for the blue LED (*see* Subheading 2.6, **item 2**) may allow for islet LED stimulation and simultaneous, rather than sequential or post-stimulus $[Ca^{2+}]i$ signal recording. However, depending on technical (the imaging system being used, its sensitivity, etc.), and biological influences (dye-loading success, etc.), already the easy-to-build solution presented above, allows for high time resolution (comparable to electrophysiological recordings) as the TTL signal switch, stimulatory illumination (1–10 ms) and image acquisition (10–100 ms) are all happening within 10–100 ms; fast enough to (sequentially) record changes in $[Ca^{2+}]i$ and perhaps membrane potential. That said, to use the $[Ca^{2+}]i$ as physiological readout is only an example. Fluorescent or near-infrared voltage-sensitive dyes may also be a possibility.

3. As alternative to the micromanipulator system holding the LED during Ca^{2+} imaging (*see* Subheading 3.7, **step 1**), miniature mounting posts and clamps (both available from e.g. Thorlabs) can be used to build a stand small enough to fit onto the microscope stage without interfering with its use (*see* Fig. 5).

4. In your breedings (*see* Subheading 3.1), keep animals hemizygous for both transgenes as animals that are both homozygous

Fig. 5 Alternative mounting of fiber-coupled LED using posts and clamps with fitted miniature collimator. Maximum illumination intensity for demonstration

ChR2-YFP and hemizygous for RipCre (ChR2-YFP$^{+/+}$RipCre$^{+/-}$) exhibit lower weight gain and may not be fertile [own observations], and should thus not be used. If the original strains have different genetic backgrounds, as is the case for the Herrera lab RipCre strain and the used ChR2-YFP strain (Jax strain no. 012569), mice should be inbred for several generations in order to produce genetically identical animals. For more information on colony management see elsewhere [15].

5. For islets that are to be used in microscopy, handpick islets in HBSS without adding BSA (*see* Subheading 3.2, **step 2**). Although BSA may increase yield [16] and facilitates easier and faster picking, it reduces the likelihood of the islets to attach to the glass-bottom dishes. Pick in sterilized glass petri dishes or plastic dishes with anti-stick surface to avoid attachment of islets during the process. You may also rinse the dish surface you pick on shortly in HBSS with BSA while proceeding with actually picking in HBSS without BSA.

6. Although animals have been genotyped, the presence of YFP in the ChR2-YFP islets (beta-cells) should be verified by fluorescence microscopy (*see* Fig. 1a). Use an YFP or a GFP filter set, the latter should work as good.

7. The voltage applied to the LEDs (*see* Subheading 3.4, **step 13**) should be initially tested for by using a multimeter.

8. Instead of incubating the islets in the waterbath during the batch incubations (*see* Subheading 3.5, **step 5**), a humidified 37 °C, 5 % CO_2 incubator may be used instead but the temperature equilibration is considerably slower and results may thus differ. The wiring length of the illumination system (*see* Subheading 3.4, **step 2**) must also be amended accordingly.

9. The optionally available software for many the laboratory power sources allows for remote control of the LEDs. This feature can be used to create pulsed-light protocols rather than using constant illumination (*see* Subheading 3.5, **step 10** and Subheading 3.9, **step 2** and Fig. 6). However, the minimum duration is 500 ms long for the PS2000B used here, so you may consider alternative devices that allow more flexible settings.

10. Controlling for the islets to behave normally is essential (*see* Subheading 3.9, **step 4**). Healthy islets do exhibit Ca^{2+} oscillations when stimulated with high glucose [17]. Using high potassium allows estimating if the acquisition settings are within a dynamic range.

11. To more precisely assess Ca^{2+} signals within single cells of the islet while allowing for stimulation of single cells out of the subpopulation rather than the whole population (*see* Subheading 3.9, **step 4**), a confocal laser-scanning microscope

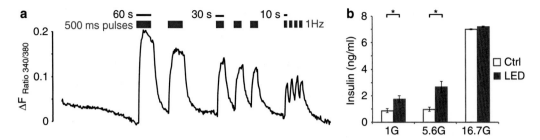

Fig. 6 (**a**) Example trace of changes in Fura-2 fluorescence ratio ($\Delta F_{340/380}$) in ChR2-YFP expressing islets stimulated with 500 ms at 1 Hz for different durations (60, 30, and 10 s) with pauses of the same duration in between. Experiment performed at 2.8 mM glucose. (**b**) Constant illumination for 1 h (LED) compared to control (Ctrl) and effects on glucose-induced insulin release at different glucose concentrations (G, in mM). Adapted from [8]

to image single cells [18] with combined fluorescence recovery after photobleaching (FRAP) to stimulate them may also be an option, although with a considerably higher price tag than the system presented here.

Acknowledgements

The work described here was made possible through financial support by the Albert Påhlsson foundation, Fredrik and Ingrid Thurings foundation, Royal Physiographic Society, and the Swedish Diabetes Association.

References

1. Chow BY, Boyden ES (2013) Optogenetics and translational medicine. Sci Transl Med 5(177):177ps175. doi:10.1126/scitranslmed.3003101

2. Fenno L, Yizhar O, Deisseroth K (2011) The development and application of optogenetics. Annu Rev Neurosci 34:389–412. doi:10.1146/annurev-neuro-061010-113817

3. Blouet C, Schwartz GJ (2010) Hypothalamic nutrient sensing in the control of energy homeostasis. Behav Brain Res 209(1):1–12. doi:10.1016/j.bbr.2009.12.024

4. Kong D, Tong Q, Ye C et al (2012) GABAergic RIP-Cre neurons in the arcuate nucleus selectively regulate energy expenditure. Cell 151(3):645–657. doi:10.1016/j.cell.2012.09.020

5. Rorsman P, Braun M, Zhang Q (2012) Regulation of calcium in pancreatic alpha- and beta-cells in health and disease. Cell Calcium 51(3-4):300–308. doi:10.1016/j.ceca.2011.11.006

6. Gromada J, Franklin I, Wollheim CB (2007) Alpha-cells of the endocrine pancreas: 35 years of research but the enigma remains. Endocr Rev 28(1):84–116. doi:10.1210/er.2006-0007

7. Gopel S, Kanno T, Barg S et al (1999) Voltage-gated and resting membrane currents recorded from B-cells in intact mouse pancreatic islets. J Physiol 521(Pt 3):717–728

8. Reinbothe TM, Safi F, Axelsson AS et al (2014) Optogenetic control of insulin secretion in intact pancreatic islets with beta-cell-specific expression of Channelrhodopsin-2. Islets 6(1):e28095. doi:10.4161/isl.28095

9. Madisen L, Mao T, Koch H et al (2012) A toolbox of Cre-dependent optogenetic transgenic mice for light-induced activation and silencing. Nat Neurosci 15(5):793–802. doi:10.1038/nn.3078

10. Herrera PL (2000) Adult insulin- and glucagon-producing cells differentiate from

two independent cell lineages. Development 127(11):2317–2322

11. Carter JD, Dula SB, Corbin KL et al (2009) A practical guide to rodent islet isolation and assessment. Biol Proced Online 11:3–31. doi:10.1007/s12575-009-9021-0

12. Li DS, Yuan YH, Tu HJ et al (2009) A protocol for islet isolation from mouse pancreas. Nat Protoc 4(11):1649–1652. doi:10.1038/nprot. 2009.150

13. Grynkiewicz G, Poenie M, Tsien RY (1985) A new generation of Ca2+ indicators with greatly improved fluorescence properties. J Biol Chem 260(6):3440–3450

14. Mahdi T, Hanzelmann S, Salehi A et al (2012) Secreted frizzled-related protein 4 reduces insulin secretion and is overexpressed in type 2 diabetes. Cell Metab 16(5):625–633. doi:10.1016/j.cmet.2012.10.009

15. Brennan K (2011) Colony management advanced protocols for animal transgenesis. Springer, Heidelberg. doi:10.1007/978-3-642-20792-1

16. Bertera S, Balamurugan AN, Bottino R et al (2012) Increased yield and improved transplantation outcome of mouse islets with bovine serum albumin. J Transpl 2012:856386. doi:10.1155/2012/856386

17. Tengholm A, Gylfe E (2009) Oscillatory control of insulin secretion. Mol Cell Endocrinol 297(1-2):58–72

18. Hodson DJ, Mitchell RK, Bellomo EA et al (2013) Lipotoxicity disrupts incretin-regulated human beta cell connectivity. J Clin Invest 123(10):4182–4194. doi:10.1172/JCI68459

Chapter 9

Optogenetics in Plants: Red/Far-Red Light Control of Gene Expression

Rocio Ochoa-Fernandez, Sophia L. Samodelov, Simon M. Brandl, Elke Wehinger, Konrad Müller, Wilfried Weber, and Matias D. Zurbriggen

Abstract

Optogenetic tools to control gene expression have many advantages over the classical chemically inducible systems, overcoming intrinsic limitations of chemical inducers such as solubility, diffusion, and cell toxicity. They offer an unmatched spatiotemporal resolution and permit quantitative and noninvasive control of the gene expression. Here we describe a protocol of a synthetic light-inducible system for the targeted control of gene expression in plants based on the plant photoreceptor phytochrome B and one of its interacting factors (PIF6). The synthetic toggle switch system is in the ON state when plant protoplasts are illuminated with red light (660 nm) and can be returned to the OFF state by subsequent illumination with far-red light (760 nm). In this protocol, the implementation of a red light-inducible expression system in plants using Light-Emitting Diode (LED) illumination boxes is described, including the isolation and transient transformation of plant protoplasts from *Arabidopsis thaliana* and *Nicotiana tabacum*.

Key words Plant synthetic biology, Plant optogenetics, Red light-inducible gene expression system, Plant leaf protoplasts, *Arabidopsis thaliana*, *Nicotiana tabacum*

1   Introduction

Inducible gene expression systems in plants are essential to study cellular processes, to control target gene expression with minimal or no interference to developmental or growth processes, and for efficient large-scale biopharmaceutical production. Spatial control of gene expression in plants has traditionally been achieved by the use of tissue-specific promoters. This leads to highly specific spatial gene expression, however, once such an expression cassette has been implemented, the promoters can no longer be exogenously controlled [1]. Likewise, classical chemically inducible systems offer temporal control over gene expression in plants (such as ethanol- or dexamethasone-inducible systems [2]) but do not fulfill key requirements of inducible gene expression systems. This is due to the intrinsic limitations of chemical inducers like solubility, diffusion, inability

Arash Kianianmomeni (ed.), *Optogenetics: Methods and Protocols*, Methods in Molecular Biology, vol. 1408, DOI 10.1007/978-1-4939-3512-3_9, © Springer Science+Business Media New York 2016

to revert induction without washing steps, inducer-removal in sample processing and pleiotropic effects, limiting their application in vitro and in vivo and their use in long-term treatments [1, 3]. Light as a *stimulus* overcomes these limitations, offering advantages such as reversibility, fast reactivity, and minimal cell toxicity, therefore allowing a precise control of gene expression in a quantitative and noninvasive manner, with both high spatial and temporal resolution.

Several light-responsive gene expression systems have been developed for gene control with UVB, blue, or red light and adapted for use in mammalian cell culture and in vivo in animals (reviewed in [4, 5]). However, the application of these optogenetic tools in plants has not yet taken root, mainly due to the fact that light is essential for plant growth and development, therefore having pleiotropic effects. Thus far, only a red/far-red light-inducible system has been applied to plants [6], in principle due to its ability to revert between ON and OFF states with two different wavelengths. In this sense, this toggle switch system is unique and differs from the rest of the optogenetic tools based on photoreceptors which can be activated by light of one wavelength but can only revert to the basal, inactive state nonphotochemically, with shut off kinetics depending on their photobiological properties (dark reversion). The red/far-red light-inducible system is based on the photoreceptor Phytochrome B (PhyB) and phytochrome-interacting factor 6 (PIF6) from *Arabidopsis* thaliana. This system is a split transcription factor in which the components interact in a light-dependent manner. It is based on three constructs: (1) PIF6 (amino acids 1–100) fused to the mphR(A) (macrolide repressor DNA-binding protein E) and a nuclear localization sequence (NLS); (2) PhyB (amino acids 1–650) fused to the *Herpes simplex* VP16 transactivation domain and an NLS; and (3) multiple repetitions of an etr motif (cognate binding site of the E protein), placed upstream of a CMV minimal promoter followed by a reporter gene, e.g. firefly luciferase (Fig. 1a, Table 1). Upon exposure to red light, PhyB changes its conformation by photoisomerization of the covalently bound chromophore, phytochromobilin (PΦB). The activated form of PhyB (P$_{fr}$) binds to PIF6 and the VP16 domain is then recruited to the etr motif in close proximity to the minimal promoter, activating transcription of the reporter gene. The PhyB-PIF6 association is readily reversed upon exposure to far-red light, when PhyB changes its conformation to the inactive form (P$_r$) resulting in the termination of reporter gene expression (Fig. 1b).

Here we describe a protocol for a light-inducible expression system that is activated by red light and deactivated by far-red light to control gene expression in leaf protoplasts of *Nicotiana tabacum* and *Arabidopsis thaliana*. The control of gene expression with high resolution in time and space overcomes intrinsic limitations of existing systems and facilitates novel applications including the precise interrogation of complex biological signaling processes in a quantitative and noninvasive manner.

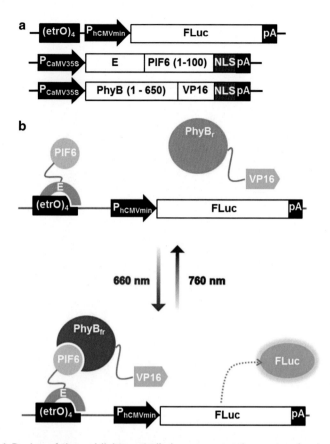

Fig. 1 Design of the red light-controlled gene expression system in plants. (**a**) Configuration of the vectors. (**b**) Mode of function. Upon exposure to red light (660 nm), PhyB changes its conformation to its active form (P_{fr}) that allows the binding to PIF6 and therefore recruitment of the transactivator to the minimal promoter, firefly luciferase is expressed as a consequence. Under far-red (760 nm) light illumination, PhyB is converted to its inactive form (P_r), PhyB-PIF6 dis-associates, thus ceasing the transcription of the reporter gene

2   Materials

Prepare all solutions using double distilled water and p.a. purity grade chemicals. Use plant cell culture tested reagents for plant growth and protoplast isolation media. Prepare and store all reagents at 4 °C unless indicated otherwise.

2.1   Plant Growth

1. SCN (Seedling Culture Nicotiana) (modified from [7]): 0.32 % (w/v) Gamborg B5 basal salt powder with vitamins (bioWORLD, GeneLinx International, Inc., USA), 4 mM $MgSO_4 \cdot 7H_2O$, 58.4 mM sucrose and 0.15 % (w/v) gelrite. Mix and adjust to pH 5.8 and autoclave. After autoclaving, add 0.1 % (v/v) of Gamborg B5 Vit Mix (bioWORLD) and pour

Table 1
Description of the plasmids encoding the red light-controlled gene expression system for plants used in this protocol

Vector	Description	References
pROF100	(etrO)₄-P_hCMVmin-FLuc-pA Vector encoding firefly luciferase (Fluc) under the control of the human cytomegalovirus minimal promoter (P_hCMVmin); placed downstream of multiple repetitions of an operator sequence for the E protein (etrO)₄.	This work
pMZ827	P_CaMV35S-E-*AtPIF6*(1-100)-NLS-pA Vector comprising the macrolide repressor DNA-binding protein (E) fused to the N-terminal 100 amino acids of *AtPIF6* under control of the cauliflower mosaic virus 35S promoter (P_CaMV35S). The fusion protein is targeted to the nucleus by C-terminal fusion of a nuclear localization sequence (NLS).	[6]
pMZ828	P_CaMV35S-*AtPhyB*(1-650)-VP16-NLS-pA Vector encoding a fusion protein of the N-terminal 650 amino acids fragment of *AtPhyB* fused to the Herpes simplex-derived VP16 transactivation domain under the control of the cauliflower mosaic virus 35S promoter (P_CaMV35S). The fusion protein is targeted to the nucleus by C-terminal fusion of a nuclear localization sequence (NLS).	[6]

Abbreviations: *E* macrolide repressor DNA-binding protein, *etrO* operator sequence for E protein, *Fluc* firefly luciferase, *NLS* nuclear localization sequence, *pA* polyadenylation signal, *P_CaMV35S* cauliflower mosaic virus 35S promoter, *P_hCMVmin* human cytomegalovirus minimal promoter, *VP16 Herpes simplex* virus-derived transactivation domain

50 ml of the medium into each Magenta Plant Culture Box (*see* **Note 1**).

2. SCA (*Seedling Culture Arabidopsis*) (modified from [8]): 0.32 % (w/v) Gamborg B5 basal salt powder with vitamins (bioWORLD), 4 mM $MgSO_4 \cdot 7H_2O$, 43.8 mM sucrose and 0.8 % (w/v) phytoagar in H_2O. Mix and adjust to pH 5.8. Autoclave and add 0.1 % (v/v) Gamborg B5 Vit Mix (bioWORLD) then pour 50 ml of the medium into each Magenta Plant Culture Box; or alternatively add 1:2000 ampicillin and pour 50 ml of the medium into each 12 cm square plate (*see* **Note 1**).

3. Seed sterilization solution for *A. thaliana* (modified from [9]): 5 % (w/v) calcium hypochlorite, 0.02 % (v/v) Triton X-100 in 80 % (v/v) EtOH. Combine the chemicals in a bottle and mix for few hours at room temperature. A precipitate will form. Place the bottle to 4 °C for storage. Allow the precipitate to settle and do not agitate the bottle before use.

4. Seed sterilization solution for tobacco: 5 % active chlorine from NaOCl solution (12 % active chlorine stock solution), 0.5 % (v/v) Tween 20 in autoclaved H_2O. Sterilize with a 0.22 μm filter. Prepare fresh prior to each use.

5. Parafilm.

6. Syringe and 22 μm filter.

7. Ampicillin stock (100 mg/ml).

2.2 Protoplast Isolation and PEG Mediated Protoplast Transformation

1. MMC (*MES, Mannitol, Calcium*) [8]: 10 mM MES, 40 mM $CaCl_2 \cdot H_2O$, add mannitol until obtaining an osmolarity of 550 mOsm (ca. 85 g/l). Adjust to pH 5.8 and filter sterilize.

2. F-PIN (*Fast Protoplast Incubation Nicotiana*) (modified from [7]): 10 mM MES, 0.32 % (w/v) Gamborg B5 basal salt powder with vitamins (bioWORLD), 0.38 M sucrose. Adjust to pH 5.8 and filter sterilize.

3. Enzyme solution stock 5 % (10× concentrated): cellulase Onozuka R10 and macerozyme R10 (SERVA Electrophoresis GmbH, Germany) in F-PIN or MMC. Weigh 10 g of cellulase and 10 g of macerozyme and dissolve in F-PIN solution or MMC (preheated to 37 °C) to a total volume of 200 ml H_2O (*see* **Note 2**). Sterile filter the solution with a bottle-top filter into a sterile bottle and make aliquots of 2 ml. Store at –20 °C, avoid thaw–refreeze cycles.

4. MSC (*MES, Sucrose, Calcium*) [8]: 10 mM MES, 0.4 M sucrose, 20 mM $MgCl_2 \cdot 6H_2O$, add mannitol until obtaining an osmolarity of 550 mOsm (ca. 85 g/l). Adjust to pH 5.8 and filter sterilize.

5. W5 solution (modified from [10]): 2 mM MES, 154 mM NaCl, 125 mM $CaCl_2 \cdot 2H_2O$, 5 mM KCl, 5 mM glucose. Adjust to pH 5.8 and filter sterilize.

6. MMM (*M*ES, *M*annitol, *M*agnesium) [8]: 15 mM MgCl$_2$, 5 mM MES, mannitol to 600 mOsm (ca. 85 g/l). Adjust to pH 5.8 and filter sterilize.

7. PEG solution: Mix 2.5 ml of 0.8 M mannitol, 1 ml of 1 M CaCl$_2$ and 4 g PEG$_{4000}$ and 3 ml H$_2$O. Made fresh for each experiment. Not filtered, prepare fresh and place the tube at 37 °C for PEG dissolution, then use directly.

8. PCA (*P*rotoplast *C*ulture *A*rabidopsis) (modified from [8]): 0.32 % (w/v) Gamborg B5 basal salt powder with vitamins (bioWORLD), 2 mM MgSO$_4$·7H$_2$O, 3.4 mM CaCl$_2$·2H$_2$O, 5 mM MES, 0.342 mM l-glutamine, 58.4 mM sucrose, glucose 550 mOsm (ca. 80 g/l), 8.4 μM Ca-panthotenate, 2 % (v/v) biotin from a biotin solution 0.02 % (w/v) in H$_2$O (warm up the biotin solution to dissolve). Adjust to pH 5.8 and filter sterilize, add 0.1 % (v/v) Gamborg B5 Vitamin Mix and 1:2000 ampicillin to the PCA before use.

9. Scalpel.

10. Disposable 100 μm and 40–70 μm pore size sieve (Greiner bio-one international, Germany).

11. Petri dish 94 × 16 mm.

12. Parafilm.

13. 200 μl and 1 ml large orifice pipette tips.

14. Round-bottom 15 ml Falcon tubes.

15. Rosenthal cell counting chamber.

16. Nontreated 6-, and 12-, or 24-well plates.

2.3 Illumination Treatment

1. 660 and 760 nm light-emitting diode (LED) illumination boxes.

In brief, the LED illumination boxes are custom-made boxes of PVC that exclude external light and at the same time allow gas exchange. The light boxes contain panels of LEDs (Roithner Lasertechnik GmbH, Austria) of one or several wavelengths. In addition, the irradiation intensity and illumination schemes can be set by using a programmable control unit (for full description see [11] and [12]). As an example, such a box is shown in Fig. 2. The light box is composed of three parts: a base for placing the cell culture plate, the walls, and the lid where the LEDs of specific emission wavelengths are built-in. In this protocol, boxes equipped with either red (660 nm) or far-red (760 nm) LEDs were used.

2.4 Luminescence Reporter Assay

1. Costar® 96-well flat-bottom white plate.

2. Firefly luciferase substrate: 20 mM tricine, 2.67 mM MgSO$_4$·7H$_2$O, 0.1 mM EDTA·2H$_2$O, 33.3 mM DTT, 0.52 mM ATP, 0.27 mM acetyl-CoA, 0.47 mM d-luciferin (Biosynth AG), 5 mM NaOH, 264 μM MgCO$_3$·5H$_2$O, in H$_2$O. Prepare

Fig. 2 LED illumination box. (**a**) Illumination box for one cell culture plate. (**b**) Opened illumination box. The LEDs are located in the lid of the box. (**c**) Three components of the light box

a beaker with a magnetic stirrer and add the components in the order as above, then add the luciferin and H_2O and mix the solution, proceed with the addition of the last two components (NaOH and $MgCO_3 \cdot 5H_2O$). Adjust to pH 8, aliquot the substrate in precooled black Falcon tubes and freeze them at –80 °C (*see* **Note 3**).

3   Methods

3.1   Seed Sterilization and Plant Material

3.1.1   Arabidopsis thaliana (Wild Type, Columbia-0)

1. Seed sterilization should be done in 1.5 ml tubes in a sterile working hood. For large-scale seed sterilization, fill tubes to a maximum of approximately 250 µl volume. Avoid sterilizing a larger volume in a single tube, as results (efficiency) may vary.

2. Rinse seeds multiple times with 80 % (v/v) ethanol until all large dirt and other plant particles are removed.

3. Sterilize the seeds with 1 ml of the *A. thaliana* sterilization solution under agitation for 10 min.

4. Remove the solution and replace with 1 ml of 80 % (v/v) EtOH. Incubate 5 min under agitation.

5. Repeat **step 4** but incubating for 2 min.

6. Replace the solution with 1 ml absolute ethanol (≥99.5 %) and incubate for 1 min under agitation.

7. Remove all ethanol and let the seeds dry completely under the sterile hood.

8. Add autoclaved water and plate in a line on autoclaved filter paper strips (200–300 seeds/strip) placed on 12 cm square plates containing SCA medium and seal with parafilm. Multiple strips may be placed in one plate. Alternatively, place 1–16 seeds, evenly dispersed, in a Magenta Box containing 50 ml SCA medium.

9. Place the plates in a growth chamber with a 16 h light regime at 22 °C. Two- to three-week old plantlets from 12 cm square plates can be used for protoplast isolation. Three- to four-week old plants grown in Magenta boxes can be used for protoplast isolation.

3.1.2 Nicotiana tabacum cv Petit Havana

1. Incubate the desired number of seeds with 1 ml of seed sterilization solution for tobacco for 5 min at room temperature under agitation. Large-scale seed sterilization for *N. tabacum* has not been tested, due to the small amount of seeds necessary when growing plants in Magenta boxes.

2. Remove the solution (centrifuge if necessary to sediment the seeds) and rinse the seeds 3–4 times with 1 ml of H_2O in the same manner.

3. Place one or two seeds in the middle of a Magenta Box containing 50 ml SCN medium. When more than one seed germinates, the seedlings must be separated to different boxes (around day 4–6 after germination) in order to have only one plant per box for optimal growth.

4. Place the Magenta boxes in a growth chamber with a 16 h light regime at 22–25 °C (plants will grow faster at higher temperatures). Leaves from 2- to 3-week old plants can be used for protoplast isolation.

3.2 Protoplast Isolation and Polyethylene Glycol-Mediated Transformation

A. thaliana and *N. tabacum* protoplast isolation per flotation and polyethylene glycol-mediated transformation were performed as described before ([8] and [13], respectively) with a few alterations. All pipetting is done with wide orifice tips to avoid damaging the protoplasts. Preferentially use medium acceleration and lowest deceleration settings for the centrifugation steps (140 s acceleration and 300 s deceleration according to DIN58970).

1. Cut the tobacco leaves in 1 mm strips with the abaxial surface facing up starting from the middle lamella with a sterile scalpel

(*see* **Note 4**). Finely slice the plant leaves of *A. thaliana* with the scalpel in 2 ml of MMC (*see* **Note 5**).

2. Transfer the cut leaf material into a new Petri dish containing 9 ml F-PIN (tobacco) or 7 ml of MMC (*A. thaliana*).

3. Proceed with the enzymatic digestion of cut plant material by adding 1 ml of 10× enzyme stock solution (the final concentration of each enzyme should be 0.5 %).

4. Seal the dish with parafilm and cover it with aluminum foil. Incubate overnight (12–16 h) in the dark at 22 °C.

5. Carefully homogenize the digested leaf material by pipetting the leaf-enzyme mixture up and down to release the protoplasts from the plant material.

6. Pass through a disposable 100 μm (tobacco) or 40–70 μm (*A. thaliana*) pore size sieve.

7. Transfer the filtered protoplast solution to 15 ml round bottom Falcon tubes. One tube should be used for each plate of digested leaf material. The remaining steps should be completed in these tubes.

8. For *A. thaliana*, centrifuge the filtered protoplast solution in round bottom Falcon tubes at $100 \times g$ for 10–20 min to sediment the protoplasts. Remove supernatant and resuspend in 10 ml of MSC. For tobacco protoplasts, centrifugation is not necessary, as the flotation of protoplasts can be done directly in the F-PIN solution.

9. Very carefully overlay 10 ml of protoplast solution with 2 ml of MMM (*see* **Note 6**).

10. For *A. thaliana* protoplasts, centrifuge for 10 min at $80 \times g$ for accumulation of the protoplasts at the interphase of MSC and MMM. For tobacco protoplasts, instead of centrifugation, incubate the tubes at room temperature for 20–30 min, in which time the protoplasts will float to the interphase of F-PIN and MMM (*see* **Note 7**).

11. Collect the protoplasts at the interphase and transfer into a new Falcon tube with 7 ml of W5 solution. For each floatation tube to be used, prepare two W5-filled collection Falcon tubes. Multiple rounds of protoplast collection can be done (if necessary overlay again with MMM) until no further protoplasts float to the interphase or enough protoplasts are obtained.

12. Centrifuge the collected protoplasts for 10 min at $100 \times g$ to pellet and resuspend in a defined volume of W5 for counting (*see* **Note 8**).

13. Determine the cell density using a Rosenthal cell counting chamber.

14. Sediment the protoplasts by centrifuging for 5 min at $80 \times g$. Discard supernatant and adjust with MMM solution to a density of 5×10^5 cells/ml for tobacco and 5×10^6 cells/ml for *A. thaliana*.

15. (a) For the transformation of tobacco protoplasts, prepare 50 μg of DNA in H_2O (*see* **Note 9**) in a round bottom Falcon tube and add 1 ml of the protoplasts in MMM. Carefully mix by pipetting and incubate for 5 min.

    (b) For *A. thaliana* protoplasts, prepare 15–30 μg of DNA in H_2O (*see* **Note 9**) adjusted to a maximum volume of 20 μl (volume adjustment with MMM). Transfer the 20 μl DNA solution to the rim of a well of a 6-well culture plate (slightly tilt the plate for easier pipetting in the following steps). Dispense 100 μl of the protoplast solution to each well with DNA and mix by gentle pipetting. Incubate for 5 min.

16. (a) For tobacco protoplast transformation, add 1 ml PEG_{4000} solution to the protoplasts in a drop-wise manner with a tip-in-tip method while slowly rotating the Falcon tube (*see* **Note 10**). After 8 min (*see* **Note 11**), consecutively add 1, 2, 3, and 4 ml of W5 per minute to the tube as a stepwise dilution of the transformation, and gently tilt the tube after each step for mixing.

    (b) For *A. thaliana* protoplast transformation, gently shake the 6-well plate from side to side to distribute the protoplasts and DNA along the rim before directly adding 120 μl of PEG_{4000} solution drop-wise, tip-in-tip. Do not mix after the addition of PEG. Incubate for 8 min (*see* **Notes 11** and **12**) and quickly add 120 μl of MMM and, directly afterwards, at least 1.2 ml of PCA. Gently mix by tilting the plate after the addition of PCA (final volume should be at least 1.6 ml).

17. Only for tobacco, sediment the cells at 5 min at $80 \times g$, discard the supernatant and resuspended in at least 1.6 ml PCA.

18. After transformation, if only one condition is to be tested, leave the *A. thaliana* protoplast suspension in a well of a 6-well plate. In the case of tobacco protoplasts transfer the 1.6 ml from the tube into a well of a 6-well plate.

    If more than one condition is to be tested, split the protoplasts in different plates according to the number of light conditions to be assayed. The volume pipetted to each well in the new plates will depend on the number of replicates per condition. Considering that 25,000 protoplasts (see below) will be used for each measurement (80 μl protoplast suspension), it follows that for 6 replicates 150,000 protoplasts are needed, amounting to 480 μl protoplast suspension. Scale down to 12- or 24-well plate to avoid high evaporation rates. Seal the plate(s) with parafilm.

3.3 Illumination Treatment and Reporter Assay

1. After transformation of the protoplasts, illuminate the plates with the appropriate wavelength (i.e. 660, 760 nm) and intensity of light with LED arrays, or incubate in the dark prior to reporter quantification. The spectra of the LEDs and the radiation intensity can be determined with a spectroradiometer (e.g. AvaSpec-ULS2048-USB2 FC/PC and FC-UVIR200-2-ME-1FCPC, Avantes, Netherlands).

   As an example, Fig. 3a shows time-course and dose-response curves for the red light-inducible gene expression system. Protoplasts were isolated from *A. thaliana* plantlets and 10 μg of each plasmid (pMZ827, pMZ828 and pROF100) were transformed into the protoplasts. Several transformations were made in parallel (22 transformations) and after transformation all the protoplasts were pooled. Aliquots of 3.5 ml of the protoplasts suspension were transferred into one well of seven different 6-well plates (one plate for each illumination condition). The luminescence determination was made for each condition at different points in time (0, 6, 12, 18, and 24 h). As a dark control, 1 ml of protoplast suspension was transferred into one well of four different 24-well plates. In this way, a single plate per time point was used and accidental exposure of the plate to ambient light avoided. The results of the kinetics and expression levels of the red light-inducible system in *A. thaliana* protoplast depicted in Fig. 3a indicate between 1 and 4 μE/m²/s as optimal illumination conditions for maximum expression rates. The highest expression levels are achieved at 24 h but a better dynamic range (399 and 395 × fold induction) is obtained at 18 h of gene expression for 2–4 μE/m²/s red-light intensities (Fig. 3b). It is, however, recommended to adjust the protocol to the application of interest.

2. To determine reporter expression, first gently mix the protoplast suspension with the pipette and transfer 80 μl (25,000 protoplasts) into a Costar® 96-well flat-bottom white plate, including 4–6 replicates for each condition (*see* **Note 13**).

3. Add 20 μl of firefly luciferase substrate and monitor the luminescence in a plate reader [14]. 10 s of shaking plate for homogeneous substrate availability and directly luminescence measurement for 20 min kinetics (interval of 2 min) are advisable.

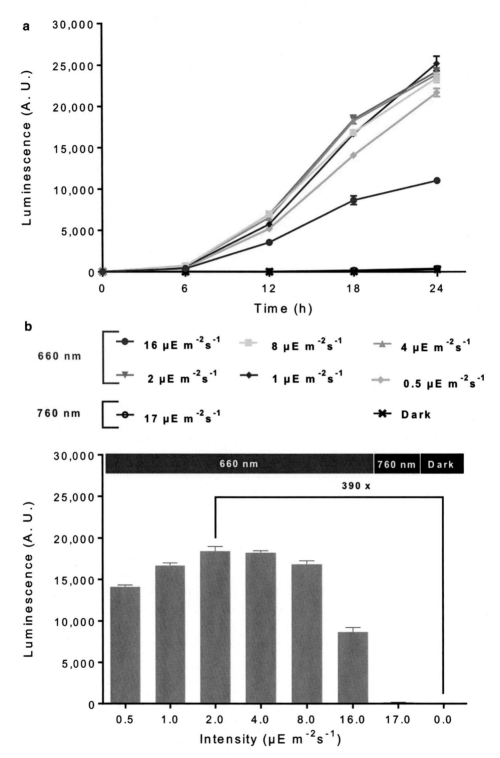

Fig. 3 Time- and dose-response curves of the red light inducible gene expression system in protoplasts of *A. thaliana*. Protoplasts from *A. thaliana* were transformed for red light–inducible firefly luciferase expression (pMZ827, pMZ828, and pROF100). After transformation, 3.5 ml aliquots of protoplast suspensions containing approximately 1.09×10^6 protoplasts, were illuminated either at different intensities of 660 nm (0.5, 1, 2, 4, 8, and 16 µE/m²/s), at 760 nm (17 µE/m²/s) light, or were kept in the dark as a control. (**a**) Samples were taken at the indicated points (0, 6, 12, 18, and 24 h after transformation) and firefly luciferase expression was determined.

4   Notes

1. Prepare the plates or Magenta Plant Culture boxes directly after autoclaving because the gelrite and phytoagar will not dissolve upon reheating.

2. Both enzyme extracts are not to be inhaled and are poorly soluble. For these reasons: solve under a fume hood by adding 10 ml of prewarmed (37 °C) MMC/F-PIN to each bottle directly, shake, and pour into beaker and rinse bottles repeatedly. Fill beaker to 200 ml. The solution will not be clear, should, however, be a clear brown after filtration.

3. For certain solutions, a stock solution can be prepared in advance; however, tricine, DTT, ATP, and acetyl-CoA should be prepared fresh. From the addition of DTT on, all steps should be performed under a fume hood. Moreover, luciferin is sensitive to light, oxygen, and high temperature so that from its addition on, the preparation should be performed in darkness and as quickly as possible. Due to the high price of acetyl-CoA, it is also preferable to purchase this substrate in small amounts (50 mg for the preparation of 200 ml of firefly luciferase substrate) and use the entire content in a single preparation of substrate to avoid freeze–thaw cycles and waste.

4. Choose healthy leaves not showing nutritional deficiency, chlorosis, or mechanical damage.

5. *A. thaliana* plant material grown in plates should be carefully cut from the plate with a scalpel in a way that avoids including roots and seeds, and should then be cut finely into small pieces. When cutting the plant material from Magenta Plant Culture Boxes, take only the leaves and either cut them in strips as described for tobacco or slice them finely. Sterile featherweight forceps are helpful in holding *A. thaliana* leaves from Magenta boxes to be cut in strips without inflicting damage to them.

6. Gentle inversion of the tube before adding the MMM solution slowly helps for a clear separation of phases. For addition of MMM use a tip-in-tip technique i.e. placing a smaller tip into the tip of a bigger tip for a slow solution dispense.

7. Collecting the first band of protoplasts at the interphase after 10–15 min increases the speed of protoplast floatation.

8. Protoplasts will not be successfully pelleted if the collection tube contains less W5 than MMM.

Fig. 3 (continued) The graph shows the reporter luminescence values at different time points and different illumination conditions. (**b**) Reporter luminescence values after 18 h expression at the indicated light intensities. Data are means ± SEM ($n = 6$ technical replicates)

9. DNA amounts mentioned in the protocol are total amounts of DNA. When more than one plasmid is used, the amounts of each plasmid must be adjusted proportionally, keeping the total DNA amount constant. Purify the plasmid DNA using midiprep or maxiprep kits and check the quality of the plasmid DNA by agarose gel electrophoresis (e.g. RNA content).

10. If the PEG has sedimented to the bottom of the tube, mixing by gently tilting the tube will be necessary.

11. The duration of PEG treatment is critical in the transformation; the suggested 8 min treatment leads to high transformation efficiency in our experience.

12. Gently shaking the plate side to side before PEG addition avoids the aggregation of protoplasts.

13. It is recommended to pipette the protoplasts in the following order: 660 nm (highest to low intensities)—760 nm—dark, as the system is rapidly activated by ambient light. Due to the sensitivity of the system, it is also recommended to work in a darkroom with green safelight emitted by LEDs (~520 nm). Green light illumination at moderate intensities does not lead to noticeable activation of PhyB.

Acknowledgments

This work was supported in part by the Excellence Initiative of the German Federal and State Governments (EXC294-BIOSS, GSC-4 Spemann Graduate School (SGBM)) and the Alexander von Humbolt Foundation (research Grant no. 1141629). We thank Susanne Knall and Frauke Bartels-Burgahn for experimental assistance. We thank J. Schmidt, D. Schächtele and J. Meßmer (University of Freiburg) for designing and constructing the illumination boxes.

References

1. Corrado G, Karali M (2009) Inducible gene expression systems and plant biotechnology. Biotechnol Adv 27(6):733–743

2. Junker A, Junker B (2012) Synthetic gene networks in plant systems. In: Weber W, Fussenegger M (eds) Synthetic gene networks, vol 813, Methods in molecular biology. Humana, New York, pp 343–358

3. Padidam M (2003) Chemically regulated gene expression in plants. Curr Opin Plant Biol 6(2):169–177

4. Zhang K, Cui B (2015) Optogenetic control of intracellular signaling pathways. Trends Biotechnol 33(2):92–100

5. Müller K, Naumann S, Weber W, Zurbriggen MD (2015) Optogenetics for gene expression in mammalian cells. Biol Chem 396(2):145–152

6. Müller K, Siegel D, Rodriguez Jahnke F, Gerrer K, Wend S, Decker EL, Reski R, Weber W, Zurbriggen MD (2014) A red light-controlled synthetic gene expression switch for plant systems. Mol BioSyst 10(7):1679–1688

7. Dovzhenko A, Bergen U, Koop HU (1998) Thin-alginate-layer technique for protoplast culture of tobacco leaf protoplasts: shoot formation in less than two weeks. Protoplasma 204(1-2):114–118

8. Dovzhenko A, Dal Bosco C, Meurer J, Koop HU (2003) Efficient regeneration from cotyledon protoplasts in Arabidopsis thaliana. Protoplasma 222(1–2):107–111

9. Luo Y, Koop H-U (1997) Somatic embryogenesis in cultured immature zygotic embryos and leaf protoplasts of Arabidopsis thaliana ecotypes. Planta 202(3):387–396

10. Menczel L, Galiba G, Nagy F, Maliga P (1982) Effect of radiation dosage on efficiency of chloroplast transfer by protoplast fusion in nicotiana. Genetics 100(3):487–495

11. Müller K, Zurbriggen MD, Weber W (2014) Control of gene expression using a red- and far-red light-responsive bi-stable toggle switch. Nat Protoc 9(3):622–632

12. Müller K, Engesser R, Metzger S, Schulz S, Kämpf MM, Busacker M, Steinberg T, Tomakidi P, Ehrbar M, Nagy F, Timmer J, Zubriggen MD, Weber W (2013) A red/far-red light-responsive bi-stable toggle switch to control gene expression in mammalian cells. Nucleic Acids Res 41(7):e77

13. Koop H-U, Steinmüller K, Wagner H, Rößler C, Eibl C, Sacher L (1996) Integration of foreign sequences into the tobacco plastome via polyethylene glycol-mediated protoplast transformation. Planta 199(2):193–201

14. Wend S, Bosco CD, Kämpf MM, Ren F, Palme K, Weber W, Dovzhenko A, Zurbriggen MD (2013) A quantitative ratiometric sensor for time-resolved analysis of auxin dynamics. Sci Rep 3:2052

Chapter 10

Enhancing Channelrhodopsins: An Overview

Jonas Wietek and Matthias Prigge

Abstract

After the discovery of Channelrhodopsin, a light-gated ion channel, only a few people saw the diverse range of applications for such a protein. Now, more than 10 years later Channelrhodopsins have become widely accepted as the ultimate tool to control the membrane potential of excitable cells via illumination. The demand for more application-specific Channelrhodopsin variants started a race between protein engineers to design improved variants. Even though many engineered variants have undisputable advantages compared to *wild-type* variants, many users are alienated by the tremendous amount of new variants and their perplexing names.

Here, we review new variants whose efficacy has already been proven in neurophysiological experiments, or variants which are likely to extend the optogenetic toolbox. Variants are described based on their mechanistic and operational properties in terms of expression, kinetics, ion selectivity, and wavelength responsivity.

Key words Optogenetics, Channelrhodopsins, Protein engineering, User guide, Microbial rhodopsin

1 Introduction

Ten years ago the first experiments using heterologous expression of Channelrhodopsin (ChR), a light-gated ion channel, triggered an avalanche of new approaches to the control of cellular function by application of light [1–3]. A new paradigm called *Optogenetics* emerged [4]. Despite diverse developments in the field of Optogenetics over the last 10 years, its tremendous penetrance into neuroscience is highly linked to the ChR molecule itself [5, 6].

And yet, after nearly a decade of protein engineering of new, more suitable ChRs for different Optogenetic applications, researchers are still largely favoring the first discovered ChR or its first mutation H134R, depicted in Fig. 1a, b [3]. Is it legitimate to

"A designer knows he has achieved perfection not when there is nothing left to add, but when there is nothing left to take."

Arash Kianianmomeni (ed.), *Optogenetics: Methods and Protocols*, Methods in Molecular Biology, vol. 1408, DOI 10.1007/978-1-4939-3512-3_10, © Springer Science+Business Media New York 2016

ask here how useful ChR engineering actually is for the field of Optogenetics? The pace at which new ChR versions are released may be blessing and curse at the same time. Some new tailored ChR variants are efficient tools for highly efficient Optogenetic applications. Whereas other variants are published for the sake of publishing, and are not useful for a standard optogenetic experiment. Inexperienced users may be overwhelmed by the large choice of tools, which is constantly increasing (*see* Fig. 1c).

This chapter highlights and reviews mechanistic properties of ChR variants which have already employed in Optogenetic studies. We want to focus on their respective intended applications, as well as their flaws. In this respect, we want to draw attention to pitfalls which arise when nonappropriate ChR variants are used for a given application.

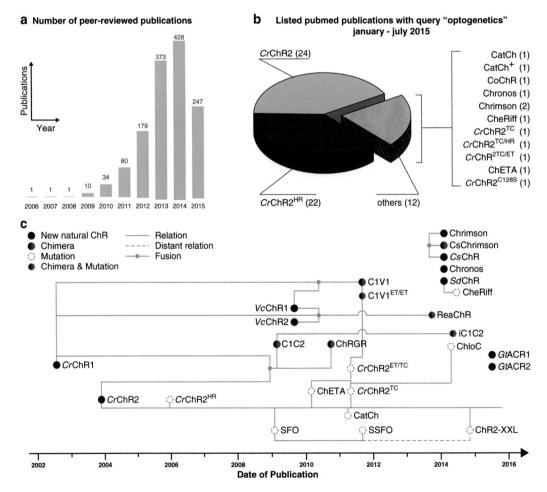

Fig. 1 Prevalence of different ChR variants. (**a**) A bar diagram exhibits the exponential increase of publications contain the keyword "*Optogenetics*" from 2005 to 2015. (**b**) Show the usage of different ChR variants in the period of 6 months. Only publications with viral injection are considered in order to avoid a bias to older variants due to the relatively slow distribution of transgenic animals. (**c**) An ancestral chart of engineered and newly discovered ChR to depict relation between variants was well showing increased paced new variants are released

Channelrhodopsins are a new class of microbial rhodopsins which were first discovered in fresh water alga *Chlamydomonas rheinhardtii* (*Cr*ChR1 and *Cr*ChR2; throughout the chapter the first two letters are the initials of the genus and species of the organism from which the rhodopsin originates, followed by the gene number and possible mutations. The exception is chimeric variants, which are given their common abbreviation) [7–9]. Like all rhodopsins, microbial rhodopsins, or so called type I rhodopsins, consist of seven transmembrane helices and utilize a retinal molecule as a chromophore for light absorption. Upon absorption of photons, the retinal molecule undergoes an isomerization from all-*trans* to 13-*cis*. This initial process efficiently triggers a cascade of conformational changes in the protein backbone that lead to the side-flip of a transmembrane helix, and opens up a continuously conductive pore allowing flux of cations across the membrane [10]. The pore closes upon reisomerization and the rhodopsin returns to its starting point. During this photocycle, the retinal molecules stay covalently bound to the seventh transmembrane helix making it a compact, self-sufficient biological photodiode (Fig. 2a) [11].

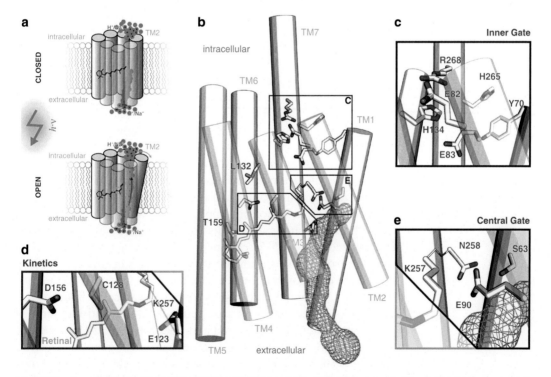

Fig. 2 Structure of Channelrhodopsin. (**a**) Closed ChR model in dark (*top*) converted by light into the open state (*bottom*) by retinal isomerization followed by structural changes, allowing transport of ions across the membrane. (**b**) Overall ChR2 model based on C1C2 crystal structure. Transmembrane helices (TM) and discussed residues are shown. The extracellular access channel (turquoise mesh) was calculated with MOLE 2.0 (**b** and **d**) [112]. A closer look into the structure showing the inner gate (**c**), residues relevant for kinetic modifications (**d**), and the inner gate (**e**)

In addition to the novelty of controlling the membrane potential of cells with light, it is also the simplicity and robustness of the ChR molecule itself that have made it such an attractive tool. A transmembrane protein which enables a unicellular alga to find optimal light conditions for photosynthesis can be readily transferred to vertebrates. Despite topological similarities between microbial and vertebrate rhodopsins, they share virtually no sequence homology. Nevertheless, the variety of electrochemical and also biochemical cascades both protein families can modulate are starting to become remarkably similar [12, 13].

2 Adapting Channelrhodopsin Properties

2.1 Functional Expression

In their native environment, ChRs act as a sensory receptor for photons enabling motile, green algae to avoid overly illuminated environments and to find optimal conditions for photosynthesis [14]. The photoreceptors are targeted to a small membrane area, the "eyespot". This localization in a membrane patch is accomplished by linking the large cytoplasmic C-terminal of ChRs to intracellular structures such as the chloroplast [15]. For heterologous expression, the C-terminal anchor can be omitted without altering channel function, enabling a more homogeneous distribution in the membrane. In that respect, truncating the C-terminal of CrChR2, and reducing the protein length from 737 to 320 amino acids can be considered as the first improvements toward rendering ChR an optogenetic tool [16]. It is therefore not surprising that the C-terminal fusion of a fluorescence protein (FP) has little or no effect on overall channel function [17].

Another, very early adaptation was to match DNA codons to the repertoire of expressed tRNAs in the host system. This proved to facilitate translation accuracy and speed, which can both become a bottle-neck when strong expression is desired [18].

In neurons an important improvement was the transformation from the GC-rich algae codons to the usage of human codons. Humanized ChR DNA sequences, which are commonly marked with *h* in front of the specific ChR variant, have been shown to give acceptable expression in most vertebrates. However, for some rhodopsins the obtained high expression levels can cause fluorescence aggregates in the endoplasmatic reticulum (ER) and blebbing of plasma membrane in hippocampal rat neurons [19, 20]. Normally, these side effects can be overcome with endogenous targeting sequences such as the ER export sequence or the membrane trafficking sequence, both of which are derived from the mammalian inward rectifying potassium channel 2.1 [21, 22]. The ER export sequence is added to the C-terminal end of a ChR–FP fusion construct, whereas the trafficking sequence is inserted between ChR and the FP. However, some commonly used fluorophores, such as mCherry, that still have a residual tendency to form oligomers can alter the ChR expression pattern and effect the viability of the host cell [23, 24]. Therefore, when considering problems with ChR expression one should always focus on the entire translated gene and not only on the ChR itself. Some ChR variants show such exceptionally good membrane targeting that it can be difficult to pinpoint the actual ChR expressing cell in an acute slice preparation since virtually any fluorescence linked to ChR is membrane bound. Here, constructs with a p2a, t2a, or IRES sequences, which separate the ChR and FP open reading frame, have proven to be useful [17].

Nonetheless, ChR also intrinsically contains endogenous targeting sequences which can facilitate its functional expression in vertebrates. For example, the first reported ChR with a red-shifted absorption spectrum was identified in a multicellular alga, *Volvox carterii,* and termed *Vc*ChR1 [25, 26]. It was speculated that improper protein folding and trafficking accounts for the low photocurrent of *Vc*ChR1 in hippocampal neurons [26]. Subsequently, several studies exemplified that the N-terminus of *Cr*ChR1, a rhodopsin with low expression in vertebrates, can increase *Vc*ChR1 photocurrents ninefold in the same expression system [27–29]. Also recently, a red-light absorbing ChR coined Chrimson from the fresh water alga *Chlamydomonas noctigama* has shown improved expression upon codon optimization and addition of trafficking sequences [22]. Nevertheless, for high expression in *Drosophila melanogaster*, a model organism known for small photocurrents due to the low endogenous concentration of all-*trans* retinal, the fusion of a the N-terminus sequence of ChR from *Chloromonas subdivisa* increased expression to an adequate level (*Cs*Chrimson) [22].

Therefore it appears as if the efficacy of endogenous trafficking sequences not only depends on the species or kingdom from which the microbial opsins originate, but also on the specific organism in which it is applied [20, 22].

Taken together, the first decade of tailoring ChRs to optimize expression has equipped protein engineers with a toolbox of various trafficking sequences to achieve high functional expression in virtually all commonly used laboratory organism such as nematode, mouse, rat, zebra fish, drosophila, or nonhuman primates [16, 30–33].

So far, the effect of high expression of microbial rhodopsins on membrane parameters and overall cell health were mostly mentioned as a side note [34, 35]. Careful monitoring is important if optimized ChR variants are under control of a strong promoter. To what extent neurons are capable of supporting the increased energy consumption, and how the increased workload effects protein folding machinery is largely unexplored. To date, standard practice has been to confirm that a given ChR variant neither alters membrane integrity (membrane resistance, resting potential, and cell capacitance), nor changes cell morphology, i.e., spine density [1, 36, 37]. But it is questionable if those complex parameters are a sufficiently sensitive readout, or if, perhaps, the integrity control should rather start at the RNA level with a full RNA sequence comparison of cells expressing a new ChR variant to control cells.

At least membrane capacitance seems to be a sensitive parameter for estimating the extent of membrane crowding. High expression of *Cr*ChR2 in HEK cells causes thinning of the lipid membrane and hence an increase in membrane capacitance [38]. In neurons, a fast cycling ChR chimeric variant called ChIEF, showed a change in membrane capacitance compared to the control (*see* Fig. 1c and Table 1) [39]. Therefore, rhodopsin expression should be carefully adjusted for a given application in respect to the necessary photocurrent.

2.2 Photocurrent Properties

The photocurrent is the sum of moved charges through the open ChR molecule at a given membrane potential. It is a key parameter for most Optogenetic applications.

As a first approximation, photocurrent of dark-adapted ChRs (I) at a given time point is defined as the product of the unitary

Table 1
List of ChR variants and there typical feature as well as their potential handicaps. The last row gives specific references to their first description as well as their first application

Name	Features	Struggle/obstacles	References
*Cr*ChR2	• Widely used ChR variant • Apparent off-kinetic ~10 ms • Peak activation 470 nm • Moderate photocurrent	• Moderate photocurrent	[1, 113]
*Cr*ChR2-H134R	• Widely used ChR variant • Reduced inactivation of to 60 % • Slightly higher sodium/proton ratio (1.2) • Larger photocurrent compared to *Cr*ChR2	• Apparent off-kinetics slightly slower than *Cr*ChR2 (~20 ms)	[3, 39, 114]
*Cr*ChR2-T159C	• Larger photocurrents (tissue dependent) • Peak activation 470 nm • Higher affinity to retinal • Longer turnover • Most efficient for low frequency stimulation	• Strong inactivation • Moderate apparent off-kinetics (~ 45 ms)	[58]

(continued)

Table 1
(continued)

Name	Features	Struggle/obstacles	References
ChETAs CrChR2-E123T CrChR2-E123A CrChR2-E123T–T159C	• Ideal for fast stimulation rates (50–200 Hz) (apparent τ_{off} 4–5 ms) • Fast peak recovery (<3 s) • Peak activation ~490 nm • Voltage independent off-kinetics • Larger photocurrent • 8–10 ms apparent off-kinetics • Fast peak recovery (<3 s) • High retinal binding affinity	• Small photocurrent for single E123X mutations • Less light sensitive • E123A enhance proton conductivity	[39, 104, 115]
Step function rhodopsins (SFR) CrChR2-C123T/A/S CrChR2-D156A/N CrChR2-C128S–D156A (stabilized)	• Prolonged open state lifetime of 2, 50, and 100 s • Yellow/red and UV light to trigger conversion to close state • High light sensitivity • Prolonged open state lifetime 300 s • Moderate photocurrent • Prolonged open state lifetime 29 min • Yellow/red and UV light to trigger conversion to close state • Moderate photocurrents • High light sensitivity	• Low expression • Subthreshold depolarization • Danger of acidifying cell when channel is open for long time	[30, 76, 109]
Chloride conducting ChRs (ChloCs) (Slow) ChloC iC1C2 iChloC GtACR-1/2	• Conducting chloride ions • Inhibiting of neurons • Slow and moderate kinetics • Large photocurrents • Reversal potential like GABA$_A$ ion channels • In vivo silencing • Large photocurrent • Reversal potential like GABA ion channels	• Still depolarizing at resting potential (E_{rev} ~55 mV) • Small photocurrents • No in vivo experiment so far	[52, 69, 70]

(continued)

Table 1
(continued)

Name	Features	Struggle/obstacles	References
Calcium translocating Channelrhodopsin (*CatCh*) *Cr*ChR2-L132C *Cr*ChR2-L132C–T159C	• 1.6× larger Ca^{2+} conductance • Higher conductance for Mg^{2+}	• Conducted calcium can trigger synaptic release and second messenger pathways • Slow kinetics (~150 ms) • Slow kinetics (~1000 ms)	[52, 69, 70]
*Cr*ChR2-L132C-T159S	• Large photocurrent • Large photocurrent • Conducts Mg^{2+} and Ca^{2+} • Large photocurrent • Conducts Mg^{2+} and Ca^{2+}		
ChIEF C1C2(25)-I170V	• Small inactivation (20 %) • Apparent off-kinetic 10 ms	• High expression; change in membrane capacitance	[53, 116]
GreenReciever (ChRGR)	• Peak absorption ~505 nm • Fast kinetics (τ_{off} 4–5 ms) • Low inactivation	• Moderate photocurrents	[57]
Chronos	• Large photocurrent • Apparent off-kinetics 5 ms • High operational light intensity • Peak absorption ~500 nm	• Very high expression	[22]
C1V1(25)	• Large photocurrent • Peak absorption 535 nm • Moderate apparent off-kinetics ~160 ms	• Increased conductance for Ca^{2+}/Mg^{2+}	[27, 117]
C1V1$^{ET/ET}$ C1V1(25)-E122T–E162T	• Large photocurrent, λ_{max} 540 nm • Peak absorption 540 nm • Moderate apparent off-kinetics ~40 ms • Large two-photon cross section	• Increased conductance for Ca^{2+}/Mg^{2+}	[30, 90, 91]
ReaChR C1V1V2V1-N511	• Large photocurrent • Peak absorption 540 nm • Can be inactivated with orange light too	• Slower kinetics with orange light	[28, 118]
(*Cs*)Chrimson	• Large photocurrent • Max absorption 590 nm • Apparent off-kinetics 20 ms	• Absorption in the blue range of spectrum	[22]

conductance (g), number of open channels (N) and membrane voltage (U) [40]:

$$I = N(f,E) \cdot g(U,C)$$

The number of open channels depends on photon flux, or in other words, on the probability of the rhodopsin to absorb a photon (Φ) and the number of functional ChRs in the plasma membrane (C). Different strategies to increase functional expression (E) have been summarized in the previous section. A higher photon flux increases the probability of a ChR molecule to convert from a closed, non-conducting state, to an open, conducting state. This relation is linear until the photon flux is so high, that virtually a close state rhodopsin is immediately reexcited. This light intensity is a variant-specific kinetic dependent parameters and causes saturation of photocurrent [41]. For example, CrChR2 has an off-kinetics of 20 ms and saturates at around 10 mW/mm^2. Whereas slow variants, with off-kinetics in the range of 100 s, are already saturated at 0.1 mW/mm^2 [27, 39]. The quantum efficiency, meaning the likelihood of an absorbed photon to successfully trigger the conversion from a closed to an open state, is between 0.3 and 0.8 for rhodopsins and is a difficult parameter to further optimize [42].

This renders the unitary conductance the salient parameter to increase photocurrents. But what are the elements controlling unitary conductance, and the mechanism of ion permeation? Most protein engineers have speculated that ChR in its open state forms a continuous pore for ion permeation [16]. Seven years of diligent mutational, spectroscopic, and modeling studies have revealed only limited insights into the mechanism of this putative ion channel pore. A major leap toward understanding ion conductivity and permeation was taken when the high-resolution crystal structure of the closed state C1C2 chimera was published [43]. Based on this atomic model, the hydrophilic pores is located between transmembrane helices TM1, TM2, TM3, and TM7 and is controlled by two main gates which are shown in Fig. 2a, b [10]. As with other ion channels, the closed ChR pore is partially filled with water molecules from the extra- and intracellular site. While the interior of this pore, between the inner gate (R268, H265, H134, Y70, E82, and E83) and the central gate (S63, N258, and E90), remains free of water (Fig. 2c, e) [10].

Based on such mechanism, the pore diameter of the central gate is estimated to be around 3–5 Å [44]. Similar diameters have been experimentally determined for sodium channels (4.6 Å), potassium channels (4 Å), and even the nicotinic acetylcholine receptor (3.5 Å) [45–47]. However, these voltage and ligand-gated ion channels have a 500–1000× larger unitary conductance then the estimated 20–50 fS for a CrChR2 molecule [48, 49].

The reasons for this difference are not known. On one hand, understanding these differences might lead to the development of extremely efficient ChR variants. On the other hand, the low unitary conductance could also indicate a less classical pore with a strong ion binding site. The successful implementations of enzyme kinetic models that describe ion permeation through ChR support this hypothesis [50, 51].

A strategy that has already been applied to increase the overall conductance is to confine the activated ChR molecules to pass only through a high performance photocycle. Based on the canonical model of a four-state photocycle of CrChR2, which will be described in more detailed in subheading 4, ChR molecules possess two open states, O1 and O2. Both can be populated from two different closed states, C1 and C2 (*see* Fig. 3a). Both open states have distinct properties: O1 is more selective for sodium, has a shorter lifetime and a larger unitary conductance compared to O2 [51, 52]. Also, the conversion from C1 to O1 exhibits slightly higher quantum efficiency [41]. Mutations hindering the interconversion to the low-performance photocycle (C2/O2) render ChR more conductive [53]. The degree of conversion between both photocycles is reflected as the ratio of peak photocurrent (I_p) to stationary photocurrent (I_s) (*see* Fig. 3b). The ratio of both photocurrent amplitudes is commonly referred as inactivation [54]. The kinetic from I_p to I_s is the deactivation kinetics (τ_{in})

Recently, a carefully performed spectroscopic analysis suggested different retinal isoforms for retinal Schiff base in both photocycles [55]. For the high-performance photocycle, a conversion from 13-*trans*, 15-*anti* to 13-*cis*, 15-*anti* is proposed, whereas low-performance photocycle converts 13-*cis*, 15-*syn* to 13-*trans*, 15-*syn* retinal upon absorption of a photon (*see* Fig. 3a). Variants such as the C1C2 chimera (ChIEF) or Channelrhodopsin GreenReciever (ChRGR) exhibit reduced inactivation, which is explained by the low probability of interconversion between the photocycles (*see* Table 1) [40, 56, 57]. These variants, with low inactivation are more suitable when repetitive pulse stimulation or constant depolarization is needed.

The discovery that protein stability and photocurrent are intermingled revealed an unexpected means of increasing photocurrents. Mutating threonine 159 to cysteine doubles the photocurrent amplitude in rat hippocampal neurons [58]. It is hypothesized that T159C increases the binding affinity of the chromophore, and therefore stabilizes protein structure, subsequently decreasing the turn-over rate of functional ChRs [59]. This mutation seems beneficial for most variants even though its impact depends on the organism in which it is expressed. Nevertheless, the efficiency with which optimized, large-photocurrent ChR depolarizes cells also depends on the interplay of electrochemical gradient of the cell with the ion selectivity of a given ChR variant.

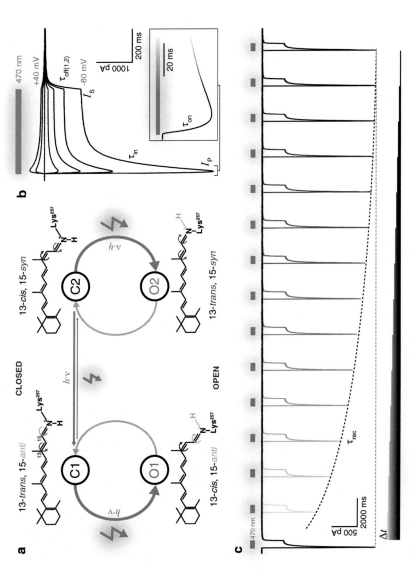

Fig. 3 Photocycle and photocurrent properties of ChRs. (**a**) Photocycle model of ChR according with two closed states (C1, C2) and two open states (O1, O2) [105]. Possible reactions are indicated by arrows. The dedicated retinal isoforms for each state are shown. (**b**) Typical photocurrents of *Cr*ChR2-T159C expressed in HEK293 cells, measured in voltage clamp mode ranging from −80 to +40 mV in 20 mV steps. Typical current parameters and kinetic constants are shown. The inset shows a closer look at the on-kinetic. (**c**) Recovery kinetic of *Cr*ChR2-T159C, depending on the time interval Δ*t* between two applied light pulses. I_p recovers with increasing Δ*t*, whereas I_s is unaffected

2.3 Ion Selectivity   First described as a light-gated proton channel, *Cr*ChR1 later turned out to also transport mono- and divalent cations across the cell membrane [7, 54]. More recent studies showed that under physiological conditions, photocurrents are carried half by protons, and half by sodium, potassium, and, to a lesser extent, calcium, and magnesium [52].

However, in neuronal tissue, extracellular sodium and chloride concentrations are high, while potassium concentration is high inside the cell. The resting membrane potential of a typical neuron expressing *Cr*ChR2 in the absence of light is approximately -70 mV. Under these conditions, a brief pulse of light triggers the influx of protons and sodium, as these ions are the major contributors of the photocurrent into the cell [60]. If this photocurrent is strong enough to recharge the membrane, the transported charges are translating into membrane depolarization and will recruit more and more voltage-gated sodium channels and therfore eventually trigger an action potential. Outward flux of potassium can be neglected due to the inward rectification of *Cr*ChR2.

Commonly, short light pulses between 2 and 10 ms are used to activate ChRs in a neuronal context. Longer illumination time can change ion composition in the vicinity of the membrane. For instance, it has been shown that exposing ChR expressed in glial cell too long periods of light can trigger glutamate release via intracellular acidification or intracellular calcium release [61–63].

Finally, extended depolarization due to inappropriate, slow off-kinetics of ChRs can result in significant activation of voltage-gated calcium channels or metabotropic glutamate receptors [64, 65].

To overcome these side effects, many efforts have been made to engineer ChRs that are more selective for sodium or potassium than protons. This has been challenging due to lack of knowledge about the exact geometry of the selectivity filter.

One small step towards a more ion-selective ChR with less proton conductance is the replacement of histidine 134 with arginine. This was one of the first introduced modifications of *Cr*ChR2 and caused lower inactivation of the photocurrent and a mild increase in the Na^+/H^+ ratio [3, 50].

In contrast, ion selectivity can also be modified so that more protons are conducted than alkali ions. The ChETA variants (named after the first version *Ch*annelrhodopsin *E123T* Accelerated), in particular E123A, have been shown in mechanistic studies to mainly conduct protons across the cell membrane [50, 58]. Under physiological conditions these ChETA variants are used when high-frequency stimulation is demanded. Nevertheless, it should be considered that long stimulation paradigms can cause a stronger proton influx in these variants when compared to *wild-type Cr*ChR2. Furthermore, to avoid acidification, variants like E123T or E123Q should be used when prolonged and fast stimulation is applied.

The consequences of a prolonged illumination not only depend on the specific cell type, but also on the illuminated subcellular structures. In axons, where the *surface area/volume* ratio is high, large and long photocurrents can significantly change intracellular and extracellular ion composition, and therefore change the reversal potential of endogenous ion channels [66, 67]. This is particularly important for intracellular calcium, since small concentration differences can trigger second messenger cascades [64]. On the other hand Ca^{2+} entry through prolonged illumination of ChR can be intentionally used to increase the calcium level, for example in cardiomyocytes [68]. The most suitable ChRs for this purpose are the so called *ca*lcium *t*ranslocating *ch*annelrhodopsins (CatCh and CatCh⁺). Exchange of leucine 132 in *Cr*ChR2 with a cysteine causes an increased Ca^{2+} selectivity by a factor of 1.6 [27, 69]. The combination of L132C with T159C or T159S (CatCh⁺ mutations) increased calcium mediated photocurrents even further [52, 70]. Also the red-shifted C1V1 chimera has an increased calcium conductance compared to *wild-type Cr*ChR2 [27, 50].

Nevertheless, the mechanism that underlies the increased calcium selectivity is still poorly understood. Leucine 132 is in close vicinity to histidine 134 which has a key role in the formation of the inner gate (Fig. 2c). However, in the crystal structure of C1C2, L132 is facing away from the inner gate toward TM5 [43]. More work is required to understand the complex interactions that lead to higher Ca^{2+} selectivity. A more detailed insight into the mechanisms that govern ion selectivity in general might help to guide engineering of ChRs with high selectivity for a desired ion.

In 2014, a milestone toward that direction was reached. Two independent groups converted ChRs from a cationic to an anionic conductance based on two independent approaches [37, 71]. The first conversion into a chloride conducting ChR (termed ChloC) introduced a positively charged arginine at the position of the negatively charged central gate glutamate (E90R). Two additional mutations were used, one to improve photocurrents (T159C), and a second to prolong the open state (D156N). These mutations together lead to a decrease in the light power necessary for silencing [37].

The second approach introduced nine mutations on the putative channel surface of C1C2 to render the electrostatic environment more positive [71]. This "inhibiting" C1C2 (iC1C2) as well as ChloC can suppress action potentials in organotypic slice preparations. However, under identical conditions, both variants show the reversal potential of around –55 mV; this reversal potential is 15 mV higher than pure chloride permeability would predict (Jonas Wietek, personal communication). Therefore, these first variants depolarize the cell at resting potential rather than hyperpolarizing it. In a recent follow-up study, Wietek and colleagues show that the first version of ChloC is not sufficient to silence unperturbed neurons at resting potential, but rather can lead to the triggering of action potentials [37].

Introducing a serine for glutamate 101, which is known to lower cation conductance, further increase the chloride selectivity of ChloC [71–73]. Additionally, the inner gate residue E83 was replaced by glutamine to eliminate the negative charge at this position without disturbing the overall geometry. Thus, by charge neutralization at amino acid residues on both entrance sides of the CrChR2, the improved ChloC (termed iChloC) exhibits the same reversal potential as endogenous GABA$_A$ receptors. This improved variant iChloC reliably suppressed visually evoked spike activity in the primary visual cortex of mice upon light stimulation [37].

Naturally occurring, anion-conducting ChRs were recently found in the alga *Guillardia theta* using an screening approach [74]. The two identified anion selective ChRs termed GtACR1 and GtACR2 (anion-conducting rhodopsin) show no dependence on pH, exhibiting a reversal potential similar to that of endogenous chloride reversal potential, and feature very large photocurrents in hippocampal neurons.

Interestingly, the central gate of the ACRs is identical to *wild-type* CrChR2 and possesses no positive charged residue at the E90 position like the ChloC variants. Compared with iC1C2, most of the acidic residues that have been mutated are missing in ACRs. Moreover, the inner gate seems to be differently arranged, as well as the extracellular access channel.

The new group of chloride conducting ChRs (below referred as ChloCs) is likely to become the new standard for light induced inhibition of neuronal activity. So far, pumping microbial rhodopsins NpHR and Arch, as well as their decedent variants have been applied. Recently, a study showed that 500 ms illumination of NpHR expressing cells shifts the reversal potential of GABA$_A$ channels by 2 mV in organotypic slice culture under moderate expression [67]. This study therefore shows that inhibition with NpHR changes synaptic responses and should only be used with the respective controls.

ChloCs, in contrast to chloride pumps, which hyperpolarize the cell by constantly pumping Cl$^-$ into the cell, have a fundamentally different mode of action. They shunt the cell by driving the membrane potential toward the reversal potential of Cl$^-$, and since the membrane resting potential is normally close to the Cl$^-$ reversal potential, only act on a nascent depolarization with an inward flux of Cl$^-$. In that respect, ChloCs are mimicking endogenous GABA$_A$ channels, and can be considered to be more physiological [75]. Additionally ChloCs exhibit several advantages compared to ion pumps. They do not require high light intensities since they transport many charges per photocycle, and thus allow for the reduction of the overall delivered energy. Another successful protein engineering approach is the design of more red-shifted ChRs. This will be discussed in the next section.

2.4 Spectral
Properties

All naturally occurring rhodopsin molecules utilize a retinal derivative to absorb light. The spectral fraction of the light they are sensitive to is defined by (1) the planarity of the retinal ring-chain system, (2) the electrostatic interaction of residues within the protein binding pocket, and (3) the counterion complex [79]. The tuning of these parameters has resulted in a tremendously expanded color palette for ChR peak absorption, spanning the spectral range from 440 to 600 nm (*see* Fig. 4).

The full width at half maximum of an action spectrum can differ between variants (compare C1V1 and *Cr*ChR2 in Fig. 4, not energy scaled axis). Often, a broad action spectrum resembles a superposition of different protonated isoforms with different absorption properties [26, 54].

The peak absorption and width of the spectra area is a crucial selection criterion for most applications. Nevertheless, despite all the different color variants, independent activation of two rhodopsins with high fidelity appears only to be possible when variants from both ends of the spectral palette are chosen, e.g., *Cr*ChR2 together with Chrimson (*see* Fig. 4). Any application where three or more independent stimulation wavelengths are desired is extremely challenging, and requires carefully adjusted light intensities and stimulation paradigms.

Since the newer generations of ChRs are all optimized for high expression, the ability to select ChRs according to peak absorption is mainly limited by equipment availability (e.g., lasers of the appropriate wavelengths, optical filters) and the spectral properties of an

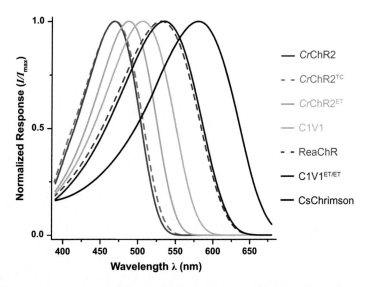

Fig. 4 Action spectra of different ChRs, recorded in HEK293 cells with equal photon flux for all wavelengths. Data was fitted with a three parametric Weibull function

intended optical activity reporters such as GCaMP, RCaMP, or pHluorin [80–82].

Important selection criteria for activation wavelength for in vivo studies are penetration depth, light cone, and potential overheating of the tissue [83]. Implications of high intensity illumination or continuous light application should be always considered. Temperature changes of >1 °C can lead to decreased firing threshold, reduced half-width of action potentials, and decreased action potential amplitude [84]. It was estimated that brain tissue temperature rises by around 0.4 °C during a typical illumination protocol for ChRs, in which light pulses of 5 ms are delivered at 20 Hz for 30 s (470 nm, 5 mW) [85].

Light absorption, and therefore heating of brain tissue, is estimated to be tenfold lower at 600 nm compared to 440 nm [86]. Furthermore, longer wavelengths show a twofold lower scattering coefficient over the same spectral range [86]. Taking these two properties together and assuming isotropic transmission, the delivered intensity of blue light of around 470 nm decays 10× more than red light (600 nm) over the same distance [85, 87]. In an in vivo experiment, this can result in a 500× larger activation cone for red light than for blue light. Certainly the volume of activation also depends on the light sensitivity of a particular ChR variant. But taking into consideration just the light traversing through brain tissue, long-wavelength sensitive variants such as C1V1, ReaChR, and, most notably, CsChrimson can recruit more neurons from a larger volume when illuminated with wavelength above 600 nm [22, 27, 28]. C1V1 as well as ReaChR are chimeric ChRs based on CrChR1, VcChR1, and VcChR2. C1V1 is a composite protein consisting of the first two helices of CrChR1 and the last five TM from VcChR1. Whereas ReaChR is mainly based on VcChR1, except for the short N-terminus of CrChR1 and TM6 from VcChR2 [28]. Both variants, C1V1 and ReaChR, show good membrane targeting and have photocurrents with a peak absorption of ~540 nm (Fig. 4). Nevertheless, so far CsChrimson has the most red-shifted absorption spectrum and it produces large photocurrents in cultured hippocampal neurons [22].

If deep brain penetration together with a small volume of activation or even single cell activation is desired, two-photon excitation is the best choice. Variants with red-shifted absorption are also beneficial since their two-photon excitation spectra are also red-shifted, and therefore longer wavelengths, (>1200 nm) with deeper tissue penetration, can be used. Nevertheless, for a two-photon approach, channel off-kinetics and the spatial pattern of stimulation gain greater importance. The small two-photon excitation volume only excites a small number of channels at once, causing a small, subthreshold depolarization. Scanning over the entire cell membrane leads to accumulation of open ChRs, eventually generating an action potential. Therefore, ChRs with moderate closing rates (30–50 ms) contribute for longer periods to the overall photocurrent. Photocurrent amplitude is therefore a function of both the total scanning time and the decay constant of a given ChR variant [88].

Theoretical studies show that certain mutations such as T159C, E83T and E123T can increase the two-photon absorption cross section [89]. This is especially true of C1V1$^{ET/ET}$ (C1V1-E122T–E162T) which incorporates all three mutations, has moderate off-kinetics, has a red-shifted absorption spectrum, and has been now frequently used for two-photon activation [90, 91].

The new freedom to choose different spectrally tuned ChRs and the improved performance of optical activity reporters makes an all-optical approach readily usable [92–94]. To get a clear optical separation between the activation and the monitoring light path, reporters with longer wavelength absorption than a chosen actuator seem at first preferable. But all ChRs, even the most red-shifted variants like CsChrimson, show a substantial absorption in the blue part of the spectrum (*see* Fig. 4). Along this line of argument, several studies have shown that action potentials can still be evoked in neurons expressing red-shifted ChRs at wavelengths above 600 nm, in which light absorption is already below 15 %, by either increasing light intensity or pulse duration [22, 30, 95]. Yet, the same holds true for the blue side of the absorption spectra, illuminating here with longer and stronger blue light pulses can trigger spikes even in the most red-shifted variant. This still limits the usefulness of red-shifted variants in conjunction with optical reporters that require blue excitation light. The thoughtful design of more blue-shifted variants might alleviate the current limitations for all-optical experiments, but will not overcome them. Mutations such as T159G or G163A have been shown to cause a substantial a blue shift of more than 20 nm [96]. Nevertheless, in a recent study, a blue-shifted ChR, termed CheRIFF ($\lambda_{max} \sim 460$ nm) from *Sherffelia dubia,* was shown to work in conjunction with a red-shifted reporter [36]. Nevertheless, blue-shifted ChRs will not be right approach for all optogenetic experiments due to the decrease of tissue penetration of light.

Two-photon stimulation is one methods to solve the problem of cross-activation between actuator and reporter. Due to the optical sectioning capability of this method, a prescanned field of view can be subdivided into regions of interest (ROI) which can be either stimulated or imaged, since fast scanning across the imaging ROI is assumed to only cause subthreshold activation. Even though those techniques have a subcellular resolution, most ROIs are based on soma detection algorithms [90]. At high light power and strong expression, axons, or dendrites in close proximity to stimulated soma can get coactivated and add unwanted noise to the experiments. Here, restricting ChR expression only to the soma or axon initial segment would be beneficial, but efforts to do this have so far been unsuccessful [97].

Microbial rhodopsins are self-sufficient molecular machines. In contrast with vertebrate rhodopsins, the retinal chromophore of microbial rhodopsins stays attached during several cycles of isomerization which allows design of faster variants [98].

During one cycle, a ChR molecule passes through several distinct states, each with different spectroscopic and conductive properties. Therefore, the literature distinguishes between a spectroscopic and electrophysiological photocycle [41, 99, 100]. Here, we consider only the electrophysiological photocycle. Excellent reviews of the spectroscopic photocycle can be found in Schneider et al. and Stehfest et al. [101, 102].

The first opsin related event upon retinal isomerization is the change in the dissociation constant of the protonated retinal Schiff base, which subsequently donates its proton to the counterion pair D253/E123. This process of deprotonation and subsequent reprotonation defines the open dwell time and off-kinetics of ChR variants [6, 103]. The ChETA mutants (E123T, E123Q, or E123A)—which favor protonation of the second counterion, D253—exhibit accelerated off-kinetics of ~5 ms for CrChR2 [104]. These single mutations, alone or in combination with other variants, are used whenever fast off-kinetics are desired. Nevertheless, the fast off-kinetics of ChETA$_T$ (CrChR2-E123T) and ChETA$_A$ (CrChR2-E123A) come at the cost of only moderate photocurrents due to their short dwell time in the open states. A good compromise is the combination of with T159C, CrChR2$^{ET/TC}$ (CrChR2-E123T-T159C). This combination leads to slightly slower off-kinetics (τ_{off} 10 ms), but large photocurrents and is maybe the most versatile and robust ChR [39].

The second event after isomerization is the restructuring of the protein side chains of the inner gate. In the dark-adapted state, it is assumed that N258 of TM7 holds TM2 in close proximity by forming two intrahelical hydrogen bonds with E90. Descriptively speaking, the conversion from the *trans* to the *cis* isoform exerts a pulling force which traverses from the retinal carbon chain to the site chain of K257, and finally initiates a movement of N258 (Fig. 1a, c). This movement flips the hydrogen bonds toward E90, and the entire helix TM2 slides 4 Å aside, leading to the opening of the ion conducting pore [10]. Based on the canonical four-state photocycle given in Fig. 4a, all common kinetic parameters as well as light sensitivity, can be modeled to a satisfactory level [105]. In the dark, ChRs can occupy one of two different closed states: the dark-adapted state C1, and the light adapted state C2. Molecules absorbing a photon from C1 pass through the high-conductivity open state, O1; whereas molecules activated from C2 go into the low-conductivity, or low-performance, state O2. Pulsed excitation of molecules in the dark-adapted state leads the photocurrent to rise to the peak photocurrent within 1–2 ms, during which time all of the molecules have to pass through the O1 state. For continuous illumination in which molecules undergo several rounds of activation, I_p decays into a stationary photocurrent constituted of a mixture of O1 and O2. In consequence, inactivation decreases the level of depolarization in neurons during constant or high-frequency application of light. Therefore, variants with low inactivation should be considered when stable, sustained depolarization is required. For example, the *wild-type* CrChR2 has an inactivation of 70 %; whereas variants such as ChIEF, CatCh, and ChRGR exhibit an inactivation of less than 20 % [39].

The recovery kinetic is the conversion of a molecule from C2 to C1 and is normally in the range of seconds (*see* Fig. 4c) [27, 100]. If molecules get reexcited from C2, they only take the path through the low-performance O2 state, resulting in a smaller overall peak current. This leads to decreased fidelity when neurons are repetitively stimulated with short light pulses.

In a typical double pulse experiment shown in Fig. 3c, the recovery from C2 to C1 can be probed. Variants which convert faster from C2 to C1 are for example ChETA$_T$, ChETA$_{TC}$, or ChIEF [39].

The apparent off-kinetics is of similar importance for a high fidelity of repetitive stimulation for a given variant. After illumination stops, molecules start to accumulate in their respective close states with time constants τ_1 or τ_2. The weighted linear combination of these two time constants is called the apparent off-kinetics, which is a single value found in most literatures.

Fast variants such as the ones based on ChETA not only exhibit accelerated channel-closing rates, but also possess decreased open state dwell times [106]. Both features lead to a reduced number of ions transported per absorbed photon. As a result, the intrinsic light sensitivity of these fast variants is reduced, and high light intensities are required to explore the full potential of fast variants as high-frequency actuators.

On the other extreme, variants with strongly decelerated closing rates have been engineered [30, 76, 107]. Two residues, C128 in TM3 and D156 in TM4, have been shown to form a hydrogen bond that is crucial for channel gating. Variants with mutations in these residues are commonly referred to as the step-function rhodopsins (SFR) (*see* Fig. 2d). For example, the two SFR mutations C128S or D156A give rise to an apparent off-kinetic of 100 s and ~300 s, respectively [76, 99, 108]. Both mutations can be synergistically joined to a form stabilized SFR, which exhibits off-kinetics of >29 min in hippocampal cultured neurons [30]. All SFRs display a drastically increased light sensitivity since activated ChRs are essentially stucked in the open state. Therefore, saturated photocurrents can be reached with 1000× lower light intensities when light pulse duration is appropriately adjusted. Due to their slow off-kinetics, SFRs remain in the opens state after a short blue light pulse, which can be used to induce a long-lasting depolarization step without constant light application—this property of SFRs can be useful for behavioral experiments [109]. Carefully titrated switching of a fraction of SFR into the open state can be used to increase the excitability of genetically defined neuronal population [30]. However, despite their indisputable usefulness, one needs to be aware that prolonged activation periods of SFRs can change the intracellular pH or Ca^{2+} levels. This can eventually lead to unintended side effects, and negatively influence cell viability.

Similar problems can occur when ChR variants are stimulated faster than the channel-closing rate. This causes a stable fraction of ChR molecules to be continuously open, and depolarizes neurons to the so called plateau potential [104]. At the plateau potential, the inactivation of sodium channels results in a depolarization block of action potential firing [110, 111]. In general, a good practice is to titrate the light intensity to find the lowest light intensity that will evoke a robust behavioral response.

3   Conclusion

The first 10 years of Channelrhodopsin tailoring equipped users with new, unforeseeable possibilities to control membrane potential with light. The ChR toolbox has been widely expanded from the initial, medium-fast, blue light sensitive excitatory rhodopsin, to a range of highly efficient excitatory and inhibitory rhodopsins with different color sensitivities and kinetics. This variety can be overwhelming to users.

In this chapter, we have summarized new variants which have been tested and approved by users in a neurophysiological context. We tried to underline their respective advantages as well as their failings. This new possibility of having variants for a specific application will further help to established Optogenetics as the standard method to functionally study neurobiological circuitry.

Acknowledgments

We thank our colleagues for providing action spectra of selected ChRs: Franziska Schneider (C1V1, C1V1-E122T–E162T), Christiane Grimm (ReaChR), and Johannes Vierock (CsChrimson). We are also in debt to Mathias Mahn, Simon Wiegert, Yoav Printz, Kirstin Eisenhauer, and Tess Oram for proofreading the manuscript and fruitful discussion.

References

1. Boyden ES, Zhang F, Bamberg E et al (2005) Millisecond-timescale, genetically targeted optical control of neural activity. Nat Neurosci 8:1263–1268

2. Li X, Gutierrez DV, Hanson MG et al (2005) Fast noninvasive activation and inhibition of neural and network activity by vertebrate rhodopsin and green algae channelrhodopsin. Proc Natl Acad Sci U S A 102:17816–17821

3. Nagel G, Brauner M, Liewald JF et al (2005) Light activation of channelrhodopsin-2 in excitable cells of Caenorhabditis elegans triggers rapid behavioral responses. Curr Biol 15:2279–2284

4. Deisseroth K, Feng G, Majewska AK et al (2006) Next-generation optical technologies for illuminating genetically targeted brain circuits. J Neurosci 26:10380–10386

5. Ziegler T, Möglich A (2015) Photoreceptor engineering. Front Mol Biosci 2:1–25

6. Hegemann P, Möglich A (2010) Channelrhodopsin engineering and exploration of new optogenetic tools. Nat Methods 8:39–43

7. Nagel G, Ollig D, Fuhrmann M et al (2002) Channelrhodopsin-1: a light-gated proton channel in green algae. Science 296:2395–2398

8. Sineshchekov OA, Jung K-H, Spudich JL (2002) Two rhodopsins mediate phototaxis to low- and high-intensity light in Chlamydomonas reinhardtii. Proc Natl Acad Sci U S A 99:8689–8694

9. Suzuki T, Yamasaki K, Fujita S et al (2003) Archaeal-type rhodopsins in Chlamydomonas: model structure and intracellular localization. Biochem Biophys Res Commun 301:711–717

10. Jens Kuhne FB, Eisenhauer K, Ritter E, Hegemann P, Gerwert K (2015) Early formation of the ion-conducting pore in channelrhodopsin-2. Angew Chem 54:4953–4957

11. Zhang F, Vierock J, Yizhar O et al (2011) The microbial opsin family of optogenetic tools. Cell 147:1446–1457

12. Man D, Wang W, Sabehi G et al (2003) Diversivication and spectral tuning in marine proteorhodopsins. EMBO J 22:1725–1731

13. Spudich JL, Jung K-H (2005) Handbook of photosensory receptors. Wiley-VCH, Weinheim

14. Kateriya S, Nagel G, Bamberg E et al (2004) "Vision" in single-celled algae. News Physiol Sci 19:133–137

15. Mittelmeier TM, Berthold P, Danon A et al (2008) C2 domain protein MIN1 promotes eyespot organization in Chlamydomonas reinhardtii. Eukaryot Cell 7:2100–2112

16. Nagel G, Szellas T, Kateriya S et al (2005) Channelrhodopsins: directly light-gated cation channels. Biochem Soc Trans 33:863–866

17. Prakash R, Yizhar O, Grewe B et al (2012) Two-photon optogenetic toolbox for fast inhibition, excitation and bistable modulation. Nat Methods 9:1171–1179

18. Fuhrmann M, Hausherr A, Ferbitz L et al (2004) Monitoring dynamic expression of nuclear genes in Chlamydomonas reinhardtii by using a synthetic luciferase reporter gene. Plant Mol Biol 55:869–881

19. Adamantidis AR, Zhang F, Aravanis AM et al (2007) Neural substrates of awakening probed with optogenetic control of hypocretin neurons. Nature 450:420–424

20. Gradinaru V, Zhang F, Ramakrishnan C et al (2010) Molecular and cellular approaches for diversifying and extending optogenetics. Cell 141:154–165

21. Zhao S, Cunha C, Zhang F et al (2008) Improved expression of halorhodopsin for light-induced silencing of neuronal activity. Brain Cell Biol 36:141–154

22. Klapoetke NC, Murata Y, Kim SS et al (2014) Independent optical excitation of distinct neural populations. Nat Methods 11:338–346

23. Asrican B, Augustine GJ, Berglund K et al (2013) Next-generation transgenic mice for optogenetic analysis of neural circuits. Front Neural Circuits 7:160

24. Madisen L, Mao T, Koch H et al (2012) A toolbox of Cre-dependent optogenetic transgenic mice for light-induced activation and silencing. Nat Neurosci 15:793–802

25. Kianianmomeni A, Stehfest K, Nematollahi G et al (2009) Channelrhodopsins of Volvox carteri are photochromic proteins that are specifically expressed in somatic cells under control of light, temperature, and the sex inducer. Plant Physiol 151:347–366

26. Zhang F, Prigge M, Beyriere F et al (2008) Red-shifted optogenetic excitation: a tool for fast neural control derived from Volvox carteri. Nat Neurosci 11:631–633

27. Prigge M, Schneider F, Tsunoda SP et al (2012) Color-tuned channelrhodopsins for multiwavelength optogenetics. J Biol Chem 287:31804–31812

28. Lin JY, Knutsen PM, Muller A et al (2013) ReaChR: a red-shifted variant of channelrhodopsin enables deep transcranial optogenetic excitation. Nat Neurosci 16:1499–1508

29. Lin B, Koizumi A, Tanaka N et al (2008) Restoration of visual function in retinal degeneration mice by ectopic expression of melanopsin. Proc Natl Acad Sci U S A 105: 16009–16014

30. Yizhar O, Fenno LE, Prigge M et al (2011) Neocortical excitation/inhibition balance in information processing and social dysfunction. Nature 477:171–178

31. Ito HT, Zhang S, Witter MP et al (2015) A prefrontal–thalamo–hippocampal circuit forgoal-directed spatialnavigation. Nature 522(7554):50–55

32. Inagaki HK, Jung Y, Hoopfer ED et al (2014) Optogenetic control of Drosophila using a red-shifted channelrhodopsin reveals experience-dependent influences on courtship. Nat Methods 11:325–332

33. Diester I, Kaufman MT, Mogri M et al (2011) An optogenetic toolbox designed for primates. Nat Neurosci 14:387–397

34. Chow BY, Han X, Dobry AS et al (2010) High-performance genetically targetable optical neural silencing by light-driven proton pumps. Nature 463:98–102

35. Zou P, Zhao Y, Douglass AD et al (2014) Bright and fast multicoloured voltage reporters via electrochromic FRET. Nature Commun 5:4625

36. Hochbaum DR, Zhao Y, Farhi SL et al (2014) All-optical electrophysiology in mammalian neurons using engineered microbial rhodopsins. Nat Methods 11:1–34

37. Wietek J, Beltramo R, Scanziani M, Hegemann P, Oertner TG, Simon WJ (2015) An improved chloride-conducting channelrhodopsin for light-induced inhibition of neuronal activity in vivo. Sci Rep 5:14807. doi:10.1038/srep14807

38. Zimmermann D, Zhou A, Kiesel M et al (2008) Effects on capacitance by overexpression of membrane proteins. Biochem Biophys Res Commun 369:1022–1026

39. Mattis J, Tye K, Ferenczi E (2011) Principles for applying optogenetic tools derived from direct comparative analysis of microbial opsins. Nature 18:159–172

40. Grossman N, Nikolic K, Toumazou C et al (2011) Modeling study of the light stimulation of a neuron cell with channelrhodopsin-2 mutants. IEEE Trans Biomed Eng 58:1742–1751

41. Nikolic K, Grossman N, Grubb MS et al (2009) Photocycles of channelrhodopsin-2. Photochem Photobiol 85:400–411

42. Ernst OP, Lodowski DT, Elstner M et al (2014) Microbial and animal rhodopsins: Structures, functions, and molecular mechanisms. Chem Rev 114:126–163

43. Kato HE, Zhang F, Yizhar O et al (2012) Crystal structure of the channelrhodopsin light-gated cation channel. Nature 482: 369–374

44. Richards R, Dempski RE (2012) Re-introduction of transmembrane serine residues reduce the minimum pore diameter of channelrhodopsin-2. PLoS One 7(11):E50018

45. Doyle DA, Cabral JM, Pfuetzner RA et al (1998) The structure of the potassium channel: molecular basis of K+ conduction and selectivity. Science 280:69–77

46. Miyazawa A, Fujiyoshi Y, Unwin N (2003) Structure and gating mechanism of the acetylcholine receptor pore. Nature 423:949–955

47. Payandeh J, Scheuer T, Zheng N et al (2011) The crystal structure of a voltage-gated sodium channel. Nature 475:353–358

48. Feldbauer K, Zimmermann D, Pintschovius V et al (2009) Channelrhodopsin-2 is a leaky proton pump. Proc Natl Acad Sci U S A 106:12317–12322

49. Govorunova E, Sineshchekov O, Li H (2013) Characterization of a highly efficient blueshifted channelrhodopsin from the marine alga Platymonas subcordiformis. J Biol Chem 288:29911–29922

50. Gradmann D, Berndt A, Schneider F et al (2011) Rectification of the channelrhodopsin early conductance. Biophys J 101:1057–1068

51. Berndt A, Prigge M, Gradmann D et al (2010) Two open states with progressive proton selectivities in the branched channelrhodopsin-2 photocycle. Biophys J 98:753–761

52. Schneider F, Gradmann D, Hegemann P (2013) Ion selectivity and competition in channelrhodopsins. Biophys J 2:91–100

53. Lin JY, Lin MZ, Steinbach P et al (2009) Characterization of engineered channelrhodopsin variants with improved properties and kinetics. Biophys J 96:1803–1814

54. Tsunoda SP, Hegemann P (2009) Glu 87 of channelrhodopsin-1 causes pH-dependent color tuning and fast photocurrent inactivation. Photochem Photobiol 85:564–569

55. Ritter E, Piwowarski P, Hegemann P et al (2013) Light-dark adaptation of channelrhodopsin C128T mutant. J Biol Chem 288:10451–10458

56. Tian L, Hires SA, Mao T et al (2009) Imaging neural activity in worms, flies and mice with improved GCaMP calcium indicators. Nat Methods 6:875–881

57. Wen L, Wang H, Tanimoto S et al (2010) Opto-current-clamp actuation of cortical neurons using a strategically designed channelrhodopsin. PLoS One 5:e12893

58. Berndt A, Schoenenberger P, Mattis J et al (2011) High-efficiency channelrhodopsins for fast neuronal stimulation at low light levels. Proc Natl Acad Sci U S A 108:7595–7600

59. Nagel G, Ullrich S, Gueta R (2013) Degradation of channelopsin-2 in the absence of retinal and degradation resistance in certain mutants. Biol Chem 394:271–280

60. Nikolic K, Jarvis S, Grossman N et al (2013) Computational models of optogenetic tools for controlling neural circuits with light. Conf Proc IEEE Eng Med Biol Soc 2013:5934–5937

61. Beppu K, Sasaki T, Tanaka KF et al (2014) Optogenetic countering of glial acidosis suppresses glial glutamate release and ischemic brain damage. Neuron 81:314–320

62. Figueiredo M, Lane S, Stout RF et al (2014) Comparative analysis of optogenetic actuators in cultured astrocytes. Cell Calcium 56:208–214

63. Perea G, Yang A, Boyden ES et al (2014) Optogenetic astrocyte activation modulates response selectivity of visual cortex neurons in vivo. Nat Commun 5:3262

64. Caldwell JH, Herin GA, Nagel G et al (2008) Increases in intracellular calcium triggered by channelrhodopsin-2 potentiate the response of metabotropic glutamate receptor mGluR7. J Biol Chem 283:24300–24307

65. Zhang Y-P, Oertner TG (2007) Optical induction of synaptic plasticity using a light-sensitive channel. Nat Methods 4:139–141

66. Ferenczi E, Deisseroth K (2012) When the electricity (and the lights) go out: transient changes in excitability. Nat Neurosci 15:1058–1060

67. Raimondo JV, Kay L, Ellender TJ et al (2012) Optogenetic silencing strategies differ in their effects on inhibitory synaptic transmission. Nat Neurosci 15:1102–1104

68. Bruegmann T, Malan D, Hesse M et al (2010) Optogenetic control of heart muscle in vitro and in vivo. Nat Methods 7:897–900

69. Kleinlogel S, Feldbauer K, Dempski RE et al (2011) Ultra light-sensitive and fast neuronal activation with the Ca?+-permeable channelrhodopsin CatCh. Nat Neurosci 14:513–518

70. Pan ZH, Ganjawala TH, Lu Q et al (2014) ChR2 mutants at L132 and T159 with improved operational light sensitivity for vision restoration. PLoS One 9(6):e98924

71. Berndt A, Lee SY, Ramakrishnan C et al (2014) Structure-guided transformation of channelrhodopsin into a light-activated chloride channel. Science 344:420–424

72. Tomita H, Sugano E, Fukazawa Y et al (2009) Visual properties of transgenic rats harboring the channelrhodopsin-2 gene regulated by the thy-1.2 promoter. PLoS One 4:e7679

73. Watanabe HC, Welke K, Schneider F et al (2012) Structural model of channelrhodopsin. J Biol Chem 287:7456–7466

74. Govorunova EG, Sineshchekov OA, Janz R et al (2015) Natural light-gated anion channels: a family of microbial rhodopsins for advanced optogenetics. Science 349:647–650

75. Jentsch TJ, Introduction I, Stein V et al (2002) Molecular structure and physiological function of chloride channels. Physiol Rev 82:503

76. Berndt A, Yizhar O, Gunaydin LA et al (2009) Bi-stable neural state switches. Nat Neurosci 12:229–234

77. Hososhima S, Sakai S, Ishizuka T et al (2015) Kinetic evaluation of photosensitivity in bi-stable variants of chimeric channelrhodopsins. PLoS One 10:e0119558

78. Stehfest K, Ritter E, Berndt AB et al (2010) The branched photocycle of the slow-cycling channelrhodopsin-2 mutant C128T. J Mol Biol 398:690–702

79. Hoffmann M, Wanko M, Strodel P et al (2006) Color tuning in rhodopsins: the mechanism for the spectral shift between bacteriorhodopsin and sensory rhodopsin II. J Am Chem Soc 128:10808–10818

80. Chen T-W, Wardill TJ, Sun Y et al (2013) Ultrasensitive fluorescent proteins for imaging neuronal activity. Nature 499:295–300

81. Inoue M, Takeuchi A, Horigane S et al (2014) Rational design of a high-affinity, fast, red calcium indicator R-CaMP2. Nat Methods 12(1):64–70

82. Miesenböck G, De Angelis DA, Rothman JE (1998) Visualizing secretion and synaptic transmission with pH-sensitive green fluorescent proteins. Nature 394:192–195

83. Stujenske JM, Spellman T, Gordon JA (2015) Modeling the spatiotemporal dynamics of light and heat propagation for in vivo optogenetics. Cell Rep 12:525–534

84. Heitler WJ, Goodman CS, Rowell CHF (1977) The effects of temperature on the

threshold of identified neurons in the locust. J Comp Physiol 117:163–182

85. Yizhar O, Fenno LE, Davidson TJ et al (2011) Optogenetics in neural systems. Neuron 71:9–34

86. Yaroslavsky N, Schulze PC, Yaroslavsky IV et al (2002) Optical properties of selected native and coagulated human brain tissues in vitro in the visible and near infrared spectral range. Phys Med Biol 47:2059–2073

87. Chuong A, Miri M, Acker L et al (2014) Non-invasive optogenetic neural silencing. Nat Neurosci 17:1123–1129

88. Rickgauer JP, Tank DW (2009) Two-photon excitation of channelrhodopsin-2 at saturation. Proc Natl Acad Sci U S A 106:15025–15030

89. Sneskov K, Olsen JMH, Schwabe T et al (2013) Computational screening of one- and two-photon spectrally tuned channelrhodopsin mutants. Phys Chem Chem Phys 15:7567–7576

90. Packer AM, Russell LE, Dalgleish HWP et al (2014) Simultaneous all-optical manipulation and recording of neural circuit activity with cellular resolution in vivo. Nat Methods 12(2):140–146

91. Rickgauer JP, Deisseroth K, Tank DW (2014) Simultaneous cellular-resolution optical perturbation and imaging of place cell firing fields. Nat Neurosci 17:1816–1824

92. Wiegert JS, Oertner TG (2013) Long-term depression triggers the selective elimination of weakly integrated synapses. Proc Natl Acad Sci U S A 110:E4510–E4519

93. St-Pierre F, Marshall JD, Yang Y et al (2014) High-fidelity optical reporting of neuronal electrical activity with an ultrafast fluorescent voltage sensor. Nat Neurosci 17:884–889

94. Akerboom J, Carreras Calderón N, Tian L et al (2013) Genetically encoded calcium indicators for multi-color neural activity imaging and combination with optogenetics. Front Mol Neurosci 6:2

95. Lin J, Sann S, Zhou K et al (2013) Optogenetic inhibition of synaptic release with chromophore-assisted light inactivation (CALI). Neuron 79:241–253

96. Kato HE, Kamiya M, Sugo S et al (2015) Atomistic design of microbial opsin-based blue-shifted optogenetics tools. Nat Commun 6:7177

97. Grubb MS, Burrone J (2010) Channelrhodopsin-2 localised to the axon initial segment. PLoS One 5(10):e13761

98. Oesterhelt D, Hess B (1973) Reversible photolysis of the purple complex in the purple membrane of Halobacterium halobium. Eur J Biochem 37:316–326

99. Bamann C, Kirsch T, Nagel G et al (2008) Spectral characteristics of the photocycle of channelrhodopsin-2 and its implication for channel function. J Mol Biol 375:686–694

100. Ernst OP, Murcia PAS, Daldrop P et al (2008) Photoactivation of channelrhodopsin. J Biol Chem 283:1637–1643

101. Grimm C, Schneider F, Hegemann P (2015) Biophysics of channelrhodopsin. Annu Rev Biophys 44:167–186

102. Stehfest K, Hegemann P (2010) Evolution of the channelrhodopsin photocycle model. Chemphyschem 11:1120–1126

103. Watanabe HC, Welke K, Sindhikara DJ et al (2013) Towards an understanding of channelrhodopsin function: simulations lead to novel insights of the channel mechanism. J Mol Biol 425:1795–1814

104. Gunaydin LA, Yizhar O, Berndt A et al (2010) Ultrafast optogenetic control. Nat Neurosci 13:387–392

105. Bruun S, Stoeppler D, Keidel A et al (2015) Light-dark adaptation of channelrhodopsin involves photoconversion between the all-trans and 13-cis retinal isomers. Biochemistry 54(35):5389–5400

106. Lórenz-Fonfría VA, Schultz B-J, Resler T et al (2015) Pre-gating conformational changes in the ChETA variant of channelrhodopsin-2 monitored by nanosecond IR spectroscopy. J Am Chem Soc 137:1850–1861

107. Bamann C, Gueta R, Kleinlogel S et al (2010) Structural guidance of the photocycle of channelrhodopsin-2 by an interhelical hydrogen bond. Biochemistry 49:267–278

108. Nack M, Radu I, Gossing M et al (2010) The DC gate in Channelrhodopsin-2: crucial hydrogen bonding interaction between C128 and D156. Photochem Photobiol Sci 9:194–198

109. Miyazaki KW, Miyazaki K, Tanaka KF et al (2014) Optogenetic activation of dorsal Raphe serotonin neurons enhances patience for future rewards. Curr Biol 24:2033–2040

110. Herman AM, Huang L, Murphey DK et al (2014) Cell type-specific and time-dependent light exposure contribute to silencing in neurons expressing Channelrhodopsin-2. eLife 2014:1–18

111. Schmitt BM, Koepsell H (2002) An improved method for real-time monitoring of membrane capacitance in Xenopus laevis oocytes. Biophys J 82:1345–1357

112. Sehnal D, Vařeková RS, Berka K et al (2013) MOLE 2.0: advanced approach for analysis of biomacromolecular channels. J Cheminform 5:1–13

113. Nagel G, Szellas T, Huhn W et al (2003) Channelrhodopsin-2, a directly light-gated cation-selective membrane channel. Proc Natl Acad Sci U S A 100:13940–13945

114. Arroyo S, Bennett C, Aziz D et al (2012) Prolonged disynaptic inhibition in the cortex mediated by slow, non-α7 nicotinic excitation of a specific subset of cortical interneurons. J Neurosci 32:3859–3864

115. Jego S, Glasgow SD, Herrera CG et al (2013) Optogenetic identification of a rapid eye movement sleep modulatory circuit in the hypothalamus. Nat Neurosci 16:1637–1643

116. Sharp AA, Fromherz S (2011) Optogenetic regulation of leg movement in midstage chick embryos through peripheral nerve stimulation. J Neurophysiol 106:2776–2782

117. Erbguth K, Prigge M, Schneider F et al (2012) Bimodal activation of different neuron classes with the spectrally red-shifted Channelrhodopsin chimera C1V1 in Caenorhabditis elegans. PLoS One 7

118. Hooks BM, Lin JY, Guo C et al (2015) Dual-channel circuit mapping reveals sensorimotor convergence in the primary motor cortex. J Neurosci 35:4418–4426

Chapter 11

Optogenetics in *Drosophila* Neuroscience

Thomas Riemensperger, Robert J. Kittel, and André Fiala

Abstract

Optogenetic techniques enable one to target specific neurons with light-sensitive proteins, e.g., ion channels, ion pumps, or enzymes, and to manipulate their physiological state through illumination. Such artificial interference with selected elements of complex neuronal circuits can help to determine causal relationships between neuronal activity and the effect on the functioning of neuronal circuits controlling animal behavior. The advantages of optogenetics can best be exploited in genetically tractable animals whose nervous systems are, on the one hand, small enough in terms of cell numbers and to a certain degree stereotypically organized, such that distinct and identifiable neurons can be targeted reproducibly. On the other hand, the neuronal circuitry and the behavioral repertoire should be complex enough to enable one to address interesting questions. The fruit fly *Drosophila melanogaster* is a favorable model organism in this regard. However, the application of optogenetic tools to depolarize or hyperpolarize neurons through light-induced ionic currents has been difficult in adult flies. Only recently, several variants of Channelrhodopsin-2 (ChR2) have been introduced that provide sufficient light sensitivity, expression, and stability to depolarize central brain neurons efficiently in adult *Drosophila*. Here, we focus on the version currently providing highest photostimulation efficiency, ChR2-XXL. We exemplify the use of this optogenetic tool by applying it to a widely used aversive olfactory learning paradigm. Optogenetic activation of a population of dopamine-releasing neurons mimics the reinforcing properties of a punitive electric shock typically used as an unconditioned stimulus. In temporal coincidence with an odor stimulus this artificially induced neuronal activity causes learning of the odor signal, thereby creating a light-induced memory.

Key words Optogenetics, Neuronal circuits, *Drosophila melanogaster*, Learning and memory, ChR2-XXL, Dopamine, Mushroom body

1 Introduction

The dissection and analysis of neuronal networks that orchestrate and control animal behavior, and the question how these networks change and mediate adaptive learning represents a key topic in modern neuroscience. This quest can be approached with two complementary experimental strategies: On the one hand, behavioral actions and reactions as well as adaptive changes in behavior can be correlated with neuronal activity or molecular processes monitored through electrophysiological or optophysiological recordings. Thereby, the computational processing performed by

Arash Kianianmomeni (ed.), *Optogenetics: Methods and Protocols*, Methods in Molecular Biology, vol. 1408,
DOI 10.1007/978-1-4939-3512-3_11, © Springer Science+Business Media New York 2016

specific neuronal circuits and the underlying mechanisms can be analyzed. On the other hand, manipulative interference with constitutive elements forming the neuronal circuits, e.g., individual neurons or specific molecules, can uncover whether these elements are necessary and/or sufficient for a particular behavior [1]. *Drosophila melanogaster* provides a favorable model organism for both approaches for three reasons. First, the development of several binary expression systems and the generation of large fly strain collections make it possible to target many individually identifiable neurons or populations of neurons selectively [2, 3]. Second, the *Drosophila* brain is comprised of only ~100,000 neurons, which represents much lower complexity than in most mammals, e.g., the mouse brain [3]. However, the fruit fly's behavioral repertoire and the brain circuits controlling it is rich enough to pose interesting questions, e.g., for the mechanisms underlying circadian rhythms [4], for complex sensory stimulus processing (e.g., 5–7), or for learning and memory formation [8, 9]. Third, genetically encoded fluorescence sensors designed to detect physiological parameters reflecting neuronal activity, e.g., Ca^{2+} dynamics, synaptic vesicle release, or second messenger synthesis, have been improved to a high degree in recent years [10]. In *Drosophila*, optical Ca^{2+} imaging nowadays represents a commonly used standard technique [11]. In addition, a variety of genetically encoded proteins have been introduced to manipulate neuronal activity, e.g., through temperature shifts. Synaptic transmission can be blocked reversibly by expressing a temperature-sensitive variant of the *Drosophila* dynamin homologue, shibire[(ts)] [12], and membrane depolarization can be induced by expressing the temperature-sensitive cation channel dTRPA1 [13].

Optogenetic techniques to depolarize or hyperpolarize neuronal membrane potential through light to affect neuronal activity through illumination represent an elegant strategy, i.e., with minimally invasive side effects [14]. In *Drosophila* larvae, wild type ChR2 [15] and the first gain-of-function mutant ChR–H134R [16] function reliably and represent very useful tools [17–21]. However, due to low levels of membrane integration and limited stability, which depends on the affinity of channelopsin-2 to its chromophore *all-trans*-retinal [22], these ChR2 versions fail to efficiently depolarize central brain neurons in adult *Drosophila*. The application of ChR2 was therefore restricted to research on larvae [17–21] or to peripheral sensory neurons of adult *Drosophila* (e.g., 23) that typically are extremely sensitive to slight membrane depolarization. Only recently, significant progress has been made by developing spectrally shifted ChR variants with high cellular expression that are readily functional in adult *Drosophila* [24–26]. In terms of photostimulation, however, the most powerful modification of ChR2 was achieved through the D156C mutation (Fig. 1b), which strongly increases the affinity of the apoprotein to *all-trans*-retinal and greatly extends the channel's open state lifetime. This leads to highly elevated expression and drastically increased photocurrent amplitudes of blue light-activated ChR2–D156C, which was therefore termed Channelrhodopsin-2-XXL (*e*xtra high e*x*pression and *l*ong open state) [27]. Using ChR2-XXL, we here describe how to optogenetically control neuronal activity in the *Drosophila* brain in order to manipulate the fly's behavior. We exemplify this with a well described aversive olfactory learning paradigm. Fruit flies can be trained to associate an odor stimulus with a temporally coinciding punitive electric shock; thereby, they learn to avoid the odor temporally paired with the punishment in a subsequent test situation [28]. A class of dopamine-releasing neurons has been shown to respond to an electric shock stimulus [29] and blocking synaptic transmission from these neurons impairs aversive associative odor learning [30]. Artificial activation of these neurons using temperature-sensitive dTRPA1 [31] or uncaging of ATP together with the expression of an ATP-sensitive cation channel [32] in coincidence with an odor leads to an induction of aversive learning. It is well established that the reinforcing properties of the punitive electric shock stimulus are mediated by small subsets of these dopaminergic neurons [31, 33]. We have previously demonstrated that ChR2-XXL can also be used in this paradigm to mimic the reinforcing properties of an aversive stimulus [27]. Here we provide a detailed protocol for this type of experiment.

2   Material

2.1   Materials for Optogenetic Induction of Aversive Odor Learning

2.1.1   Transgenic Drosophila Strains

1. The "driver strain" *Th-Gal4* [34].

2. The "optogenetic effector strain" *UAS:chop2-XXL* [27].

   To drive the expression of ChR2-XXL specifically in dopaminergic neurons, we employ the most commonly used bipartite expression system, the *Gal4–UAS* system [35]. Crossing the *UAS:chop2-XXL* effector strain to the well described driver line *Th-Gal4* [34] will result in F1 progeny expressing ChR2-XXL in a large number of dopaminergic neurons except for most of those neurons located in the protocerebral anterior medial (PAM) cluster. Dopaminergic neurons covered by *Th-Gal4* encompass cell clusters for which a contribution to the reinforcing properties of aversive stimuli has been shown, in particular the cluster PPL1 (Fig. 1a).

2.1.2   Materials for the Optogenetic Training Apparatus

Olfactory training of adult fruit flies using electric shocks is typically done using an apparatus that consists of a training tube, lined inside with an electric grid, closed with fine plastic mesh permeable to air on both ends, and equipped with a holder for an odorant reservoir attached to one end. Air provided by a pump is guided over the surface of the odorant and through the training tube [28].

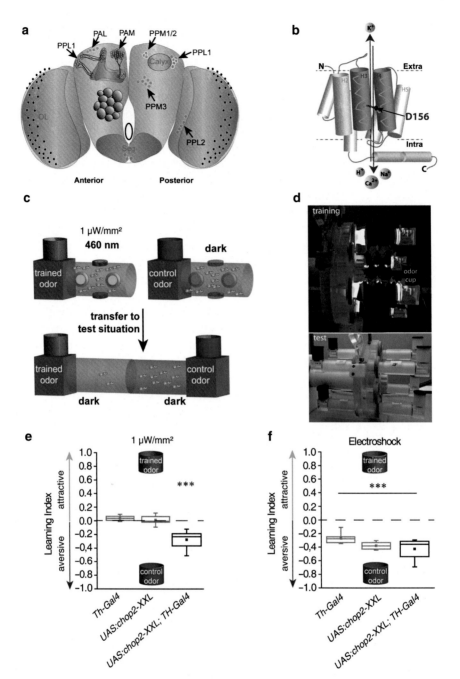

Fig. 1 Optogenetic induction of aversive olfactory learning in *Drosophila*. (**a**) Schematic illustration of a *Drosophila* brain with clusters of dopaminergic neurons shown in color. Those neurons covered by *Th-Gal4* are highlighted in *red*. Neurons located in the PAL, PAM, and PPL1 clusters innervate the mushroom body, a central brain structure critical for associative olfactory learning. (**b**) Schematic illustration of the structure of ChR2-XXL, with the mutated aspartic acid, D156C, indicated. (**c**) Schematic illustration of the training and test apparatus. (**d**) Photograph of the training and test apparatus. The test situation is pictured at room light, but note that the experiment should be performed under red light conditions. (**e**) Representative example of an aversive learning index optogenetically induced by activating ChR2-XXL in *Th-Gal4*-positive neurons in comparison with genetic control strains. (**f**) Representative example of an aversive learning index induced by electric shocks in flies expressing ChR2-XXL in *Th-Gal4*-positive neurons and in genetic control strains. All genotypes associated odors with electric shock punishment. *Box plots* represent median, interquartile ranges, and 10–90 percentiles. $n = 8$ for each group. ***$p < 0.001$ (Student's *t*-test with Bonferroni correction). Data from [27]

We used a modified, barrel-shaped version of this machine as reported in [30], which contains four training tubes in each of which ~100 flies can be trained simultaneously. In addition, the training apparatus contains an "elevator section" through which the flies can be transferred to a T-maze section. Here, two equivalent tubes, but without electric grids, are positioned oppositely such that the animals can distribute across both tubes. These tubes are equipped with different odorants, one with the odorant presented together with the electric shock, the other one with a control odorant not presented with punishment. In order to substitute the electric shock with blue light, the training tubes containing the electric grids are replaced by transparent training tubes of 8.6 cm length and 1.9 cm diameter each equipped with 12 blue light diodes with a peak wavelength of 468 nm and a power of ~1 $\mu W/mm^2$ that are evenly inserted into each tube's surface. Light intensity of the diodes is regulated using an adjustable power supply reaching 50 mA max. The light intensity can be adjusted using a regular lux-meter. A vacuum pump is connected to the apparatus to apply a constant airflow of ~167 ml/min. Environmental conditions should be kept constant at 25 °C and 65–75 % humidity.

1. Barrel-type olfactory training apparatus [30].

2. Vacuum pump (airflow adjusted to ~167 ml/min in each training tube).

3. Odorants: 4-methylcyclohexanol (diluted 1:750), 3-octanol (diluted 1:500), mineral oil as diluent.

4. Plastic odorant cups of 5 mm diameter, with a capacity of 60 μl for the diluted odorants.

5. Blue-light diodes with peak wavelength of 468 nm and a power of ~1 $\mu W/mm^2$, evenly inserted into the tube surface of transparent training tubes (Fig. 1d).

6. Power supply (50–150 V, 50 mA).

7. Pulse-generator (12–15 V DC, 250 mA).

8. For training using electric shock stimuli: training tubes lined with an electrifiable grid, as described in [28].

9. Humidifier.

10. Adjustable heating system.

---

3 Methods

1. Flies expressing ChR2-XXL in neurons of interest can be generated by crossing the appropriate Gal4 line with the transgenic *UAS:chop2-XXL* line. In our experiments, we crossed *UAS:chop2-XXL* [27] with *Th-Gal4* [34], which drives expression in the majority of dopaminergic neurons (*see* **Note 1**).

2. After egg laying parental flies are removed from the vials. During development and until the day of the experiment the F1 generation is kept in constant darkness (*see* **Note 2**). As ChR2-XXL is not activated by long-wavelength light, alternatively, red foil can be wrapped around the vials to prevent undesired photostimulation [27].

3. To adjust for innate odor preferences the exact odor concentrations need to be equilibrated. In our experiments, the dilutions given above result in approximately equal distributions when both odorants are presented in a T-maze situation. However, a new setup may deviate from this, depending on the air flow and the exact odorant concentrations in the tubes. Therefore, transfer ~100 flies directly into the T-maze part of the apparatus and present each odor alone against pure mineral oil as control. Calculate the preference index by subtracting the number of flies on the side of the odorant-containing tube from the number of flies on the opposite side divided by the total number of flies. The odor preference indices of both odors should be comparable. If not, adjust odor concentrations accordingly. In a second step, both odorants are presented simultaneously and the preference index is calculated. The odorant concentrations should be adjusted such that the flies are distributed equally between the two arms of the T-maze, i.e., the preference index should be ~0 (*see* **Note 3**).

4. Ten minutes before each experiment, flies are transferred under red light from the food vials into empty fly culture vials.

5. The odorants 4-methylcyclohexanol (CAS 589-91-3; Sigma) and 3-octanol (CAS 589-98-0; Sigma) need to be diluted in mineral oil (CAS 8042-47-5; Sigma) to the final concentration by vortexing thoroughly.

6. Groups of approximately one hundred 4–7 days old flies are transferred under red light from the empty culture vials into the training tubes. Training starts 1 min after transferring the flies.

7. Apply 60 μl of the diluted odorant into the odorant cups. During associative training each odor will be presented consecutively for 1 min, with a 1 min break between two odorant stimulations. The odor used as the conditioned stimulus (CS+) is temporally paired with the blue light. The odor used as control (CS–) is presented after a 1 min break in complete darkness.

8. To optogenetically train the flies, blue light with a peak wavelength of 468 nm and intensity of ~1 μW/mm^2 is administered via 12 diodes evenly inserted in the tube surface. For training 12 illumination pulses with duration of 1.25 s and separated by 3.75 s are applied (Fig. 1c, d).

9. After the training sequence is complete transfer the flies to the T-maze part of the apparatus where both odors are presented from each side and let the flies distribute for 2 min (Fig. 1c, d).

10. The experiment is subsequently repeated with new flies in a reciprocal manner, i.e., the odorant that has served as CS+ in the first experiments will now be presented as CS–.

11. Count the flies and calculate the preference indices as described under **step 3**. Learning indices are calculated by averaging the preference indices of the two reciprocal experiments using each odor as CS+ or CS–, respectively (*see* **Note 4**). Representative data are shown in Fig. 1e.

---

4   Notes

1. It is not advisable to create flies that are homozygous for both the *UAS* and the *Gal4* constructs. Raising the animals at higher temperatures (e.g., 29 °C) may help to increase expression levels of ChR2-XXL, if necessary.

2. In contrast to other ChR variants, *all-trans*-retinal administration to the food medium is not required when using ChR2-XXL [27].

3. ChR2-XXL is highly light sensitive, and neuronal membrane depolarization can potentially be induced even at regular ambient room light. Therefore, all subsequent steps must be conducted under red light conditions.

4. In order to ensure that all fly lines used are capable of learning, it is advisable to perform the classical odor-shock learning paradigm, as described by Tully and Quinn (1985) [28] as a positive control. Here, one odorant serves as the conditioned stimulus (CS+) and is temporally paired with 12 electric shocks of 90 V DC, of 1.25 s duration and 3.75 s interpulse interval, administered through an electrifiable grid covering the inside of the tubes. A control odor (CS–) is presented without shocks. Representative data are shown in Fig. 1f.

Acknowledgement

This work was supported by the Deutsche Forschungsgemeinschaft (FI 821/3-1 and SFB 889/B4 to A.F., and KI 1460/1-1, SFB 1047/A5, and FOR 2140/TP3 to R.J.K.).

References

1. Fiala A (2013) Optogenetic approaches in behavioral neuroscience. In: Hegemann P, Sigrist S (eds) Optogenetics. De Gruyter, Berlin/Boston, pp 91–97

2. Wimmer EA (2003) Innovations: applications of insect transgenesis. Nat Rev Genet 4(3):225–232

3. Venken KJ, Simpson JH, Bellen HJ (2011) Genetic manipulation of genes and cells in the nervous system of the fruit fly. Neuron 72(2):202–230

4. Helfrich-Förster C (2005) Neurobiology of the fruit fly's circadian clock. Genes Brain Behav 4(2):65–76

5. Behnia R, Desplan C (2015) Visual circuits in flies: beginning to see the whole picture. Curr Opin Neurobiol 34:125–132

6. Wilson RI (2013) Early olfactory processing in Drosophila: mechanisms and principles. Annu Rev Neurosci 36:217–241

7. Albert JT, Göpfert MC (2015) Hearing in Drosophila. Curr Opin Neurobiol 34:79–85

8. Fiala A (2007) Olfaction and olfactory learning in Drosophila: recent progress. Curr Opin Neurobiol 17(6):720–726

9. Guven-Ozkan T, Davis RL (2014) Functional neuroanatomy of Drosophila olfactory memory formation. Learn Mem 21(10):519–526

10. Looger LL, Griesbeck O (2012) Genetically encoded neural activity indicators. Curr Opin Neurobiol 22(1):18–23

11. Riemensperger T, Pech U, Dipt S et al (2012) Optical calcium imaging in the nervous system of Drosophila melanogaster. Biochim Biophys Acta 1820(8):1169–1178

12. Kitamoto T (2001) Conditional modification of behavior in Drosophila by targeted expression of a temperature-sensitive shibire allele in defined neurons. J Neurobiol 47(2):81–92

13. Hamada FN, Rosenzweig M, Kang K et al (2008) An internal thermal sensor controlling temperature preference in Drosophila. Nature 454(7201):217–220

14. Fiala A, Suska A, Schlüter OM (2010) Optogenetic approaches in neuroscience. Curr Biol 20(20):897–903

15. Nagel G, Szellas T, Huhn W et al (2003) Channelrhodopsin-2, a directly light-gated cation-selective membrane channel. Proc Natl Acad Sci U S A 100(24):13940–13945

16. Nagel G, Brauner M, Liewald JF et al (2005) Light activation of channelrhodopsin-2 in excitable cells of Caenorhabditis elegans triggers rapid behavioral responses. Curr Biol 15(24):2279–2284

17. Schroll C, Riemensperger T, Bucher D et al (2006) Light-induced activation of distinct modulatory neurons triggers appetitive or aversive learning in Drosophila. Curr Biol 16(17):1741–1747

18. Pulver SR, Pashkovski SL, Hornstein NJ et al (2009) Temporal dynamics of neuronal activation by Channelrhodopsin-2 and TRPA1 determine behavioral output in Drosophila larvae. J Neurophysiol 101(6):3075–3088

19. Hwang RY, Zhong L, Xu Y et al (2007) Nociceptive neurons protect Drosophila larvae from parasitoid wasps. Curr Biol 17(24):2105–2116

20. Bellmann D, Richardt A, Freyberger R et al (2010) Optogenetically induced olfactory stimulation in Drosophila larvae reveals the neuronal basis of odor-aversion behavior. Front Behav Neurosci 4:27. doi:10.3389/fnbeh.2010.00027

21. Ljaschenko D, Ehmann N, Kittel RJ (2013) Hebbian plasticity guides maturation of glutamate receptor fields in vivo. Cell Rep 3(5):1407–1413

22. Ullrich S, Gueta R, Nagel G (2013) Degradation of channelopsin-2 in the absence of retinal and degradation resistance in certain mutants. Biol Chem 394(2):271–280

23. Suh GS, Ben-Tabou de Leon S, Tanimoto H et al (2007) Light activation of an innate olfactory avoidance response in Drosophila. Curr Biol 17(10):905–908

24. Lin JY, Knutsen PM, Muller A et al (2013) ReaChR: a red-shifted variant of channelrhodopsin enables deep transcranial optogenetic excitation. Nat Neurosci 16(10):1499–1508

25. Inagaki HK, Jung Y, Hoopfer ED et al (2014) Optogenetic control of Drosophila using a red-shifted channelrhodopsin reveals experience-dependent influences on courtship. Nat Methods 11(3):325–332

26. Klapoetke NC, Murata Y, Kim SS et al (2014) Independent optical excitation of distinct neural populations. Nat Methods 11(3):338–346

27. Dawydow A, Gueta R, Ljaschenko D et al (2014) Channelrhodopsin-2-XXL, a powerful optogenetic tool for low-light applications. Proc Natl Acad Sci U S A 111(38):13972–13977

28. Tully T, Quinn WG (1985) Classical conditioning and retention in normal and mutant Drosophila melanogaster. J Comp Physiol A 157(2):263–277

29. Riemensperger T, Völler T, Stock P et al (2005) Punishment prediction by dopaminergic neurons in Drosophila. Curr Biol 15(21):1953–1960

30. Schwaerzel M, Monastirioti M, Scholz H et al (2003) Dopamine and octopamine differentiate between aversive and appetitive olfactory

memories in Drosophila. J Neurosci 23(33): 10495–10502

31. Aso Y, Siwanowicz I, Bräcker L et al (2010) Specific dopaminergic neurons for the formation of labile aversive memory. Curr Biol 20(16):1445–1451

32. Claridge-Chang A, Roorda RD, Vrontou E et al (2009) Writing memories with light-addressable reinforcement circuitry. Cell 139(2):405–415

33. Aso Y, Herb A, Ogueta M et al (2012) Three dopamine pathways induce aversive odor memories with different stability. PLoS Genet 8(7):e1002768

34. Friggi-Grelin F, Coulom H, Meller M et al (2003) Targeted gene expression in Drosophila dopaminergic cells using regulatory sequences from tyrosine hydroxylase. J Neurobiol 54(4):618–627

35. Brand AH, Perrimon N (1993) Targeted gene expression as a means of altering cell fates and generating dominant phenotypes. Development 118(2):401–415

Chapter 12

Optogenetic Control of Mammalian Ion Channels with Chemical Photoswitches

Damien Lemoine, Romain Durand-de Cuttoli, and Alexandre Mourot

Abstract

In neurons, ligand-gated ion channels decode the chemical signal of neurotransmitters into an electric response, resulting in a transient excitation or inhibition. Neurotransmitters act on multiple receptor types and subtypes, with spatially and temporally precise patterns. Hence, understanding the neural function of a given receptor requires methods for its targeted, rapid activation/inactivation in defined brain regions. To address this, we have developed a versatile optochemical genetic strategy, which allows the reversible control of defined receptor subtypes in designated cell types, with millisecond and micrometer precision. In this chapter, we describe the engineering of light-activated and -inhibited neuronal nicotinic acetylcholine receptors, as well as their characterization and use in cultured cells.

Key words Optogenetic pharmacology, Photoswitches, Optogenetics, Chemical–optogenetics, Ligand-gated ion channels, Receptor, Nicotinic acetylcholine receptor

1 Introduction

Neuronal mechanisms governing cell excitability, synaptic transmission, and plasticity are tightly regulated by a myriad of ion channels and receptors, each having a very well-defined regional specificity, cellular targeting, and compartmentalization. Having a deep molecular understanding of the ion channels involved in these neural mechanisms would be extremely valuable in providing new targets for the treatment of neuropsychiatric disorders. Such advances, however, will only come with significant progress in the development of novel techniques for activating/inhibiting specific subtypes of ion channels and receptors in defined pathways. Traditional pharmacological agents may show some receptor-subtype selectivity but, even when locally microinjected into brain nuclei, activate or inhibit a large, heterogeneous group of neurons. They lack temporal specificity as well, which is required to link fast events such as synaptic transmission with modulation of neural circuits and behavior. In addition, pharmacological tools lack

Arash Kianianmomeni (ed.), *Optogenetics: Methods and Protocols*, Methods in Molecular Biology, vol. 1408,
DOI 10.1007/978-1-4939-3512-3_12, © Springer Science+Business Media New York 2016

functional specificity, in the sense that they cannot be targeted to genetically defined cell types, such as pre- versus postsynaptic terminals. To overcome the issues of classical pharmacology, we have developed a strategy for photosensitizing specific ion channel or receptor subtypes in designated neurons. This strategy combines the power of genetics and photochemistry:

– Light provides a strategic solution to control biological processes with high accuracy. First, light can be directed with exquisite spatial and temporal precision (e.g., subcompartments of a cell) and at time scales compatible with fast neuronal mechanisms (e.g., action potential propagation or synaptic transmission). Second, light can be projected from afar and is therefore a relatively noninvasive stimulus. Third, light can be used as an orthogonal, highly specific stimulus; indeed, since mammalian proteins are for the most part not light responsive, bestowing light sensitivity to a particular protein ensures a specific control of that very protein with light.

– Genetics provides two additional levels of control: a specificity of receptor subtype (only the designer receptor will respond to light) and a cellular specificity (expression of the transgene can be restricted to specific tissue using molecular genetic techniques).

In the past decade, a number of strategies have been developed to render proteins light sensitive [1]. Here, we focus on one chemical–optogenetic strategy enabling cell-surface proteins (*see* **Note 1**) to be photocontrollable [2–4]. The strategy is based on site-specific protein labeling with photoswitchable tethered ligands (PTLs). PTLs are made of three elements: (1) a reactive moiety (maleimide) for covalent attachment to the thiol group of an ideally located cysteine-substituted amino-acid residue (*see* **Note 2**); (2) a central azobenzene photoswitch, which can be reversibly photoconverted between a *trans*, elongated, and a *cis*, bent configuration; and (3) a pharmacologically active ligand, such as an agonist [5–7], an antagonist [5, 7, 8], a pore blocker [9–11], or an allosteric modulator [12]. Two different wavelengths of light (classically violet and green) are used to toggle the ligand in and out of its binding pocket, resulting in an optical control of the activity of the target protein. Multiple classes of cell-surface proteins have been made photosensitive this way, including ligand-gated ion channels [5, 6, 8], G-protein coupled receptors [7], and voltage-gated ion channels [9–11], showing the versatility of this approach.

In this chapter, we describe the design and characterization of light-activated and -inhibited neuronal nicotinic acetylcholine receptors (LinAChRs). Nicotinic acetylcholine receptors (nAChRs) are widely expressed in different organs and especially in the brain [13].

Nicotinic neuromodulation plays a key role in many physiological functions such as sleep or hunger but also in high-level behaviors such as learning, motivation, decision-making, or reward processes [14–16]. Moreover they play a key role in development, neuronal plasticity, and aging. nAChRs are ligand-gated ion channels from the cys-loop family. In the central nervous system, 12 different subunits can be found (α2–9 and β2–4), which self-assemble into homo- (e.g., α7) or heteropentameric (e.g., α4β2, α3β4) combinations, giving rise to numerous receptor subtypes with different pharmacological and kinetic properties [15]. Heteromeric receptors have two binding sites for ACh, located at the interface between a principal (α) and a complementary (β) subunit. The ability to control a specific nAChR subtype using PTLs may enable to assess its various roles within the brain.

The following concepts for designing light-controlled nAChRs may be used as a guide to develop in principle any new photoswitchable cell-surface protein. Engineering a light-controlled receptor is a three-step process. First one needs to design the PTL itself. The ligand (agonist, antagonist, pore blocker, etc.) requires to be derivatized with a relatively bulky azobenzene photoswitch without compromising its biological function. Extensive knowledge of the features required for the molecular recognition of the ligand in its active site is therefore highly beneficial. Second, one needs to identify potential amino acid candidates on the surface of the target protein for substitution to cysteine. This step is greatly aided by three-dimensional structures of the ligand-binding domain. Finally, one needs to screen for a cysteine mutant–PTL pair that gives robust photocontrol over the protein activity. This is commonly done using electrophysiology in heterologous expression systems.

2 Materials

2.1 Chemicals

1. The PTLs MAACh and MAHoCH were synthesized as previously described [5]. All compounds are stored in dry DMSO at –80 °C, at a concentration of at least 2 mM, to ensure a final DMSO content ≤1 % during bioconjugation (*see* **Note 3**).

2. Anhydrous DMSO as solvent for stock solutions (*see* **Note 4**).

3. Drierite desiccant (calcium sulfate) with indicator (*see* **Note 4**).

2.2 Molecular Biology

1. Clones for the α3, α4, β2, and β4 nAChR subtypes in a vector for expression in oocytes (e.g., pNKS2) and in a vector for mammalian cell expression (e.g., pCDNA3.1 or pIRES, *see* **Note 5**).

2. Plasmid DNA mini- and maxi-prep kits.

3. Site-directed mutagenesis kits.

4. High yield capped RNA transcription kits with the adequate promoter (SP6 in case of the pNKS2 vector).

5. UV–visible spectrophotometer.

2.3 Xenopus Oocyte Preparation and mRNA Injection

1. *Xenopus laevis* oocytes. Freshly prepared oocytes can be ordered directly from companies such as Ecocyte Bioscience (*see* **Note 6**).

2. Microinjector.

3. Micromanipulator.

4. Pipette puller.

5. Glass capillaries.

6. Mineral oil.

7. Binocular dissecting microscope.

8. Incubator (18 °C).

2.4 Cell Culture

1. Neuro-2A cells: derived from neuroblastoma in an albino strain A mouse (*see* **Note 7**).

2. Culture medium: Modified Essential Medium (MEM), supplemented with 10 % Foetal Bovine Serum (FBS), 1 % nonessential amino-acids, 100 U/ml penicillin, 100 μg/ml streptomycin, and 2 mM Glutamax (*see* **Note 8**).

3. 6-well plates.

4. 10 ml serological pipettes.

5. 10 mm glass coverslips.

6. Tweezers.

7. 25 cm^2 culture flasks with filter cap.

2.5 Calcium Phosphate Transfection

1. Twice concentrated HEPES buffered salt solution (HeBS 2×): 280 mM NaCl, 10 mM KCl, 1.5 mM NaH$_2$PO$_4$·7H$_2$O, 50 mM HEPES, 13 mM glucose. pH adjusted to 7.05–7.12 with NaOH. Filter-sterilize and store at 4 °C.

2. 2 mM CaCl$_2$, filter-sterilize and store at 4 °C.

3. 1 N HCl solution. An acidic transfection medium solution (pH 6.3) helps dissolving DNA–Ca^{2+} phosphate precipitates.

4. Culture medium (*see* **Note 8**).

5. Minimal essential medium without glutamine.

6. Transfection Medium: MEM plus 14 mM glucose (e.g., 0.13 g in 50 ml). Filter-sterilize and store at 4 °C.

7. Acidic transfection medium: transfection medium, with pH adjusted to 6.3 with HCl. Filter-sterilize and store at 4 °C. Important: readjust to pH 6.3 the day of transfection.

2.6 Solutions for Electrophysiology

1. Oocyte ringer solution (ORI): 96 mM NaCl, 2 mM KCl, 1.8 mM $CaCl_2$, 1 mM $MgCl_2$, and 5 mM HEPES, pH 7.4 with NaOH. We usually prepare ORI from a 10× stock solution. Store at 4 °C up to a month.

2. 3 mM KCl, 100 ml, stored at room temperature.

3. Patch-clamp external solution: 140 mM NaCl, 2.8 mM KCl, 2 mM $CaCl_2$, 2 mM $MgCl_2$, 10 mM HEPES, 12 mM glucose. NaOH for pH adjustment (pH 7.3), filter-sterilize and store at 4 °C up to a week.

4. Patch-clamp internal solution: 140 mM KCl, 2 mM $MgCl_2$, 5 mM EGTA, 5 mM HEPES. KOH for pH adjustment (pH 7.3), filter-sterilize and store 0.5 ml aliquots at –20 °C.

2.7 Two-Electrode Voltage-Clamp (TEVC) Recordings

1. TEVC amplifier.

2. Digitizer.

3. Acquisition software.

4. Two micromanipulators.

5. Faraday cage (*see* **Note 9**).

6. Air table.

7. Stereomicroscope.

8. Recording chamber.

9. Perfusion system.

10. Vacuum pump.

11. Light source (*see* **Note 10**).

12. Optic fiber (1 mm diameter) terminated with a collimator.

13. Handheld optical power meter with a UV–visible detector.

14. Glass capillaries.

15. Pipette puller.

16. Nonmetallic syringe needle for filling micropipettes (28 gauge).

2.8 Patch-Clamp Recordings

1. Faraday cage (*see* **Note 9**).

2. Air table.

3. Upright microscope (*see* **Note 11**) with a bright field oil immersion condenser and a long working distance 40× objective (*see* **Note 12**).

4. Transmission light source (e.g., 850 nm LED) associated with a Dodt contrast tube (*see* **Note 12**).

5. Standard CCD camera for both fluorescence detection and cell imaging.

6. High power LED light source for photoswitching.

7. Patch-clamp amplifier with head stage.

8. Digitizer.

9. Acquisition software.

10. Micromanipulator.

11. Glass pipette puller.

12. Thin wall borosilicate pipette glass with filament (e.g., G150TF-3, Warner Instruments).

13. Nonmetallic syringe needle for filling micropipettes (28 gauge).

14. A fast-step solution delivery system (SF-77B, Warner Instruments) coupled with a digitally controlled multichannel perfusion system.

15. Vacuum pump.

3   Methods

3.1   Rational Design of LinAChRs

In this section, we describe the experimental strategy we followed for the design of light-activated and -inhibited nAChRs. Similar strategies may be adapted to other cell-surface receptors.

3.1.1   Design of the PTLs

As mentioned earlier, a PTL is composed of three elements:

1. A cysteine-reactive moiety: Maleimide is an excellent thiol-reactive group because it is very selective for cysteines (*see* **Notes 1** and **2**), it reacts with thiols rapidly (a few minutes) and the covalent adduct is virtually irreversible (as opposed to methanethiosulfonates for instance).

2. A central photoswitch: Azobenzene is a small photoswitch with excellent photostability and robust photoisomerization properties, which makes it an ideal candidate for the PTL approach [2]. The two photoisomers (*cis* and *trans*) have well-defined geometries. In darkness, azobenzenes exist predominantly in the *trans*, extended configuration. Illumination with 360–400 nm light promotes isomerization to the *cis*, bent form which is around 7 Å shorter. The *cis* isomer relaxes back to its *trans* form slowly in the dark (minutes to hours, but *see* **Note 13**), while illumination with 480–540 nm greatly accelerates the process. As much as 90–95 % of the *cis* or *trans* isomer can be obtained with proper illumination conditions (*see* **Note 14**).

3. A biologically active ligand. In the case of the nAChR, two different PTLs have been engineered: an agonist MAACh (Maleimide-Azobenzene-Acetylcholine, Fig. 1a) and a competitive antagonist MAHoCh (Maleimide-Azobenzene-HomoCholine, Fig. 1b), resulting in nAChRs that could be either photoactivated (Fig. 1c) or photoinhibited (Fig. 1d).

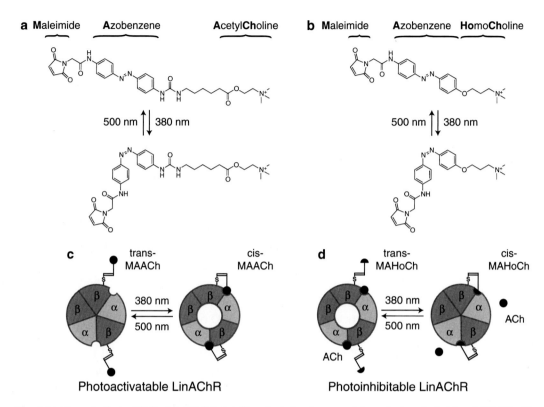

Fig. 1 Light-controlled nAChRs (LinAChRs). (**a**) Chemical structures and *trans* (*up*) and *cis* (*down*) MAACh. (**b**) Chemical structures of *trans* (*up*) and *cis* (*down*) MAHoCh. (**c**) Photoactivatable nAChR tethered with MAACh. In the *trans* configuration, MAACh extends away from the agonist-binding pocket. Illumination with 380 nm light photoisomerizes MAACh to *cis* and opens the receptor. (**d**) Photoinhibitable nAChR tethered with MAHoCh. In the *trans* configuration, MAHoCH extends away from the agonist-binding pocket and ACh can bind to activate its receptor. Illumination with 380 nm light photoisomerizes MAHoCh to *cis*, which competes with ACh and antagonizes the receptor

When designing a PTL, the trickiest part is to retain the biological activity of the ligand after derivatization with a relatively bulky group such as an azobenzene. Our ability to do so relies heavily on extensive research in the fields of pharmacochemistry and protein crystallography. Fortunately, in the case of the nAChR, several fluorescent or photoaffinity probes were developed with efficient agonist properties [17, 18]. The chemical structures of these compounds provided excellent scaffolds for engineering MAACh. In particular, a flexible spacer was positioned in between the acetylcholine moiety and the aromatic ring of the azobenzene (Fig. 1a). This spacer was shown to be crucial for enabling the activation mechanism of nAChRs, which involves a molecular motion of a receptor loop around the agonist-binding pocket [17–19]. This motion might be seen as a door closing after the agonist has entered. For designing the antagonist, we predicted that positioning the azobenzene closer to the ligand would prevent loop closure (like a foot in the door mechanism), thereby antagonizing the receptor. We designed the PTL MAHoCh, where the azobenzene is directly connected to the oxygen atom of the ligand homocholine (Fig. 1b).

3.1.2 Cysteine Mutant Selection

The choice of residues to be mutated to cysteine is greatly aided by three-dimensional structures of ligand-binding domains, preferably docked with agonists and antagonists. In the case of the nAChR, we used the structure of acetylcholine-binding protein (AChBP) in complex with carbamylcholine (CCh) [19]. AChBP is a homopentameric, soluble protein from snail that is structurally homologous to the ligand-binding domain of nAChRs.

1. Download the pdb file of the AChBP–CCh complex (pdb code 1UV6) from http://www.rcsb.org/pdb/explore/explore.do?structureId=1uv6. This structure provides valuable information as how CCh docks in the agonist-binding pocket.

2. Visualize the structure with an open source molecular visualization software (e.g., VMD or Pymol). We identified an access tunnel at the interface between two subunits, in which the flexible spacer of MAACh may fit. We anticipated the maleimide part of MAACh and MAHoCh to projects toward the complementary subunit, which is composed of β sheets (Fig. 2a).

3. Identify amino-acids that are candidates for cysteine replacement. In our case, we identified 10 residues, located on the complementary subunit of AChBP (Fig. 2a).

4. Find the homologous positions on the nAChR β2 and β4 subunits. In heteromeric nAChRs, the α and β subunits correspond to the principal versus complementary subunits in

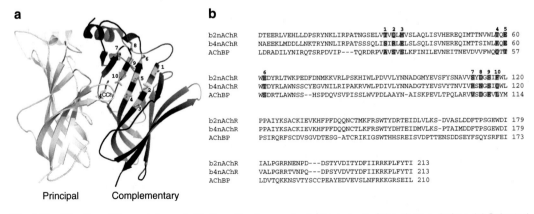

Fig. 2 Identification of the cysteine substitution sites for covalent attachment of the photoswitches. (**a**) Selected residues are depicted in *orange sticks* on the AChBP structure in cartoon representation (1UV6, PDB) in complex with carbamylcholine (CCh). The CCh-binding site is located at the interface between the principal subunit (*left*) and the complementary one (*right*). (**b**) Sequence alignment between AChBP and the ligand-binding domain of the mouse β2 and β4 nAChR subunits. Residues identified in (**a**) are highlighted in *orange*

AChBP, respectively. We aligned the sequences of the ligand-binding domain (first 213 residues) of the *Rattus norvegicus* β2 and β4 subunits with that of AChBP (210 residues) using the online multiple sequence alignment tool CLUSTAL W2 (available at http://www.ebi.ac.uk/Tools/msa/clustalw2/). The result is shown in Fig. 2b. The sequence homology between the two proteins is high (e.g., identity 19.7 % and similarity 38.7 % between AChBP and the rat β2 nAChR subunit), enabling us to identify the positions to be mutated on the nAChR.

3.2 Molecular Biology

1. Ten amino acids in the identified region of contact (*see* residues labeled 1–10 in Fig. 2b) are mutated to cysteine residues, one at a time, both on β2 and β4, using a site-directed mutagenesis kit (*see* **Note 2**).

2. Isolate five colonies and make a mini-prep for each.

3. Send plasmid DNA for sequencing to check for the presence of the desired mutation.

4. Make a DNA plasmid maxi-prep for one of the positive constructs.

3.3 Expression in Oocytes

1. Prepare mRNAs using the appropriate kit (*see* **Note 15**). Store at –20 °C.

2. Pull a glass pipette to make two glass needles with long tapers. Use a pair of fine scissors to trim the tip. Tip should preferentially be sharp, with an oblique opening (about 20–40 μm diameter, measured under a dissecting microscope).

3. Transfer the oocytes together with ORI solution into a small petri dish using a sucking glass pipette (with fire-polished opening). To prevent oocytes from moving, a polypropylene mesh glued at the bottom of the dish may be used.

4. Back-fill a glass needle with mineral oil. Ensure no air bubble is present and insert the plunger of the microinjector all the way into the needle. The plunger will force the oil toward the tip of the needle.

5. Drop 1–2 μl of mRNA solution (e.g., a 1/1 mixture of α4 and β2 mRNAs at a concentration of 40–400 ng/ml) into the cap of a 500 μl microtube. Draw the drop into the glass needle with the injector under the microscope. Use a new needle for each sample to prevent cross-contamination.

6. Position the tip of the glass needle close to the surface of the oocyte, at the border between the animal and the vegetal pole. Impale the oocyte using a micromanipulator. A resistance should be seen when oocytes are healthy.

7. Inject 50 nl of mRNA per oocyte. Slight swelling of the oocyte may be seen. Remove the glass needle slowly to avoid yolk from exiting the oocyte.

8. After injection is finished, transfer the oocytes into a petri dish containing ORI supplemented with gentamicin (100 μg/ml, to prevent bacterial contamination) and incubate at 18 °C for 1–3 days. Check oocytes everyday, and transfer healthy cells into freshly filtered ORI solution supplemented with gentamicin.

3.4 Screening of Cysteine Mutants Using Two-Electrode Voltage Clamp

1. Chlorinate the electrodes (same procedure applies to patch clamp electrodes). Silver wires should be coated with AgCl once a week. Clean the electrode with fine sandpaper, ethanol, and rinse it with ultrapure water. Immerse the electrode in newly purchased or well-sealed chlorine bleach for 20–30 min. Electrode will appear darker where chlorinated. Rinse well with distilled water (no ethanol at this stage).

2. Check the perfusion system and adjust the flow rate to about 2 ml/min. Make sure all perfusion lines have the same flow using flow regulators.

3. Incubate oocytes 15 min in a freshly prepared solution of 25 μM MAACh or MAHoCh (diluted in ORI from a 10 mM DMSO stock) in a small petri dish (*see* **Note 16**). Because attachment is covalent, several oocytes can be conjugated simultaneously.

4. Transfer oocytes in a PTL-free ORI solution.

5. Place one healthy oocyte into the recording chamber, with the animal (dark) pole facing up.

6. Back-fill the glass pipettes with 3 M KCl.

7. Position the pipette into the bath and check resistance (should be in the range of 0.7–1 MΩ). Make sure KCl does not leak out.

8. Impale the oocyte under the microscope, first with the voltage electrode and then with the current electrode. Check membrane resistance (should be in the –50 to –60 mV range).

9. Position the optic fiber above the oocyte, so that the spot of collimated light covers the entire animal pole. We have noticed that nAChRs express preferentially in the animal pole. Check light intensity at 380 and 500 nm with the power meter (should be in the 20–100 mW/cm^2 range, *see* **Note 14**).

10. Switch to voltage-clamp mode (set at –80 mV). Acquire data at 1 kHz and filter at 10 Hz.

11. Perfuse subsaturating (e.g., 10 or 100 μM) and saturating (e.g., 1 mM) concentration of the agonist acetylcholine (ACh) or carbamylcholine (CCh) to check for receptor expression and functionality. This step should also be performed without MAACh or MAHoCh treatment, because PTL covalent attachment may alter the function of the receptor dramatically

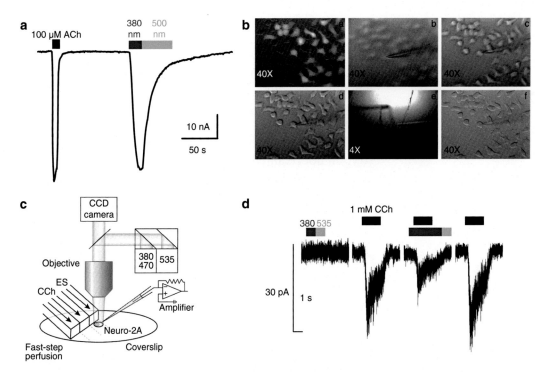

Fig. 3 Electrophysiology with LinAChRs. (**a**) Photoactivation with MAACh tethered onto α3β4E61C expressed in *Xenopus laevis* oocytes. Illumination at 380 nm triggers photoinduced current and at 500 nm shuts it off. (**b**) Whole cell patch clamp experiment. (a) Looking for an isolated cell, selected using GFP fluorescence. (b–d) Pipette positioning as close as possible to the selected cell without touching it. (e) Positioning of the fast perfusion system above the selected cell. (f) Contact with cell and gigaohm seal formation. (**c**) Schematic representation of a patch-clamp setup customized for optogenetic experiments, showing the fast-perfusion and illumination LED systems, mounted on an upright microscope equipped with a 40× immersion objective. (**d**) Photoinhibition with MAHoCh tethered onto α4β2E61C expressed in Neuro-2A cells. Illumination alone (380 or 535 nm) does not produce any detectable current. In the dark, 1 mM CCh induces about 30 pA current. This current is photoinhibited under 380 nm light, and this effect is fully reversible upon illumination with 535 nm light. Data is acquired at 2 kHz and filtered at 1 kHz, R-series resistance and whole cell capacitance are compensated. Post-acquisition filtering at 280 Hz with an eight-pole Bessel low-pass filter

(unpublished observations). Because nAChRs desensitize in the continuous presence of agonists, it is important to let the receptor recover from desensitization 1–2 min in between two agonist applications.

12. For experiments with the photoswitchable agonist MAACh, apply different pulses of 380 and 500 nm light, and compare the amplitude of the response to that of ACh (Fig. 3a) (*see* **Note 6**).

13. For experiments with the photoswitchable antagonist MAHoCh, apply different concentrations of ACh in the dark, or under 380 or 500 nm light illumination. Compare the amplitude of the responses under the different illumination conditions.

3.5   Cell Culture

1. Culture plate preparation. Place 5–6 sterile glass coverslips (10 mm) in each well of a 6-well plate. Add one drop (~50 µl) of poly-l-lysine (70,000–150,000, sigma) on top of each coverslip. Incubate 1 min, add 2 ml of PBS, and remove all the liquid using suction. Coverslips should be dry.

2. Dissociate cells. Remove medium from a 25 mm^2 flask (90 % confluence) and add 5 ml of fresh culture medium (preheated at 37 °C). Detach cells by triturating gently with a pipette (trypsin not required).

3. Place 2 ml of a 90,000 cells/ml cell suspension in each well. Leave cells to adhere at least an hour before transfecting.

3.6   Transfection Protocol

1. Replace culture medium with transfection medium (2 ml).

2. Prepare DNA mixture. Put 12.4 µl of a 2 mM CaCl$_2$ solution together with 2 µg of DNA and add sterile ultrapure water up to 100 µl. Homogenize with vortex.

3. Prepare DNA–Ca^{2+} phosphate precipitates. Add in a drop wise fashion 16 µl of the DNA mixture to 100 µl HeBS 2× solution. Triturate gently with pipette (*see* **Note 17**). Repeat procedure 6×, or until 100 µl of DNA mixture has been added to the mix.

4. Place precipitate mixture at 4 °C for 30 min.

5. Add 200 µl DNA mixture (or 10 % v/v) in a drop wise fashion on top of cells, and incubate at 37 °C for 3 h.

6. To remove precipitates, replace medium by 2 ml of acidic transfection medium and incubate at 37 °C for 20 min.

7. Replace acidic transfection medium by culture medium and incubate at 37 °C for at least 24 h before performing patch clamp electrophysiology.

3.7   Patch-Clamp Recordings in Neuro-2A Cells

1. Patch pipettes are pulled using a filament-based puller. Settings should be adjusted to get a 3–6 MΩ resistance.

2. Fill one perfusion line with a standard extracellular solution and another line with a saturating solution (1 mM) of CCh.

3. Stick a coverslip at the bottom of the recording chamber (using silicon grease) and start perfusion (should be 0.5–2 ml/min). Select a healthy and adherent transfected cell using the GFP fluorescent filter cube (Fig. 3b, *see* **Note 18**).

4. Back-fill a glass pipette with intracellular solution. Remove any air bubble by flicking the pipette gently. Place the pipette in the electrode holder and apply a positive pressure, then clamp the pressure tubing with a stopcock.

5. Using the micromanipulator, place the pipette in the bath (resistance should be in the 3–6 MΩ range) with its tip positioned just above the selected cell (Fig. 3b).

6. Place the fast perfusion system over the cell (approximately 50–100 μm distance). The fast perfusion system contains a three-square barrel tube. A stepper motor allows fast lateral motion of the perfusion, so to rapidly switch between the three tubes (noted t0, t1, and t2, *see* Fig. 3c). Note that the resting state of the perfusion is t0; it is thus preferable to use t0 for external solution and t1 and t2 for CCh.

7. Using a membrane test function (voltage pulse of –5 mV during 10 ms), adjust the pipette offset to 0 mV. Then slowly lower the pipette using the micromanipulator until the resistance increases slightly and a dimple can be seen at the cell surface. Release the positive pressure. A contact between the tip of the pipette and the cell membrane should form. The gigaohm seal formation is usually facilitated by a gentle and constant suction, while holding the potential at –40 mV.

8. When the gigaohm seal is reached, open the stopcock to release the suction. Then apply a brief and gentle suction to break the membrane. This step corresponds to the switch from the "cell-attached" to "whole-cell" configuration. Breaking of the membrane is associated with change in amplitude of the capacitive currents, which can be compensated with the amplifier. Measurements of high currents can be error minimized by compensating the series resistance.

3.8 Photoswitching Experiments

1. Prepare the PTL labeling solution (final photoswitch concentration 20 μM) by adding 0.5 μl of 2 mM MAACh or MAHoCh (DMSO stock) to 49.5 μl of external solution (ES). This solution should be prepared just prior to the patch clamp experiment (*see* **Note 4**).

2. Place a glass coverslip in the recording chamber and drop 50 μl of the freshly prepared labeling solution directly onto the glass coverslip.

3. After 20 min, wash the excess of unreacted photoswitch by adding 2 ml of ES. Turn perfusion on and place suction in the bath.

4. Patch an eGFP-positive cell (*see* Subheading 3.7).

5. Measure the current for a saturating concentration of CCh (e.g., 1 mM). Repeat 2–3× to check for reproducibility. Wash with ES for at least 1 min between two CCh applications to avoid current run-down due to receptor desensitization.

6. Photoactivation experiments with MAACh. Record photocurrents induced by 2 s light pulses. Illumination period should be adjusted to a time that is sufficient to fully activate and deactivate receptors (*see* **Note 14**). Photoelicited currents are normalized to currents evoked by saturating CCh (1 mM).

7. Photoinhibition experiments with MAHoCh (Fig. 3d). Since MAHoCH is an antagonist, apply 380 and 535 nm light pulses (1 s) to verify that light alone does not elicit any current. Measure the CCh-evoked currents (1 mM) in the dark. Then apply 380 nm light to isomerize MAHoCh to *cis*, at least 500 ms before applying CCh (*see* **Note 14**). Because MAHoCh is relatively thermostable ($t_{1/2} \approx 75$ min in solution), it is required to apply 535 nm light for at least 500 ms to convert MAHoCh back to its *trans* state. This is done in Fig. 3d just after CCh has been washed out, but alternatively it could be done just prior to the next CCh application. Then record the CCh-induced current in the dark (*trans* state) or alternatively under 535 nm light. CCh currents under the different light illuminations are normalized in relation to CCh currents in absence of light for calculating the percent photoinhibition.

4 Notes

1. Only cell-surface proteins can be rendered light-sensitive using this technology. Indeed, the cytoplasm of cells is a highly reductive environment rich in free thiols, precluding maleimide-thiol chemistry.

2. The amino acid cysteine has attractive chemistry for attaching chemical linkers, with minimal perturbation of the target protein (single amino-acid substitution). Despite the fact that cysteines are relatively abundant amino acids, our strategy has proven highly specific, since the PTL has to be anchored at a very well defined distance from a ligand-binding pocket to bestow photosensitivity to the target protein. In addition, due to the strong nucleophilicity of cysteines, other nucleophilic amino acids (i.e., lysines, histidines, etc.,) won't react with maleimides which are weak electrophiles.

3. Because azobenzene photoswitches do not photobleach or photodecompose, it is not necessary to handle them in the dark.

4. Maleimides can react with water, especially at alkaline pH, resulting in a product that is no longer cysteine reactive. It is therefore of utmost importance to use anhydrous DMSO for dissolving PTLs. Anhydrous DMSO can be purchased as small (5 ml) sealed ampoules. Aliquots of PTL stock solutions (classically 10 μl volume) should be kept at −80 °C in a box that contains drierite in order to trap water molecules.

5. The plasmid pIRES enables the expression of two genes, for example a nAChR subunit and eGFP, from the same bicistronic transcript.

6. The *Xenopus* oocyte is a large (~1 mm diameter) and opaque cell. It is an excellent expression system for membrane proteins, making it the ideal choice for screening cysteine mutants for PTL attachment. However, its big size limits perfusion exchange kinetics, which may be problematic in case of ligand-gated ion channels such as the nAChR. In addition, its opacity precludes photostimulation of the entire membrane surface. A few strategies exist for either photostimulating the entire oocyte using mirrors [20] or for restricting perfusion of the agonist to the photostimulated area [21]. For a proper quantification of the photoinduced agonism or antagonism, however, it is recommended to run experiments in transparent, small mammalian cells.

7. Neuro-2A cells are well suited to study nAChRs [22]. They are easy to maintain and allow for robust expression of membrane proteins. Their little size and transparency allows a more precise quantification of photoinduced currents compared to oocytes.

8. Prefer Glutamax (l-alanyl-l-glutamine dipeptide) over glutamine because it is more stable. Glutamine degrades over time, producing ammonia that alkalizes culture media.

9. For proper optical photostimulation, the TEVC and patch clamp setups should be either placed in a dark room with low ambient light (our preferred option), or protected from light using an opaque faraday cage (or a curtain).

10. Ultra bright light-emitting diodes (LEDs) have become a cheap and advantageous alternative over traditional Xenon arc lamps or lasers. They generate sufficient light intensity to provide a useful illumination source. They exist in many different colors, eliminating the needs of using bandpass filters. They generate minimal heat. They can be switched on and off very rapidly, eliminating the needs of using a fast shutter. They also have a high emission stability and very long life span.

11. An inverted microscope offers a better platform for cultured cells but we classically use an upright microscope, which we also use for experiments on brain slices. Similarly, a standard 100-W halogen lamp housing with classical oblique contrast system is sufficient for cultured cells, but we prefer using an infra-red LED together with a Dodt contrast system for brain slices.

12. All the optical components should have a good transmittance in the near-UV range due to the photochemical properties of azobenzene photoswitches.

13. Azobenzene substituents (in the *ortho*, *meta*, and *para* positions) can greatly influence the spectral characteristics as well as the thermal relaxation rates of the photoswitch [23–25].

For example, it is possible to design photoswitches that relax back to the *trans* configuration very rapidly (ms) in the dark, or that can be isomerized to the *cis* configuration with blue light [26].

14. Azobenzenes exist in two states (*trans* and *cis* isomers), and their interconversion can be controlled using different wavelengths of light. For a given wavelength and a given light intensity, there will be a photostationary state, i.e., a given mixture of both isomers at a particular ratio. We found that 380–390 nm light gives us the highest percent of *cis* MAACh or MAHoCh (>90 %), while 520–540 nm light gives the highest percent of *trans* isomer. However, because *cis* MAACh and MAHoCh do not absorb very much for $\lambda > 500$ nm [5], efficient photoconversion requires intense light. A powerful light source is thus required to enable both fast photoswitching (ms range) and maximization of the accumulation of a given isomer.

15. Special precautions should be taken when preparing and handling RNAs, because of the ubiquitous presence of RNases. Allocate a designated bench space (or a small room) to the preparation and handling of RNAs. Always wear gloves. Use sterile plasticware. Clean all equipment and bench space with RNase inactivating agent. Use DEPC-treated water only.

16. The pKa of the thiolate anion of cysteine amino acids depends on its environmental context, but is classically in the 8.3 range, which implies that it will be a better nucleophile (i.e., more reactive) at pH above 8.3. However, because the stability of maleimides decreases with increasing pH, a neutral pH of 7–7.4 is classically used for thiol modification with PTLs.

17. Avoid fast vortex agitation, which produces large and inhomogeneous precipitate sizes [27].

18. Healthy Neuro-2A cells are adherent with numerous dendrites. Cells in apoptosis are turgescent, e.g., spherical, and some time show surface bubbles. Some dead cells stay adherent with granular surface, and often display extremely bright fluorescence.

Acknowledgements

We wish to thank Richard H. Kramer (UC Berkeley) and Dirk Trauner (LMU Munich) for continuous support. This work was supported by the Agence Nationale de la Recherche (ANR-JCJC 2014), by a NARSAD Young Investigator Grant from the Brain & Behavior Research Foundation, and by the Fondation pour la Recherche Médicale (FRM). Romain Durand-de Cuttoli was supported by a Ph.D. fellowship from the DIM Cerveau & Pensée program of the Région Ile-de-France.

References

1. Gautier A et al (2014) How to control proteins with light in living systems. Nat Chem Biol 10:533–541

2. Fehrentz T, Schönberger M, Trauner D (2011) Optochemical genetics. Angew Chem Int Ed Engl 50:12156–12182

3. Kramer RH, Mourot A, Adesnik H (2013) Optogenetic pharmacology for control of native neuronal signaling proteins. Nat Neurosci 16:816–823

4. Mourot A, Tochitsky I, Kramer RH (2013) Light at the end of the channel: optical manipulation of intrinsic neuronal excitability with chemical photoswitches. Front Mol Neurosci 6:1–15

5. Tochitsky I et al (2012) Optochemical control of genetically engineered neuronal nicotinic acetylcholine receptors. Nat Chem 4:105–111

6. Volgraf M et al (2005) Allosteric control of an ionotropic glutamate receptor with an optical switch. Nat Chem Biol 2:47–52

7. Levitz J et al (2013) Optical control of metabotropic glutamate receptors. Nat Neurosci 16:507–516

8. Lin W-C et al (2014) Engineering a light-regulated GABAA receptor for optical control of neural inhibition. ACS Chem Biol 9:1414–1419

9. Banghart MR, Borges K, Isacoff EY, Trauner D, Kramer RH (2004) Light-activated ion channels for remote control of neuronal firing. Nat Neurosci 7:1381–1386

10. Fortin DL et al (2011) Optogenetic photochemical control of designer K+ channels in mammalian neurons. J Neurophysiol 106:488–496

11. Sandoz G, Levitz J, Kramer RH, Isacoff EY (2012) Optical control of endogenous proteins with a photoswitchable conditional subunit reveals a role for TREK1 in GABAB signaling. Neuron 74:1005–1014

12. Lemoine D et al (2013) Optical control of an ion channel gate. Proc Natl Acad Sci U S A 110:20813–20818

13. Zoli M, Pistillo F, Gotti C (2014) Diversity of native nicotinic receptor subtypes in mammalian brain. Neuropharmacology 1–10

14. Nees F (2015) The nicotinic cholinergic system function in the human brain. Neuropharmacology 96:289–301

15. Taly A, Corringer P-J, Guedin D, Lestage P, Changeux J-P (2009) Nicotinic receptors: allosteric transitions and therapeutic targets in the nervous system. Nat Rev Drug Discov 8:1–18

16. Naudé J, Dongelmans M, Faure P (2015) Nicotinic alteration of decision-making. Neuropharmacology 96:244–254

17. Mourot A et al (2006) Probing the reorganization of the nicotinic acetylcholine receptor during desensitization by time-resolved covalent labeling using [3H]AC5, a photoactivatable agonist. Mol Pharmacol 69:452–461

18. Krieger F et al (2008) Fluorescent agonists for the torpedo nicotinic acetylcholine receptor. ChemBioChem 9:1146–1153

19. Celie PHN et al (2004) Nicotine and carbamylcholine binding to nicotinic acetylcholine receptors as studied in AChBP crystal structures. Neuron 41:907–914

20. Chambers JJ, Gouda H, Young DM, Kuntz ID, England PM (2004) Photochemically knocking out glutamate receptors in vivo. J Am Chem Soc 126:13886–13887

21. Bhargava Y, Rettinger J, Mourot A (2012) Allosteric nature of P2X receptor activation probed by photoaffinity labelling. Br J Pharmacol 167:1301–1310

22. Srinivasan R et al (2012) Pharmacological chaperoning of nicotinic acetylcholine receptors reduces the endoplasmic reticulum stress response. Mol Pharmacol 81:759–769

23. Samanta S et al (2013) Photoswitching azo compounds in vivo with red light. J Am Chem Soc 135(26):9777–9784

24. Sadovski O, Beharry AA, Zhang F, Woolley GA (2009) Spectral tuning of azobenzene photoswitches for biological applications. Angew Chem Int Ed Engl 48:1484–1486

25. Beharry AA, Sadovski O, Woolley GA (2011) Azobenzene photoswitching without ultraviolet light. J Am Chem Soc 133:19684–19687

26. Mourot A et al (2011) Tuning photochromic ion channel blockers. ACS Chem Neurosci 2:536–543

27. Jiang M, Chen G (2006) High Ca2+-phosphate transfection efficiency in low-density neuronal cultures. Nat Protoc 1:695–700

Chapter 13

Optogenetic Modulation of Locomotor Activity on Free-Behaving Rats

Kedi Xu, Jiacheng Zhang, Songchao Guo, and Xiaoxiang Zheng

Abstract

The technology of optogenetics provides a new method to modulate neural activity with spatial specificity and millisecond-temporal scale. This nonelectrical modulation method also gives chance for simultaneous electrophysiological recording during stimulations. Here, we describe our locomotor activity modulation on free-behaving rats using optogenetic techniques. The target sites of the rat brain were dorsal periaqueductal gray (dPAG) and ventral tegmental area (VTA) for the modulation of defensive and reward behaviors, respectively.

Key words Optogenetics, Locomotor activity modulation, Optical-electrode array, dPAG, VTA, Virus injection

1 Introduction

Modulation of animal behaviors can be achieved by stimulations in specific areas of brain, which have been employed for disease studies and the development of bio-robots [1–3]. Both electrical and chemical methods are widely applied as effective stimulations in animal behavior modulation [4, 5]. However, with the development of neurobiology, these traditional stimulation methods are unable to meet the demands of higher resolutions both spatially and temporally. Chemical stimulation targets accurately on postsynaptic regulations, but lacks spatial and temporal resolution due to its long-term pharmacokinetic properties. The electrical stimulation, on the other hand, provides a flexible stimulation parameters control, including frequency, duration, and intensity, but the wide-spread electrical current could activate all types of neurons in region unselectively [6].

The advent of optogenetics makes it feasible to regulate the activity of specific types of cells [7]. Stimulated by laser illumination, specific neurons expressing light-sensitive opsins can be effectively activated on a time scale of milliseconds. The optogenetics

Arash Kianianmomeni (ed.), *Optogenetics: Methods and Protocols*, Methods in Molecular Biology, vol. 1408,
DOI 10.1007/978-1-4939-3512-3_13, © Springer Science+Business Media New York 2016

thus ensures the region limitation, neuron specificity, and stimulation variability. Moreover, the optical stimulation can give rise to safe excitation of neurons and allows simultaneous application of neural recording techniques in vivo. Here, we describe novel methods for locomotor activity modulation by optogenetic ways, which are based on our previous works on modulations in areas of the dorsal periaqueductal gray (dPAG) area [8, 9] for defensive behavior and in the ventral tegmental area (VTA) [10] to enhance locomotor activity in free-moving rats.

2   Materials

2.1   Surgical Components

1. Anesthetic: 1 % pentobarbital sodium solution. Dissolve 1 g pentobarbital sodium powder (Merck, Germany) with 100 mL saline solution. Mix and store at 4 °C.

2. Surgical instruments: All of the surgical instruments, including scalpel blade, ophthalmic scissors, dental spatula, fine forceps, skull screws (*see* **Note 1**) are standard types for surgeries. The instruments are sterilized by a laboratory autoclave before surgery.

3. Standard stereotaxic apparatus (Stoelting Co. Ltd., USA).

4. Operating microscope (YZ20P6, 66 Vision-Tech Co., Ltd, Suzhou, China).

5. Dental cement: Luxatemp Handmix temporary crown and bridge material (DMG Chem., Hamburg, Germany), Vertex Self Curing Cold-curing acrylic denture repair material (Vertex-Dental B.V., the Netherlands).

2.2   Virus Injection

1. Virus: AAV-hSyn-ChR2-mCherry and AAV-CaMKIIα-ChR2-mCherry were purchased from Neuron Biotech (Shanghai, China) and stored at −80 °C before use (*see* **Note 2**).

2. Micro4 microinjection pump (World Precision Instruments, USA), microliter syringe (5 μL, Model 75 RN SYR, Hamilton Co., Reno, Nevada, USA) with small hub needle (34GA RN B-TYPE 1.5″ PT 3, Hamilton Co., USA).

2.3   Electrophysiological Recording

1. Formvar-Insulated Nichrome Wires: bare 0.0010″, coated 0.0015″ in diameter for electrophysiological recording; bare 0.002″, coated 0.0026″ in diameter for "Reference" (Ref) electrode (A-M SYSTEMS, USA).

2. Silvered copper wire (10 μm in diameter) as "Ground" (GND) electrode.

3. Plexon Multichannel Acquisition Processor OmniPlex (Plexon Inc., Dallas, TX, USA) for electrophysiological signal recording.

2.4 Optical Stimulation

1. Laser device: A power adjustable 473 nm Laser device was purchased from Shanghai Laser & Optics Century Inc. (BL473T5-320FC, max power: 500 mW, Shanghai Laser & Optics Century Inc., China).

2. Dual-channel digital stimulator PG4000A (Cygnus Technology, USA) was used to generate the trigger pulse for laser.

3. Optogenetics patch cable with integrated rotary joint (with FC/PC connector and ferrule end, Thorlabs Inc., New Jersey, USA).

4. Ceramic ferrule with 50/125 Multimode glass optical fiber, defined as "fiber tail."

5. Ceramic split mating sleeves.

6. Guide cannula, 13.9 mm total length, 300/480 μm inner/outer diameter.

7. Optical power meter (LTE-1A, Chinese Academy of Sciences, Beijing, China).

8. Diamond stylus.

3 Methods

3.1 Optical-Electrode Preparation

1. Prepare the guide cannula and electrodes (see [11] for more details): Integrate the PCB skeleton with the Omnetics connector: connect a reference electrode ("Ref," formvar-insulated Nichrome wire, 0.0026″ in diameter, about 2 cm in length) and a ground electrode ("GND," silvered copper wire, about 5 cm in length) onto each side of the PCB skeleton, respectively. Bend the end of the PCB skeleton into an L-shaped substrate (Fig. 1a–c).

2. Twine eight pieces of electrode wires (0.0015″ in diameter) into four pairs, with the length of about 10 cm (see **Note 3**). Bend each electrode pair into 90° in the middle of the electrode bundle.

3. Adhere the electrode pairs at the bended point onto the root of the guide cannula with flex gel sequentially, forming a cruciform arrangement. Cut the tips of the electrode pairs to 2.5 mm extended the cannula (see **Note 4**) (Fig. 1d).

4. Adhere the electrode-bundled guide cannula onto the short arm of the L-shaped PCB skeleton with super glue (Fig. 1e).

5. Solder the top of the electrodes to the bonding pads on the PCB skeleton. An overview of the optical-electrode cannula is shown in Fig. 1f.

6. Determine the brain area and the coordinates to be stimulated. Here, we demonstrate the optogenetic modulations in the

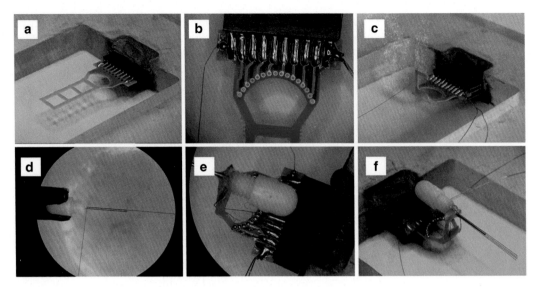

Fig. 1 Preparation of the optical-electrode array. (**a**) Integration of the PCB skeleton and the Omnetics connector. (**b**) Electronic structure of the "Ref" and "GND" channels. (**c**) The L-shaped substrate. (**d**) Connection of the electrode bundle with the guide cannula. (**e**) Assembly of the electrode-bundled guide cannula with the PCB skeleton. (**f**) The overall view of the optical-electrode array

dPAG area (7.0 mm posterior to the Bregma, 2.3 mm mediolateral, 5.5 mm dorsoventral at 16° to the surface of skull, *see* **Note 5**) to modulate defensive behavior, and in the VTA (4.8 mm posterior to the Bregma, 1.0 mm mediolateral, 8.0 mm dorsoventral) to enhance locomotor activity in free-moving rats. The coordinates are determined according to the rat brain atlas of Paxinos and Watson. Bregma and Lambda sites are identified as landmarks.

7. Cut the optic fiber tail by a diamond stylus to proper length according to the guide cannula length and the target site (*see* **Note 6**). Sand the cut smoothly with a fine grade of sandpaper, until the light output is evenly distributed.

8. Measure the light attenuation of the optic fiber tail. Change the intensity of the laser output and measure the output power in ferrule end of the fiber optic patch cable (the "INPUT" power of the optic fiber tail) and the final output from the optic fiber tail (the "OUTPUT" power of the optic fiber tail). Draw the input–output curve of the optic fiber tail and estimate the light attenuation.

3.2 Optical-Electrode Implantation and Virus Injection

3.2.1 Animal Preparation

1. Anesthetize animal with pentobarbital sodium (1 %, i.p., 50 mg/kg initial dose). Check the state of anesthetic by toe-pinch test every hour; apply an additional dose of 0.5 mL anesthetic if necessary.

2. Remove the hair around the surgical area by hair clipper, from between the eyes to the ears. Sterilize the area with iodine and 75 % ethanol solution.

3. Mount the animal in a stereotaxic apparatus, using heating pad or towel to keep animal warm. Apply atropine sulfate solution (i.p., 2 mg/kg) to avoid asphyxia. Apply a drop of glycerin in the eyes for protection.

4. Make a sagittal cut with a scalpel blade to expose the surface of the skull. Remove the connective tissue with ophthalmic scissors and dental spatula. Stop the bleeding by pressing with a sterile cotton swab for 5 min, or applying bone wax if necessary.

3.2.2 Optical-Electrode Implantation and Virus Injection

1. Mark the Bregma and Lambda sites on the skull by grease pencil, and note the coordinates. Convert the actual site coordinates according to the theoretical value determined previously in Subheading 3.1 (*see* **Note 7**). Mark the calculated target site on the skull (Fig. 2a).

2. Use tubular drill (1 mm in diameter) to remove the skull above the target site carefully (*see* **Note 8**) (Fig. 2b).

3. Use sterile needle tip to peel the dura under operating microscopy (*see* **Note 9**) (Fig. 2b inset).

4. Drill holes for fixing Ref electrode and skull screws (*see* **Note 10**).

5. Screw a silver wire bundled screw (a "GND screw") into one anterior hole, and fill the other three holes with bare skull screws (*see* **Note 11**).

6. Twine the silver wire around the screws in sequence to form a shielding ring.

7. Fix the electrode-bundled guide cannula to the cannula holding arm; make sure the cannula is vertical straight.

8. Reposition the cannula, lower it into the target site. Calculate the depth from the tips of the electrodes (*see* **Note 12**) (Fig. 2c).

9. Put the uncovered tip of the Ref electrode wire into the Ref hole horizontally between the skull and the dura.

10. Solder the GND electrode wire with the GND screw wire.

11. Seal the skull-removed hole and the Ref hole with medical adhesive.

12. Mix the Handmix temporary crown and bridge material part A and B. Apply the material quickly between the cannula and the nearest screws for a preliminary fixation of the cannula. Then mix and apply Cold-curing denture repair material for an overall secure (*see* **Note 13**) (Fig. 2d, e).

13. Prepare the micro syringe. Hold the sterile micro syringe onto the microinjection pump. Withdraw 3 µL paraffin oil at the speed of 3 µL/min, hold the needle tip in the paraffin oil for 10 min, and then infuse 1 µL paraffin oil at the rate of 6 µL/min. This step ensures a good airtightness of the injection system (*see* **Note 14**).

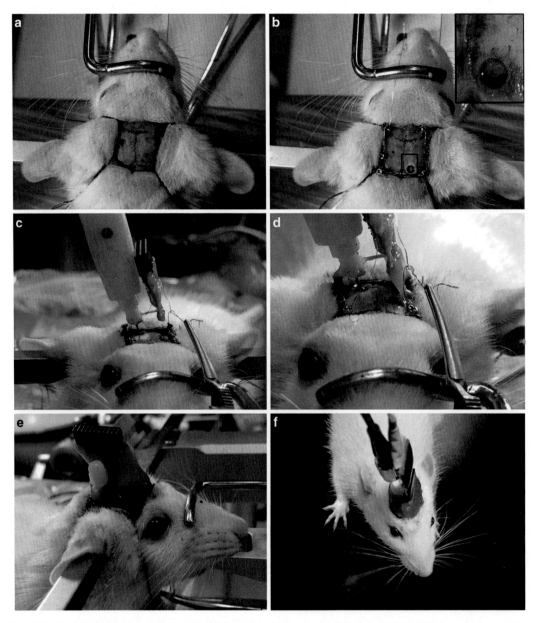

Fig. 2 Optical-electrode implantation. (**a**) Mark spots of Bregma, Lambda, and dPAG. (**b**) The positions of the skull screws and the silver wire twined around the screws. An enlargement of the dura-free window is shown in the *upper-right corner*. (**c**) Implantation of the optical-electrode array. A 16° should be applied for dPAG site. (**d**) A preliminary fixation applied by Handmix temporary crown and bridge material. Connect the guide cannula and the PCB skeleton with the skull screws. (**e**) A second dental cement applied for an overall secure. Virus injection can be applied through the guide cannula. (**f**) An overview of the completed implantation with optical stimulation on

14. Remove the virus from −80 °C freezer, dilute to the final titer (about 5×10^{12} titer) by saline and store in ice (*see* **Note 15**).

15. Withdraw 1.5 μL virus liquid at the rate of 2 μL/min. Infuse 200 nL at 200 nL/min for airtightness insurance check (*see* **Note 16**).

16. Reposition the holding arm to the target site according to landmark bregma and the coordinates calculated, and then lower the tip of the needle into the injection depth.

17. Infuse 200 nL virus at the rate of 100 nL/min each time for five repeats. The total dose of injection for each animal is 1 μL (*see* **Note 17**). Hold a 3 min interval between each injection to ensure the virus is spread and absorbed adequately. Once the final dose has been injected, leave the needle tip rest in place for 10 min before a lift of 0.5 mm; completely remove the syringe needle after another 10 min rest.

18. Connect a prepared optic fiber tail with a mating sleeve; fix the mating sleeve to the holding arm.

19. Locate the fiber tip to the cannula, and lower the fiber tip to the calculated depth. The depth should be 0.5 mm above the target site. For example, as the depth of the dPAG site in our experiment is 5.5 mm, a theoretical depth of 5.0 mm is appropriate for the fiber tip insertion.

20. Build dental cement around the cannula and the fiber mating sleeve. Leave the end of the ferrule clean so that the transmittance does not reduce.

21. House the animal separately in home cage for at least 1 week for recovery (*see* **Note 18**). Apply penicillin if necessary.

3.3 Optical Stimulation and Electrophysiological Recording In Vivo

3.3.1 Electrophysiological Recording Device Preparation

1. Setup the Plexon Multichannel Acquisition Processor OmniPlex. Connect the Plexon adapter on the optical-electrode array to Plexon headstage. Check the signals recorded to make sure that the animal recovers well.

2. Adjust the properties of different electronic recording channels individually. Make sure that for most of the channels, the connectivity is functioning well and the signal-to-noise ratio (SNR) runs high enough to detect neural electronic signals out of background noise. Then the devices and the animal are ready for electrophysiological recording (*see* **Note 19**).

3.3.2 Optical Stimulation Pathway Preparation

1. Set the pulse parameters for the stimulation (*see* **Note 20**). For most opsins, the optical stimulation parameter, including stimulation intensity, frequency, and pulse number need to be optimized with different target areas. In our experiments, we vary the stimulation pulse frequencies and train burst widths while the pulse duration is fixed with 15 ms (*see* **Note 21**).

2. Warm up the laser device. Connect the laser with the patch cable (*see* **Note 22**). Adjust the intensity to a proper value so that the final intensity illuminated onto the brain tissue is limited below 10 mW to avoid heat damage.

3. Once the patch cable and the laser devices are setup, carefully connect the Omnetics connector to the Plexon headstage, and couple the mating sleeve on rat's head to the ceramic connector on optic patch cable, respectively (*see* **Note 23**). An overview of the connections is shown in Fig. 2f.

4. Change the laser intensity (from low to high, below the limitation intensity), and observe the electrophysiological signal changes in animal. A significant increasing in spike firing rate during the photo stimulation indicates a successful expression of opsin, like Synp and ChR2. The light intensity is defined as the minimum threshold of optical stimulation to induce neuron activity changes. For individual variability and operating precision limitations, each animal should be tested its own threshold.

3.3.3 In Vivo Optical Stimulation on Various Areas of Rat Brain

1. Optical stimulation on dPAG area: Put the animal into the behavioral field (circular, 1.2 m in diameter). Leave the animal in the field exploring for about 10 min (*see* **Note 24**). The stimulation in dPAG induces defensive behaviors, from alertness to typical freeze and escape as the stimulation grows stronger (*see* **Note 25**). The reaction varies as the stimulation parameters changes.

2. Optical stimulation on VTA area: Put the animal into the field. Leave the animal for a 15-min habituation. Setup the Laser stimulation parameters with 50 Hz frequency, 15 ms pulse width, and 0.2 s duration as one single stimulation. The stimulation in VTA induces an enhancement of locomotor activity. The laser stimulation activates the animal for a long-term roaming, and far more approaches into the center region of the field.

4 Notes

1. For using on the rat's skull, one can use flat head screw as skull screw. The diameter of the screw head should be a little bigger than the drilled holes on the skull. In our experiment, we use 6 mm standard M1 flat head screw.

2. The AAV Virus can be stored at $-80\ ^{\circ}C$ for 1/2 to 1 year with higher titer ($>10^{13}$). The virus needs to be melted and diluted to about 10^{12} titer by saline solution right before injection. The unused virus diluents can be stored at $4\ ^{\circ}C$ and be injected within 2 following days. The virus will lose their effectiveness with long-term storage at $4\ ^{\circ}C$ environment or storage with lower titer.

3. The electrode bundle should be twined tight enough for tenacity assurance.

4. When adhering, make sure the tips of the electrode pairs clung close and are parallel to the cannula.

5. The dPAG site is very close to the sutura. To avoid critical bleeding during implantations, we apply an inclination angle of 16° to the sagittal plane.

6. When implanted in the rat brain, the end of the optical fiber should be 0.5 mm above the target area. In our experiment, the end of the implanted cannula is 1.0 mm above the target. Thus the length of the optical fiber tail is at least 0.5 mm longer than the cannula length.

7. Rat's brain vary in size as age and body weight change. It is necessary to apply a coefficient to the theoretical values accordingly. In our study, the coefficient Co is calculated by the landmarks of Bregma and Lambda as landmarks:

$$Co = \left(d_{measured} = \left| AP_{Bregma} - AP_{Lambda} \right| \right) / \left(d_0 = 9mm \right)$$

8. It is very important to avoid hard pressure while drilling and removing the skull above the target site. Press the drill steadily and softly on the skull till part of the skull becomes removable. Apply sterile saline to wash out the bone powder and lower the drill temperature. Use fine tipped forceps to break the remained connections and remove the skull part carefully. Clean the surroundings of the window with fine forceps and sterile saline.

9. Bend the tip of the smallest needle by pressing it against hard surface (like an iron table). The tip of the needle thus forms a small hook. While breaking the dura, gently hook the dura by drawing the needle tip on the dura surface. Never poke the dura with the needle tip to avoid bleeding and damaging of the brain tissue.

10. While drilling, hold the drill perpendicular to the skull surface and press steadily to avoid bleeding. For skull screw secure, the skull holes should be a little smaller than skull screws in size. Make sure the drilled holes are distributed evenly on the skull surface. Keep the target site have enough room for the optical-electrodes implantation.

11. The screw depth should be about 0.5 mm and the above-skull part should remain 0.5–1 mm to provide enough support for the dental cement secure.

12. Avoid blood vessels during cannula implantation if possible. Vessel rupture may cause excessive blood loss and even lead to death.

13. Be careful when building dental cement. Rickety adhesive may cause serious problems in subsequent neural activity modulation

trials as the animal may break the electrodes and the optical fiber with a sudden movement responding to stimulations, especially in defensive behavior modulations. In our experiments, we applied two different kinds of dental cement: Luxatemp Handmix temporary crown and bridge material and Vertex Self Curing Cold-curing acrylic denture repair material. The former one is an instant adhesive, which provides a fast fixation, ensuring the location of the cannula. The later one is much more viscous, and thus can anchor the cannula with skull screws steadily.

14. The preparation of the micro syringe may cost a lot of time. This step can be parallel done with cannula implantation. The virus injection should be applied in a biosafety cabinetry.

15. Prepare the virus solution by calculated the rats' number to avoid waste. Centrifuge the virus tube before use and make sure there is no liquid on the wall or the cap of the tube.

16. Leave the micro syringe tip in the air and infuse the virus out. If the airtightness is good, a small liquid drop can be observed at the syringe tip. Carefully clean the tip with kimwipe.

17. For rat weight around 250 g, 1 μL virus with titer at 5×10^{12} is most appropriate. In terms of virus volume and titer, larger volume or higher titer may cause severe tissue damage of the rat brain, while less volume or lower titer might be insufficient for the opsin expression.

18. The expression of the opsins normally takes for 3–4 weeks. House the animal singly in plastic cage. Both hSyn and CaMKIIα promoters can be used for opsin expression. The CaMKIIα promoter leads to a selective activation of dopaminergic neurons which is useful in dopaminergic system studies. The former one activates larger ranges of neurons and thus is suitable for locomotor activation controls.

19. The electrophysiological recording is optional. The electrodes insertion gives chance to a monitor on neural activity in target site in vivo, to estimate the expression of the opsins and the effects of the stimulation.

20. The laser device is triggered by high-level signals from signal generator. In our experiments, we use a dual-channel digital stimulator to generate rectangular wave train for manual/automatic laser trigger.

21. In addition to the light intensity, the stimulation duration and frequency also take part in the modulation mechanism. Generally, longer stimulation and higher frequencies leads to a stronger modulation result. However, stimulations slower than 5 Hz can hardly induce any electrophysiological or behavioral changes, and frequencies higher than 50 Hz contribute little in

modulation improvements. Thus, proper initial stimulating parameters can be a 1-s pulse train of 20 Hz, with pulse duration of 15 ms, and the output light intensity at 1 mW.

22. Use optical power meter to measure the output laser intensity from the FC end of the fiber cable. The illuminating attenuation of the optic fiber should be measured before the surgery.

23. The animal could be trained to accommodate to the connection of patch cable and headstage. One can also anesthetize the animal before the connection. Align the shafts of the ceramic ferrule on the animal's head and patch cable to ensure the light pathway is available.

24. The change of the environment will introduce extra pressure on the animals. Allow the animal move freely for a short term to get familiar with the new environment.

25. Gradually raise the laser intensity or stimulation pulse number. The application of strong laser stimulation may cause violent behavior change and introduce permanent heat damage on the brain tissue. Set recovery time (15 s in our experiment) between each stimulation to avoid tissue damage.

Acknowledgements

We thank Dr. Sicong Chen and Chaonan Yu for providing part technical supports. This work was supported by (1) International Science & Technology Cooperation Program of China, 2014DFG32580; (2) National Natural Science Foundation of China, 61305145; (3) Specialized Research Fund for Doctoral Program of Higher Education, 20130101120166; (4) Fundamental Research Funds for the Central Universities.

References

1. Umemura A, Oyama G, Shimo Y et al (2013) Update on deep brain stimulation for movement disorders. Rinsho Shinkeigaku 53:911–914

2. Talwar SK, Xu S, Hawley ES et al (2002) Behavioural neuroscience: rat navigation guided by remote control. Nature 417(6884):37–38

3. Feng Z, Chen W, Ye X et al (2007) A remote control training system for rat navigation in complicated environment. J Zhejiang Univ Sci A 8(2):323–330

4. Lim LW, Blokland A, Visser-Vandewalle V et al (2008) High-frequency stimulation of the dorsolateral periaqueductal gray and ventromedial hypothalamus fails to inhibit panic-like behaviour. Behav Brain Res 193(2):197–203

5. Brandao ML, Anseloni VZ, Pandossio JE et al (1999) Neurochemical mechanisms of the defensive behavior in the dorsal midbrain. Neurosci Biobehav Rev 23(6):863–875

6. McIntyre CC, Mori S, Sherman DL et al (2004) Electric field and stimulating influence generated by deep brain stimulation of the subthalamic nucleus. Clin Neurophysiol 115(3):589–595

7. Deisseroth K, Feng G, Majewska AK et al (2006) Next-generation optical technologies for illuminating genetically targeted brain circuits. J Neurosci 26(41):10380–10386

8. Chen S, Zhou H, Guo S et al (2015) Optogenetics based rat–robot control: optical

stimulation encodes "stop" and "escape" commands. Ann Biomed Eng 43(8):1851–1864

9. Guo SC, Zhou H, Wang YM et al (2014) A rat-robot control system based on optogenetics. Appl Mech Mater 461:848–852

10. Guo S, Chen S, Zhang Q et al (2014) Optogenetic activation of the excitatory neurons expressing CaMKIIα in the ventral tegmental area upregulates the locomotor activity of free behaving rats. Biomed Res Int 2014:687469

11. Guo S, Zhou H, Zhang J et al (2013) A multi-electrode array coupled with fiberoptic for deep-brain optical neuromodulation and electrical recording. In: Engineering in Medicine and Biology Society (EMBC), 2013 35th annual international conference of the IEEE. pp 2752–2755

Chapter 14

Combined Optogenetic and Chemogenetic Control of Neurons

Ken Berglund, Jack K. Tung, Bryan Higashikubo, Robert E. Gross, Christopher I. Moore, and Ute Hochgeschwender

Abstract

Optogenetics provides an array of elements for specific biophysical control, while designer chemogenetic receptors provide a minimally invasive method to control circuits in vivo by peripheral injection. We developed a strategy for selective regulation of activity in specific cells that integrates opto- and chemogenetic approaches, and thus allows manipulation of neuronal activity over a range of spatial and temporal scales in the same experimental animal. Light-sensing molecules (opsins) are activated by biologically produced light through luciferases upon peripheral injection of a small molecule substrate. Such luminescent opsins, luminopsins, allow conventional fiber optic use of optogenetic sensors, while at the same time providing chemogenetic access to the same sensors. We describe applications of this approach in cultured neurons in vitro, in brain slices ex vivo, and in awake and anesthetized animals in vivo.

Key words Luminopsin, Luciferase, Bioluminescence, Coelenterazine, Optogenetics, Chemogenetics, Neuron, Electrophysiology, Multielectrode array, Behavior

1 Introduction

Currently, there are two major approaches to control activity of genetically defined neurons in the brain of freely behaving animals: chemogenetic approaches that utilize diffusible small molecules [1] and optogenetic approaches that utilize externally delivered light [2]. The two methods have their own distinct merits; optogenetics offers the advantage of temporal precision while chemogenetic approaches offer scalability and ease of application. Combining these two approaches within single molecules complements each other and allows the use of either mode of interrogation in the same brain circuit. Such molecular actuators would allow acute activation with precise time resolution in defined spaces, as well as chronic and noninvasive control of entire populations throughout the brain through the same molecules.

Arash Kianianmomeni (ed.), *Optogenetics: Methods and Protocols*, Methods in Molecular Biology, vol. 1408,
DOI 10.1007/978-1-4939-3512-3_14, © Springer Science+Business Media New York 2016

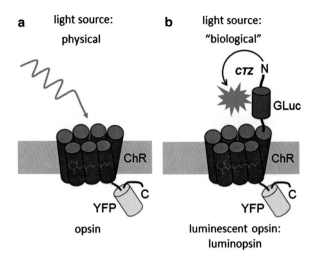

Fig. 1 Combined opto- and chemogenetic control with luminopsins. (**a**) An optogenetic element (here: channelrhodopsin, ChR) can be activated by a physical light source (LED, laser, arc lamp). (**b**) The same optogenetic element, when fused to a luciferase (here: *Gaussia* luciferase, Gluc) by a 15-amino acid linker, can also be activated by "biological" light, which is produced when the attached luciferase catalyzes oxidation of the supplied substrate coelenterazine (CTZ). Fused to the C-terminus of the optogenetic element is the fluorescent reporter YFP, allowing identification of cells expressing the element. The optogenetic element can be a channel or a pump, and depending on its biophysical properties can activate (LMO) or inhibit (iLMO) a neuron

Luminopsins (luminescent opsins or LMOs) were developed to achieve combined chemo- and optogenetic manipulation [3]. They are fusion proteins of a light-emitting luciferase and a light-sensing opsin (*see* Fig. 1). Application of the luciferase substrate, coelenterazine (CTZ), leads to emission of photons that is sufficient to activate the coupled opsin. We established proof of concept of this technology by using fusion proteins that directly link *Gaussia* luciferase (GLuc; [4, 5]) to *Chlamydomonas* channelrhodopsin 2 (ChR2; [6]) or *Volvox* channelrhodopsin 1 (VChR1; [7]). As for inhibition, we were able to harness bioluminescence from engineered *Renilla* luciferase (Nano-lantern; [8]) for activation of *Natronomonas* halorhodopsin [9] (inhibitory luminopsin or iLMO). When these fusion proteins are expressed in neurons, bright bioluminescence from GLuc and Nano-lantern was able to excite or silence a neuronal population, respectively, in cultured neurons in vitro, in brain slices ex vivo, and in awake and anesthetized animals upon application of CTZ. Moreover LMO and iLMO were able to elicit specific motor behavior in awake animals in vivo [unpublished results; also *see* Society for Neuroscience Meeting Abstracts 2013 (Tung et al., Berglund et al.) and 2014 (Clissold et al., Higashikubo et al.)].

LMOs retain the capability for conventional optogenetic control of neuronal activity. Light from a physical source (e.g. arc lamp, laser, LED) through optical fibers can activate the opsin moiety similar to when they are expressed by themselves without luciferase. As chemogenetic probes, however, LMOs have characteristics distinct from conventional chemogenetic probes such as DREADDs [10]. First, chemogenetic access is mediated by opsins instead of G-protein coupled receptors. Thus, there are no requirements for additional signaling pathways for LMOs to function. Second, while there is commonly one designer receptor each for activation and silencing, for example Gq and Gi DREADDs, respectively, LMOs can capitalize on the entire molecular palette of optogenetic actuators, which have a wide range of kinetics and sensitivity, and which can be matched with luciferases with a wide variety of emission spectra. Third, while there is a single designer drug (CNO) for activating both Gq and Gi DREADDs, different luciferases utilize different substrates, making multiplexing feasible. For example, hCTZ, a 2-deoxy derivative of native coelenterazine, which we use as a substrate for our RLuc-based inhibitory LMO (iLMO), cannot be used by Gaussia luciferase, which only emits light with native CTZ. Lastly, the approach is unique in that LMOs integrate opto- and chemogenetic approaches, and thus allow manipulation of neuronal activity over a range of spatial and temporal scales in the same experimental animal.

In this chapter, we will detail critical aspects of working with LMOs for combined opto- and chemogenetic manipulation of neuronal activity. We will address handling and application of the luciferase substrate, simultaneous bioluminescence imaging and electrophysiological and behavioral readouts in vitro and in vivo using LMOs.

2 Materials

Standard equipment and materials used for the application of interest will suffice (tissue culture, single-cell patch clamp recording, multielectrode array recording, brain slice recording, in vivo electrophysiology, behavioral testing). Detailed below are materials we routinely use for tissue culture experiments, and supplies and procedures for preparing the luciferase substrate.

2.1 Cell Culture

1. Tissue culture plasticware (4-well plates, 24-well plates, 12-well plates).

2. 12 or 18 mm poly-D-lysine-coated glass coverslips.

3. 293T human embryonic kidney fibroblasts.

4. Rat embryonic cortical tissue for preparing primary neurons.

5. suppDMEM: Dulbecco's modified Eagle's medium (DMEM) supplemented with 10 % fetal bovine serum, 1× Nonessential Amino Acids, 2 mM GlutaMax, 100 U penicillin and 0.1 mg streptomycin per milliliter. Store at 4 °C. Warm to at least room temperature before use.

6. Serum-NB: Neurobasal Medium (NB) containing 1× B27, 2 mM Glutamax, and 5 % FBS (medium for neuron plating) or 2 % FBS (medium for MEA cultures).

7. Plain-NB: Neurobasal Medium (NB) containing 1× B27, 2 mM Glutamax without any serum.

8. All-*trans* retinal: Prepare, in a dark environment, a 10 mM stock by dissolving 25 mg in 8.8 ml ETOH, aliquot, and store at –80 °C. As needed, prepare a 1:10 dilution in PBS to obtain a working stock of 1 mM, which can be kept at 4 °C and protected from light for a few days. Add to the culture medium directly or after further dilution for a final concentration of 1 μM.

2.2 Coelenterazine

Coelenterazine (CTZ), native as well as analogs (hCTZ, eCTZ, etc.), can be commercially obtained from several sources (Note: they differ in purity).

!*CTZ should always be protected from light*!

2.3 Solvents for Coelenterazine

1. Acidified alcohol (ethanol or methanol): 0.06 N HCl in alcohol; add 0.2 ml 3 N HCl to 10 ml alcohol.

2. Cyclodextrin: CTZ can be solubilized in (2-Hydroxypropyl) β-cyclodextrin, as originally described by Teranishi and Shimomura [11], and used by Naumann et al. in live zebrafish [12]. The amount of cyclodextrin used depends on the amount of CTZ being complexed and can be determined from ref. 11, Fig. 3. After dissolving cyclodextrin in PBS, filter-sterilize and add to CTZ dissolved in a small volume of ethanol. We typically dissolve 250 μg CTZ in 10 μl of ethanol and dilute it with 500 μl of 20 mM β-cyclodextrin to reach a stock concentration of 600 μM CTZ. Stocks are then kept frozen and protected from light until they are ready for use. On the day of the experiment, a stock aliquot is thawed and diluted with aqueous solvent (PBS, saline) to the desired concentration.

3. NanoFuel Solvent (NanoLight Technology): a proprietary solvent for coelenterazine. Native coelenterazine can be dissolved in a ten times higher concentration compared to 100 % ethanol or methanol. In our experience, NanoFuel-dissolved CTZ emits more bioluminescence than ethanol-dissolved CTZ when applied at the same final concentration. We dissolve CTZ at 50 mM, a concentration higher than the company's recommendation, to minimize the amount of solvent used and avoid

possible unwanted side effects. Native CTZ is dissolved in the solvent (23.6 µl solvent for 500 µg CTZ, resulting in a 50 mM solution) at room temperature. Pipette up and down and vortex to insure complete dissolving. The stock solution is stored in the same tube with a tightly closed lid at –80 °C for continued use up to several months. In the morning of the day of experiments, we make small aliquots of concentrated CTZ (typically 1 µl) and keep them on ice. Just before application, we dilute it with a buffer of choice (HEPES-buffered saline or PBS) to a specified final concentration (typically 1:500 for a 100 µM solution). Unlike the stock solution, CTZ diluted in a buffer can only last several hours at room temperature. Loss of typical yellowish color in a buffer indicates auto-oxidation and loss of activity. The solvent is also compatible with many other coelenterazine analogs that have become available in recent years.

4. Inject-A-Lume (NanoLight Technology): injectable CTZ specifically designed for in vivo use. Coelenterazine comes in sterile injection vials with a low retention volume and with sterile Fuel-Inject diluent. Fuel-Inject allows one to dissolve CTZ at high concentrations without precipitating CTZ. Inject-a-Lume is available for native CTZ, hCTZ, and eCTZ. Inject-A-Lume is typically stored in the freezer and warmed up prior to use.

5. Water-soluble coelenterazine (NanoLight Technology): CTZ (native or h) has nontoxic additives for water-solubility and is formulated to be isosmotic and easily secreted by the kidneys. It is safe for repeated intravenous injections and at high concentrations (up to 500 µg/100 µl) achieving very high coelenterazine levels within the body and leading to high bioluminescence signals [13].

3 Methods

3.1 Cell Culture

While the goal is to use tools for targeted modification of neuronal activity in the intact living brain of a behaving animal, there are several applications in vitro, specifically in cultured cells. These applications range from simply testing constructs or validating viral vector preparations for expression of bioluminescence to studying neuronal networks in long-term cultures to investigating network connections in brain slices. HEK cells provide a convenient heterologous expression system to test and verify new luminopsin constructs, while primary neurons are usually needed to test viral preparations with neuron-specific promoters.

3.1.1 HEK Cells

1. Grow HEK cells in medium of choice such as suppDMEM at 37 °C and 5 % CO_2 in a humidified atmosphere.

2. Trypsinize cells, count, and adjust to 5×10^4 cells per 0.5 ml culture medium. Using sterile forceps, place uncoated or coated (poly-D-lysine) glass coverslips into as many 24-well size wells as needed. Seed 0.5 ml per well cells onto coverslips. For plate reader assays, seed appropriate number of cells onto a 96-well plate ($1–5 \times 10^4$ cells per 0.1 ml per well) (also: *see* **Note 1**).

3. The next day, transfect each well with a luminopsin construct (*see* **Notes 2** and **3**). It is convenient to set up transfections for 5 wells per construct, and transfect 4 wells per 4-well dish. For best results, change medium 4 h after transfection.

4. (Optional) One or two days after transfection, the day before electrophysiology experiments, add all-*trans* retinal to the cultures. Opsins require retinal as a chromophore. This step is not necessary when HEK cells are cultured in serum-supplemented medium, which contains trans-retinal.

5. At the day of electrophysiology experiments (2–3 days after transfection of DNA) transfer coverslips from tissue culture wells to a recording chamber.

3.1.2 Primary Neurons

1. Isolate primary cortical or hippocampal neurons (*see* **Note 4**) and determine cell number. Adjust cells to $1–2.5 \times 10^5$ per 0.5 ml in Serum-NB. Using sterile forceps, place poly-D-lysine-coated glass coverslips (*see* **Note 5**) into as many wells as needed (24-well for 12-mm coverslips; 12-well for 18-mm coverslips). Seed 0.5 ml per well cells onto coverslips. Together with the dishes of seeded neurons place a culture tube in the incubator with its lid slightly loosened, containing Plain-NB medium sufficient to add 1 ml per seeded well (pre-equilibrated Plain-NB).

2. Neurons can be transfected with LMOs using one of the following methods:

    Lipofection: low efficiency; highest toxicity; ideal when testing constructs in single cells by patch-clamp recording.

    Electroporation: moderate efficiency; moderate toxicity; requires special equipment (e.g. Lonza Nucleofector) and a higher number of cells. Typically done before seeding.

    Viral transduction: highest efficiency; least toxic; ideal when examining activity of many neurons at a time using a multielectrode array (MEA).

    For lipofection, collect 0.5 ml of conditioned culture medium from each well 1 or 2 days after medium change-out in **step 3** below (DIV2 or DIV3) and save in a culture tube (we usually collect 2 ml per 4-well dish and save in a separate 15 ml tube for each 4-well dish). Keep tubes in incubator with lid slightly loosened to allow air exchange. Transfect cells in the remaining 0.5 ml medium per well with a luminopsin construct. For transfection of neurons to be used for patch clamping, we use Lipofectamine 2000 according to the manufacturer's recommendations, except that less of the recommended amount of Lipofectamine and DNA per well is used (per well of a 4-well or 24-well dish: 0.2 µl Lipofectamine, a tenth of the recommended amount; 200 ng plasmid DNA, a quarter of the recommended amount). Again, it is convenient to set up transfections for 5 wells per construct (1 µg DNA/250 µl OptiMEM; 1 µl Lipofectamine/250 µl OptiMEM), and transfect 4 wells per 4-well dish. Change out medium 4 h after transfection with the prewarmed and pre-equilibrated saved medium from above (*see* **Note 6**).

For electroporation, we follow the manufacturer's recommendations with slight modifications. In brief, spin down up to six million cells per reaction in a 1.5 ml tube. Resuspend the cells in 100 µl of the proprietary solution provided or in DPBS with calcium, magnesium, glucose, and pyruvate. Add and mix with 2 µg of LMO DNA. Transfer the cell suspension into a cuvette. After electroporation, using the recommended parameters for primary neurons, add 500 µl prewarmed, CO_2-equilibrated low-calcium recovery medium (i.e. RPMI) and transfer the cell suspension into the original tube. Incubate the tube in a CO_2 tissue-culture incubator for 5 min before seeding onto coverslips in a well filled with prewarmed, CO_2-equilibrated Serum-NB at a density higher than that used for lipofection.

For viral transduction, add high titer virus (>10^8 infectious units/ml lentivirus, >10^{11} vg/ml AAV) to the Plain-NB medium used in **step 1**. Use viral preparations with a multiplicity of infection (MOI)>10 to ensure near 100 % transduction efficiency (Fig. 2). Depending on the promoter, expression can typically be visualized by fluorescence microscopy 3–4 days after infection with lentivirus, and 7–10 days after infection with AAV.

3. Regardless of the method of transfection, the next day after plating (day in vitro 1, DIV1), remove Serum-NB medium completely from neurons and replace with 1 ml per well prewarmed and pre-equilibrated Plain-NB medium. Carry out the medium exchange well by well, so as to not leave neurons without medium for more than a few seconds.

4. Neurons are good for electrophysiological recordings between DIV10 and DIV14. Add ¼ volume of prewarmed and pre-equilibrated Plain-NB every 3–4 days, if needed. Make sure cultures do not evaporate by placing dishes in a humidified incubator or humidity chamber (*see* **Note 7**).

5. (Optional) The day before electrophysiological experiments, add all-*trans* retinal to the cultures for a final concentration of 1 µM. This step is not necessary if Plain-NB is supplemented with vitamin A-containing B27.

6. On the day of electrophysiological experiments transfer coverslips from tissue culture wells to the recording chamber. Coverslips should be immediately perfused with recording buffer to minimize amount of time coverslips are exposed to air.

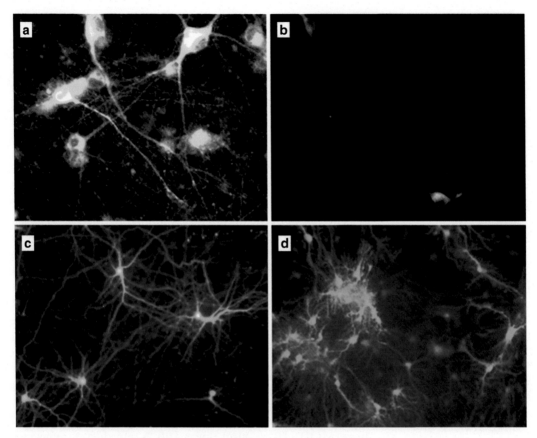

Fig. 2 Viral transduction of cultured neurons. Primary rat embryonic day 18 cortical neurons were plated on poly-ᴅ-lysine-coated coverslips at equal densities. The next day (DIV1), neurons were transduced with viral vectors. (**a, b**) The same transgene (CAG-Gluc-VChR1-EYFP) expressed from different vectors (a: AAV, b: lentivirus). (c,d) Lentivirus expressing GLuc-Mac-GFP under control of different promoters (**c**: human synapsin; **d**: CAG), with hSyn favoring expression in neurons versus glia

3.2 Bioluminescence Imaging In Vitro

For a simple bioluminescence assessment (i.e. bioluminescence intensity and emission spectrum), luminometers provide the most sensitive and quantitative measurements. They are specifically designed to measure photon emission upon time-controlled substrate injection, usually in a plate format (96-well plate in the basic models, various plate formats in the higher end models). Alternatively, we have utilized bioluminescence imagers designed for in vivo imaging (see below) and obtained satisfactory results.

If none of the above is available, a good option is imaging systems designed for chemiluminescence, such as image documenting systems for Western blots equipped with a cooled CCD camera (e.g. Fuji Film LAS-3000, Li-Cor Odyssey Fc). For high-emission luciferases, even common gel documentation systems can be used (*see* Fig. 3).

Simultaneous imaging with electrophysiological recordings is addressed in each individual section below.

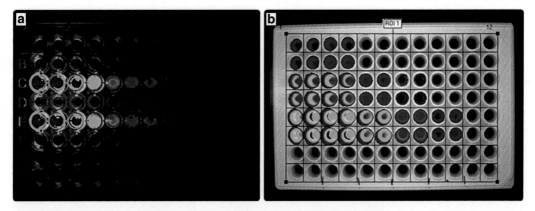

Fig. 3 Bioluminescense imaging in 96-well plate format. HEK cells (50,000 per well) transfected with LMOs were plated. Forty-eight hours later bioluminescence intensities were measured after adding different concentrations of CTZ (1–100 μM final concentrations). Images were taken with a LiCor Odyssee Fc using the chemiluminescence channel (**a**) and with a Caliper IVIS Kinetic system for in vivo imaging (**b**). Region of Interest (ROI) can be selected and bioluminescence intensity can be quantified

3.3 Intracellular Recording In Vitro

The effects of LMOs on neuronal firing can be assessed in primary neurons in culture. Concurrent imaging of bioluminescence helps to establish a causal relationship between biological light and activation of LMOs.

1. Cells are imaged and recorded using an upright or inverted epifluorescence microscope equipped with a 40× 0.8 NA water immersion objective or a 60× 1.35 NA oil immersion objective, respectively, an arc lamp, an electronic shutter, a GFP filter cube, and a cooled CCD camera or sCMOS camera with acquisition software. The recording chamber is constantly superfused with the extracellular solution at ~500 μl/min. Experiments are carried out at room temperature.

2. Locate an LMO-positive cell by its fluorescent-tag expression.

3. Obtain whole-cell recording under voltage or current clamp mode.

4. We routinely measure photocurrent to conventional wide-field photostimulation using the lamp and an appropriate filter cube (e.g. a GFP filter cube for ChR2). The measured photocurrents reflect the total functional surface expression level of LMO of the cell.

5. CTZ-induced photocurrents can be measured under voltage clamp. Effects of CTZ on spiking can be recorded under current clamp. The effect of inhibitory LMOs should be assessed under constant action potential firing, which can be evoked by a train of brief peri-threshold current injection (note: large current injection can mask subtle effects of CTZ). Immediately before application, reconstitute CTZ in the extracellular solution at 100 μM and bath-apply CTZ solution (~0.5 ml) to

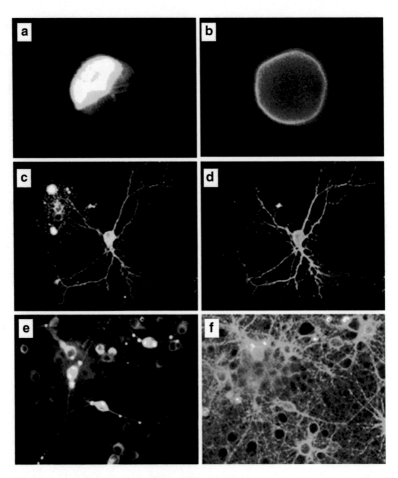

Fig. 4 Luminescence in cell culture. HEK cells (**a, b**), individual neurons (**c, d**), and neuronal cultures (**e, f**) transfected (**a–d**) or transduced (**e, f**) with LMOs were imaged under fluorescent light (**a, c, e**) and after adding coelenterazine to the culture (**b, d, f**)

a recording chamber (*see* **Note 8** for different options) (Fig. 4). Bioluminescence is imaged concurrently throughout the recording through the camera (*see* **Note 9** for specific settings). To synchronize the two recordings, an exposure signal from the camera can be used to trigger electrophysiological recording. To maximize bioluminescence collection, move the filter turret to an open position before CTZ application. For bioluminescence imaging the microscope needs to be in complete darkness (*see* **Note 10**).

6. Measure photocurrent in response to direct photostimulation by lamp after CTZ application to confirm recording conditions have not changed.

3.4 Extracellular Recording In Vitro

Recordings from multiple neurons at a time using a multielectrode array (MEA) provides not only a convenient and high-throughput method [14], but also enables analysis on change in network dynamics after activation of LMOs.

Fig. 5 MEA culture expressing iLMO1. Rat embryonic cortical neurons were plated on electrodes in a MEA chamber (Multichannel Systems) and transduced on DIV 2 with lentivirus containing Ubiquitin-NpHR-TagRFP-Rluc (iLMO1). The image was taken on DIV14

1. Prepare MEAs according to the manufacturer's recommendations (sterilize, coat) (*see* **Note 11**).

2. Seed MEAs with cortical neurons (for example: 5×10^4/array in 1-well or 1.25×10^4/array in 6-well arrays from MultiChannel Systems).

3. The next day (DIV1) transduce cultures with AAV or lentivirus with an MOI >10 to achieve close to 100 % transduction (Fig. 5).

4. Change half-volume of media every 3–4 days.

5. Spontaneous synchronous bursting activity can be recorded around DIV 14. Insert an MEA into the appropriate headstage and connect to an amplifier. Allow the MEA to equilibrate to achieve a stable baseline of activity.

6. Optical stimulation can be conducted with an external LED or lamp since the MEAs are optically clear.

7. To measure effects of CTZ, add CTZ (5 μl per 200 μl culture volume) to reach a final concentration of 10–50 μM. Recording can continue for up to several hours if cultures are kept in a humidified incubator. For simultaneous imaging of bioluminescence, use 1-well MEAs and place the headstage on an inverted microscope stage. Image bioluminescence as described above (Subheading 3.3, **step 5**).

3.5 Intracellular Recording Ex Vivo

Brain slice preparations retain innate synaptic connection to some degree. Recordings from LMO-expressing brain slices enable precise control of a given network using conventional optogenetic photostimulation [15] as well as activation of the entire network by the chemical, CTZ, in a more physiologically relevant setting than in cultured cells.

1. Inject AAV carrying the desired LMO or iLMO gene into the brain region of interest in mice or rats. Wait at least 10 days after injection to allow adequate expression before carrying out experiments.

2. Prepare acute brain slices (~300 μm thick sections) using a vibratome.

3. Whole-cell patch clamp recordings are made similar to in vitro recordings using conventional methods. Recordings can be made from cells expressing LMOs or from cells receiving synaptic inputs from LMO-expressing neurons.

4. LMOs can be activated by photostimulation with an arc lamp, with laser spots for detailed photo-mapping of synaptic inputs, or by adding CTZ to the recording chamber (see above: Subheading 3.3).

5. For simultaneous imaging of bioluminescence, a camera can be used as described above (Subheading 3.3, **step 5**). Alternatively, if working with a scanning microscope (i.e. confocal or two-photon microscope), a photodetector (i.e. a photomultiplier tube) can be used to record bioluminescence continuously. Switching to a lower magnification lens will help to collect more photons.

3.6 Animal Studies

For studying neuronal activity and/or behavior in awake animals expressing LMOs, viral vectors (generally AAV) are placed into the brain region of interest by stereotaxic injections as done for other optogenetic and chemogenetic approaches (Fig. 6). The sections below focus on specific usage of LMOs.

3.7 Administration of CTZ In Vivo

For all in vivo experiments, CTZ preparations specifically designed for in vivo applications should be used (i.e. Nanolight Inject-A-Lume or water soluble CTZ). The choice of routes of delivery of CTZ in rodents depends on the goal of the experiment. Concentrations of CTZ should be aimed at final concentrations in the animal of 100–200 μM (*see* **Note 12**).

1. Intracranial injection: Concentrated CTZ in a minute volume can be delivered to a target brain region directly using an injection cannula inserted in the guide cannula (example: 34 ng CTZ in 0.4 μl PBS). Although this causes very rapid action only near the injection site, it may cause irreversible damage to the injection site if injection rate is too fast or large volumes are injected.

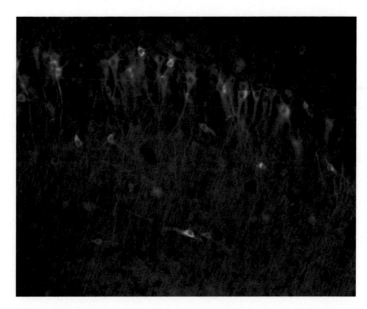

Fig. 6 Hippocampal pyramidal cells expressing iLMO2. The dorsal hippocampus in a rat was injected with AAV expressing CamKIIa-NpHR-Nano-Lantern (Venus-Rluc), i.e., iLMO2. The rat was perfused and the brain sectioned 3 weeks after viral injection. Robust expression can be seen in pyramidal cells

2. Intravenous injection: Application of substrate intravenously results in fast onset of effects (rapid rise in spike rate within several seconds of CTZ entering the bloodstream), a peak typically between 20 and 30 s after injection, and a slower decay (several minutes). Volumes for injections should be small (50–75 μl for mice; 200–250 μl for rats).

   (a) Retro-orbital injections in mice are an easy and reliable way of intravenous delivery, but usually require brief anesthesia.

   (b) Tail vein injections in mice or rats can be done without anesthesia. The animals should be warmed up by a heat lamp to improve circulation. There are single-use catheters available for tail vein injections (*see* **Note 13**).

   (c) Jugular catheters in mice or rats offer an easy access port for repeated administration of precisely timed injections with defined concentrations (e.g. obtaining a dose-response curve). Note: catheters must be kept patent with bi-weekly flushing of heparin and gentamicin (*see* **Note 14**).

3. Intraperitoneal injection: Application of substrate intraperitoneally results in a slower onset (several minutes) of measurable effects, which subside after a longer period of time (~30 up to 60 min) compared to intravenous injections. This is the route of choice for CTZ administrations repeatedly over long periods of time, for example to test the effects of chronic stimulation or

silencing of populations over several months on an animal's behavior. Example: 200 µg CTZ in 200 µl PBS intraperitoneally into a 25–30 g mouse.

3.8 Bioluminescence Imaging In Vivo

In vivo bioluminescence imaging can be conducted as a means for noninvasive confirmation of LMO expression and delivery of CTZ to a target. Imaging is conducted ideally with a dedicated small animal imager (e.g. IVIS imaging system). Depending on the species and area of expression in the brain, imaging can be conducted through the intact skull (Fig. 7).

If such an imaging system is not available, other systems equipped with a cooled CCD camera or a sCMOS camera can be utilized, although with lower sensitivity. We have had success utilizing a gel image documenting system equipped with a cooled CCD camera (i.e. Fuji Film LAS-3000). In this pre-terminal setting, the skull of the animal was removed for bioluminescence imaging from the rat striatum (*see* **Note 15**).

3.9 Extracellular Recording In Vivo

Since luminopsins modulate neuronal activity, their activity can be measured by any conventional means of in vivo extracellular recording in both acute and chronic settings. Implantable electrodes obtained commercially or custom-made electrodes/optrodes (e.g. glass pulled electrodes, cannula-electrodes, or micro-drivable tetrodes such as FlexDrive, http://www.open-ephys.org/flexdrive/; [16]) are especially useful for recording both single-unit and local field potentials of LMO-expressing cells. EEG electrodes can also be utilized depending on the extent of expression, but these approaches would be unable to detect changes at single-unit firing rates.

Luminopsins offer the ability for multiple modes of activation: by direct light stimulation and by chemical substrate. Either mode can be used as a control for the other because the same population of neurons is targeted with the same single molecule. This can be

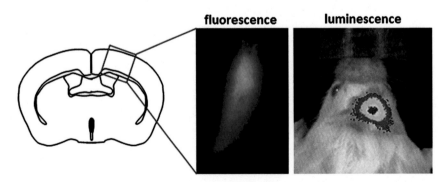

Fig. 7 Bioluminescence imaging in vivo. Lentivirus carrying synapsin-LMO1 (Gluc-ChR2-EYFP) was injected into the right premotor cortex of a mouse. A month later mice were imaged in vivo after intravenous injection of CTZ (luminescence), then perfused and brain-sectioned for fluorescence imaging (fluorescence)

achieved using a cannula-electrode, whereby photostimulation (through a fiber optic inserted through the cannula) and chemical stimulation (through injection of CTZ through the same cannula) can be restricted to only the neurons being recorded from. Cannula-electrodes can be fabricated by attaching guide cannulas to commercially available electrodes using superglue. These can be assembled before the actual implantation surgery to customize distance of the fiber tip/injection cannula from the recording electrodes (Fig. 8).

The general procedure for recording responses of LMO-expressing cells to chemical and photostimulation is as follows:

1. Inject AAV into the rodent brains as described above. The amount of time for adequate expression will be specific to the construct/vector and will need to be empirically determined. We typically wait at least 10 days for expression.

2. Commercially bought or custom-made electrodes/optrodes should be configured so that the intended target is covered with the appropriate number of recording contacts and is able to be illuminated by a fiber optic (if included).

3. Stereotaxically implant the electrodes under anesthesia. Correct placement of the electrodes can be determined in real time by optically stimulating while advancing the electrode towards the intended target depth. The cell types being recorded can be determined by analyzing single-unit firing properties. After correct placement of the electrodes, a chronic implant can be created by securing everything to the skull via

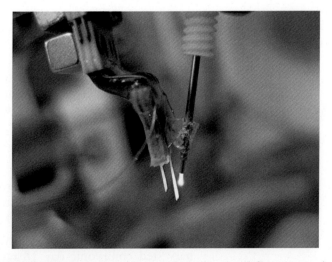

Fig. 8 Cannula-electrode. A 16-channel electrode array (*left*) was superglued to a guide cannula (*right*). An optical fiber was inserted through the cannula to illuminate cells while recording from them. Subsequent CTZ injections were also given through the same guide cannula

skull screws and dental acrylic. A chronic implant will allow for multiple injections/stimulations within the same animal: we were able to record up to 3 months after viral injection.

4. Recordings from implanted electrodes can be carried out in freely moving animals before and after administration of CTZ as described in Subheading 3.7 above. The effect of CTZ can be determined by quantifying changes in the single-unit activity and low frequency power in the local field potential.

5. Acute recordings can be conducted in head-fixed awake or lightly anesthetized animals. Wait at least 10–15 min after advancing an electrode before collecting data (*see* **Note 16**). Then establish a baseline before CTZ injection by recording at least 10–20 min without moving the probe. This is usually sufficient to confirm a stable recording and to capture characteristics of neurons at rest (but: *see* **Note 17**). Using an optrode (simultaneous recording and fiber optic light stimulation) allows stimulation before and after CTZ administration: stimulation before is used to confirm electrode location, and after to see if there are any remaining effects on driven activity as well as to confirm that the same units were recorded for the duration of the trial.

3.10 Behavioral Testing

Any behavioral test can be carried out in freely moving animals before and after administration of CTZ, acutely or to assess changes in behavior after chronic stimulation with bioluminescence. Combining behavioral assays with electrophysiological recording and/or bioluminescence imaging in a freely moving animal will offer a very unique application of LMOs.

4 Notes

1. Rather than transfect cells in 96-well plates, we usually transfect cells in 6-well or 6 cm plates. The next day, transfected cells are trypsinized, and seeded into 96-well plates.

2. Unlike neurons, the choice of a transfection agent is not so critical for HEK cells. We have used Effectene (Qiagen) or Lipofectamine 2000 (Invitrogen) with equal results.

3. Quality of DNA is critical for transfection, especially in neurons. Use DNA prepared from endotoxin-free maxi prep kits (e.g. Qiagen EndoFree Plasmid Maxi Kit) instead of mini prep kits.

4. Primary neurons of high quality can be isolated from freshly harvested embryonic day 18 rat pups, or from tissue pieces shipped the day before by BrainBits, Inc.

5. Prepare coated coverslips according to standard protocols. Pre-coated coverslips are also commercially available through NeuVitro and BD Biosciences.

6. For lipofection of neurons replacing the transfection medium with the "preconditioned" medium within 4 h (can be shortened to 2 h) is critical for obtaining healthy neurons for patch clamp recordings.

7. A humidity chamber can be a plastic Tupperware box or pyrex glass dish containing a glass plate kept just above an inch of water. The glass plate holds the culture dishes. Another glass plate put on top of the container without completely closing it serves as lid.

8. For bath-application of CTZ, we do not recommend manual pipetting, which will likely disrupt electrophysiological recordings. A better way is to fill a line of Tygon or silicone tubing of small diameter with the CTZ solution, connect a syringe at the other end, and deliver it to the recording chamber through a micro-manifold with flow control (AutoMate Scientific) by manually pushing the plunger of the syringe. We had limited success by localized application with a Picospritzer: it seems CTZ oxidizes inside the glass pipette before the recording starts.

9. Settings for bioluminescence imaging: (a) cooled CCD camera (for example, CoolSNAP-fx; Photometrics): without any filter cube with 4 by 4 binning. (b) scientific CMOS camera (for example, OptiMOS; QImaging) without binning. For *Gaussia* luciferase, exposure time is 1–5 s, for *Renilla* luciferase, 10–20 s.

10. Bioluminescence imaging should be conducted in complete darkness. This includes taping over all lights from instrument panels, and covering the microscope with light-impenetrable material (plastic, heavy cloth, felt) or keeping the entire room in the dark.

11. There are three main suppliers of MEAs: MultiChannel Systems (distributed through ALA Scientific), Axion Biosystems, and MED64 (distributed through AutoMate Scientific). We are using the MultiChannel MEAs, specifically the 6-chamber MEA with 9 electrodes per unit, and the 1-chamber MEA with 64 electrodes per unit for concurrent recording and bioluminescence imaging.

12. Example calculations based on an estimated blood volume for a 30 g mouse of 2.5 ml (molecular weight of native CTZ: 423.46): (1) Injections at 4 mg/kg of CTZ result in 120 µg for a 30 g mouse or a final concentration of ~112 µM CTZ in the bloodstream. (2) Intravenous injection of 50 µl of a 6 mM solution of CTZ results in an estimated final concentration in the bloodstream of ~120 µM. III. Using Nanolight's instructions for Inject-A-Lume, dissolve one vial CTZ (500 µg) in 150 µl NanoFuel (3.33 mg/ml); injection of 60 µl (200 µg) in a 25–30 g mouse should result in ~200 µM final concentration. Given the small volumes of liquid it is recommended to

use a syringe with as little dead space as possible (Insulin syringes in either 0.3 or 0.5 ml sizes).

13. Instech Laboratories has a selection of catheters, connectors, and vascular access ports for rats and mice (http://www.instechlabs.com/).

14. We typically push a mixture of 50 µl of heparin (30 U/ml) and 50 µl of Gentamicin (4 mg/ml) in sterile saline twice a week.

15. Whole animal imaging must be conducted with the animal under anesthesia, which may make it difficult to access peripheral veins for intravenous CTZ delivery due to vasoconstrictive effects of inhaled anesthetics like isoflurane.

16. The typical amount of wait time prior to starting baseline recording after electrode implantation takes into consideration deformation of tissue and transient depression in response to the mechanical stress. We found it important to be especially aware of this when using an optrode (custom made or commercially obtained). Even small-core fiber optics are large when compared to extracellular electrode tips, including multicontact silicon probes. As a result, the tissue movement and recovery time can be greater. Advance slowly after penetrating the cortical surface, and while looking for responsive cells. After finding light-activated units at the tip of the probe (if using a multielectrode array), wait 5–10 min to see if they are stable, and whether they drift to another contact. Overall, regardless of the recording site, let the electrode sit 20–30 min before starting baseline recordings when using a typical optrode.

17. Longer times for establishing a baseline may be required when recording in a structure with very high variability or if the intent is to capture the effect of luminopsins on neural or behavioral state transitions, as a certain number of events would be needed.

Acknowledgement

This work was in part supported by grants from NIH (NS079268, R.G.; NS086433, J.T.; MH101525, U.H.), NSF (CBET1464686, U.H.), Duke Institute for Brain Sciences (U.H.), The Michael J. Fox Foundation (C.M.), and The Brain Research Foundation (C.M.).

References

1. Sternson SM, Roth BL (2014) Chemogenetic tools to interrogate brain functions. Annu Rev Neurosci 37:387–407

2. Fenno L, Yizhar O, Deisseroth K (2011) The development and application of optogenetics. Annu Rev Neurosci 34:389–412

3. Berglund K, Birkner E, Augustine GJ, Hochgeschwender U (2013) Light-emitting channelrhodopsins for combined optogenetic and chemical-genetic control of neurons. PLoS One 8:e59759. doi:10.1371/journal.pone.0059759

4. Verhaegen M, Christopoulos TK (2002) Recombinant Gaussia luciferase. Overexpression, purification, and analytical application of a bioluminescent reporter for DNA hybridization. Anal Chem 74:4378–4385

5. Tannous BA, Kim D-E, Fernandez JL et al (2005) Codon-optimized Gaussia luciferase cDNA for mammalian gene expression in culture and in vivo. Mol Ther 11:435–443

6. Nagel G, Szellas T, Huhn W et al (2003) Channelrhodopsin-2, a directly light-gated cation-selective membrane channel. Proc Natl Acad Sci U S A 100:13940–13945

7. Zhang F, Prigge M, Beyrière F et al (2008) Red-shifted optogenetic excitation: a tool for fast neural control derived from Volvox carteri. Nat Neurosci 11:631–633

8. Saito K, Chang Y-F, Horikawa K et al (2012) Luminescent proteins for high-speed single-cell and whole-body imaging. Nat Commun 3:1262. doi:10.1038/ncomms2248

9. Schobert B, Lanyi JK (1982) Halorhodopsin is a light-driven chloride pump. J Biol Chem 257:10306–10313

10. Zhu H, Roth BL (2014) DREADD: a chemogenetic GPCR signaling platform. Int J Neuropsychopharmacol 18(1):pii: pyu007. doi:10.1093/ijnp/pyu007

11. Teranishi K, Shimomura O (1997) Solubilizing coelenterazine in water with hydroxypropyl-BETA-cyclodextrin. Biosci Biotechnol Biochem 61:1219–1220

12. Naumann EA, Kampff AR, Prober DA et al (2010) Monitoring neural activity with bioluminescence during natural behavior. Nat Neurosci 13:513–520

13. Morse D, Tannous BA (2012) A water-soluble coelenterazine for sensitive in vivo imaging of coelenterate luciferases. Mol Ther 20:692–693

14. Hales CM, Rolston JD, Potter SM (2010) How to culture, record and stimulate neuronal networks on micro-electrode arrays (MEAs). J Vis Exp. doi:10.3791/2056

15. Wang H, Peca J, Matsuzaki M et al (2007) High-speed mapping of synaptic connectivity using photostimulation in Channelrhodopsin-2 transgenic mice. Proc Natl Acad Sci U S A 104:8143–8148

16. Voigts J, Siegle JH, Pritchett DL, Moore CI (2013) The flexDrive: an ultra-light implant for optical control and highly parallel chronic recording of neuronal ensembles in freely moving mice. Front Syst Neurosci 7:8. doi:10.3389/fnsys.2013.00008

Chapter 15

Intracranial Injection of an Optogenetics Viral Vector Followed by Optical Cannula Implantation for Neural Stimulation in Rat Brain Cortex

Christopher Pawela, Edgar DeYoe, and Ramin Pashaie

Abstract

Optogenetics is rapidly gaining acceptance as a preferred method to study specific neuronal cell types using light. Optogenetic neuromodulation requires the introduction of a cell-specific viral vector encoding for a light activating ion channel or ion pump and the utilization of a system to deliver light stimulation to brain. Here, we describe a two-part methodology starting with a procedure to inject an optogenetic AAV virus into rat cortex followed by a second procedure to surgically implant an optical cannula for light delivery to the deeper cortical layers.

Key words Optogenetics, Neuromodulation, Optical cannula, Rat brain, Intracranial injection, Neural stimulation

1 Introduction

Development of brain stimulation techniques in which penetrating electrodes deliver current pulses to stimulate neural circuitries in deep brain objects introduced a new era in treating neurological diseases. Previously, the main approach in treating mental disorders has been the chemical imbalance paradigm in which we hypothesize that a mental disorder is the result of an imbalance in the concentration of chemicals, such as neurotransmitters, within the central or peripheral nervous system. Such diseases can be treated by controlling the concentration of the appropriate chemicals through pharmacologic manipulation. In contrast, interventional psychiatry is based on the idea that brain disease can be treated by directly modulating neuronal activity. One example of this new approach is the implantation of electrodes for deep brain stimulation (DBS) used to treat Parkinson's disease. Although DBS can be quite successful, electrode-based stimulation has at least three inherent deficiencies. One is the lack of practical strategies to

Arash Kianianmomeni (ed.), *Optogenetics: Methods and Protocols*, Methods in Molecular Biology, vol. 1408, DOI 10.1007/978-1-4939-3512-3_15, © Springer Science+Business Media New York 2016

target specific cell-types of interest. Typically, different cell-types are intermixed within most brain regions with each cell-type being involved in different microcircuits or data processing pathways. Injection of electrical current generally stimulates all cells in the vicinity of the electrode with little or no specificity thus potentially causing unintended side effects. Another limitation is that electrical stimulation typically causes an increase in neuronal firing while suppressive effects are minimal or difficult to control precisely. Moreover, due to these limitations the development of high density, arrays that can provide complex yet focally specific patterned stimulation/recording is problematic. Electrode arrays that are made for neuroprosthetic applications usually contain no more than few hundred electrodes at best and even for arrays with limited number of electrodes the data acquisition process is quite cumbersome.

Optogenetics is a new neuro-stimulation modality with the potential to remarkably improve upon previous technologies [1–3]. With optogenetics, specific cell-types of interest are genetically targeted to express light-sensitive ion channels or ion pumps [4, 5]. Once these proteins are produced within a cell, its activity can be increased or suppressed by exposing the cell to light of appropriate wavelengths. Cell-specific genetic targeting is achieved by controlling the gene delivery mechanism through choice of appropriate promoters and/or viral receptors.

Only the targeted cells are transfected efficiently and respond to light pulses. During the last few years, the list of light-sensitive proteins that are used for optogenetic stimulation has significantly expanded. Currently, there are proteins that function as cation channels to stimulate excitable cells or as anion pumps to hyperpolarize cells thus causing neural inhibition once exposed to light. By co-expressing different optogenetic proteins, it is possible to reversibly excite or inhibit cellular activity by simply changing the wavelength of the stimulation light [5]. Optogenetic stimulation also benefits from the inherent parallelism of optics. Using state-of-the-art technology, it is possible to generate complex patterns of stimulation with high spatial and temporal resolution making it feasible to manipulate the dynamics of extended cortical networks. As a result, optogenetics provides new opportunities for neuroscience research and has revolutionized the development of optoelectronic brain interface technologies [6].

Optogenetic tools are proteins that function as light-sensitive ion channels or ion pumps. These proteins, which are mostly categorized as microbial rhodopsin molecules, were originally isolated from microorganisms but are being used to manipulate the activity of mammalian cells such as neurons or muscle cells [5]. The most widely used member of this family is the channelrhodopsin (ChR) molecule which was isolated from the fresh water algae

Chlamydomonas reinhardtii and functions as a cation channel that opens when the protein is exposed to blue light with maximum spectral sensitivity around 445 nm [3]. When the pore of this channel opens, it allows the passage of positive ions, including Na^+ and Ca^{2+}, across the membrane along the direction of the diffusion force induced by electrochemical gradients. The influx of these cations depolarizes the cell and increases cellular activity. Another member of this family is halorhodopsin (HR) protein which is isolated from *Archaebacterial* and functions as an anion pump. When exposed to yellow light, wavelengths around 590 nm, it pumps negative ions, mostly Cl^-, into the cell which causes hyperpolarization and reduces activity [4, 5].

Light sensitivity in almost all rhodopsin molecules is achieved by isomerization of a covalently bound retinal cofactor. The electron–photon interaction in this molecule causes a conformational change from all-*trans* to 13-*cis* which reconfigures the structure of the protein so that ions transport across the membrane (*see* Fig. 1). Since the required exogenous cofactor all-trans-retinal (ATR) already exists in mammalian cells, both halorhodopsin and channel rhodopsin can function in these species without the addition of ATR; a positive aspect that facilitates the design of optogenetic experiments and simplifies the process for potential translation of optogenetics to humans in future clinical use.

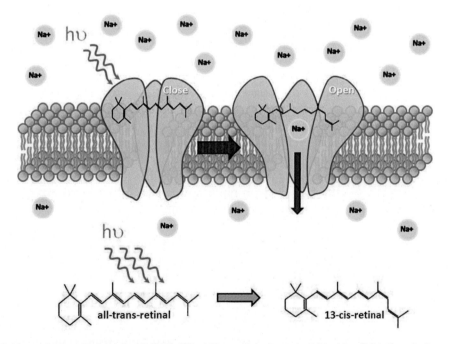

Fig. 1 Light sensitivity mechanism of ChR2: When the protein is exposed to blue light, the electron–photon interaction in this molecule causes a conformational change from all-*trans* to 13-*cis* which reconfigures the structure of the protein and opens the pore of the channel

Different strategies have been used to develop a wide range of opsins for optogenetic stimulation. In one approach, known as bio-prospecting, researchers investigated a variety of different species to discover other opsin molecules with different optical or kinetic properties [7]. For example, the protein VChR1 was isolated from *Volvox carteri* and was first discovered by searching the genome database from the US Department of Energy Joint Genome Institute [3]. The spectral response of this protein is approximately 50 nm red-shifted compared with ChR2 allowing it to stimulate cells deeper within the tissue. However, VChR1 photo currents are usually less than half of the photo current passing through the stimulated ChR2 channels. Another approach to optimizing optogenetic proteins is genetic engineering. For instance, initial tests with inhibitory microbial rhodopsins, such as NpHR, were derived from prokaryotes and were not sufficiently compatible with mammalian cells. Higher level expression of these ion pumps was required to achieve augmented inhibitory function but caused accumulation of the protein to toxic levels within the cell [8]. To address this problem, signaling peptides from endogenous ion channels were added to each end of the microbial rhodopsin, which optimized the efficiency of the membrane targeting of the NpHR molecule and improved the amplitude of the induced photocurrent.

Over the last decade, significant effort was focused on developing new optogenetic tools or optimizing protein kinetics, sensitivity, or spectral response. For instance, light-sensitive cation channels with distinct spectral sensitivities are required to target multiple cell populations intermingled within the same brain area. To reach deeper inside the brain and to minimize photodamage or phototoxicity, the spectral response of the proteins must be shifted toward red wavelengths. To make optogenetic stimulation more light efficient the kinetics of optogenetic proteins can be slowed down so that once the channel opens, it remains open for a longer period. This allows more cations to pass through each channel and reduces the number of open channels required to generate a requisite level of depolarization. This, in turn, reduces the number of effective electron-photon interactions required to generate an action potential. However, this increase in light efficiency then comes at the price of a considerable decrease in temporal resolution. If the generation of high frequency action potential bursts is required, a less efficient protein with fast kinetics may be the better choice. Ultimately, optogenetic protein structures can be optimized for a variety of different applications. It was also shown that green light exposure facilitates closing the channel and, as a result, these opsins can operate as bi-stable actuators with blue light triggering the ON state and green light triggering the OFF-state (closed channel) [1, 3]. Moreover, some point mutations can alter the selectivity of the ion channel increasing the conductance of cations such as Ca^{2+} for example. These proteins can then be used to control the intracellular concentration of specific ions.

Detailed information regarding the structure of ChR2 has been revealed recently through a sequence of protein crystallography experiments [9]. This information will help protein engineers to tailor their approaches to precisely design new variants of ChR2 with diverse spectral response, optical sensitivity, biological compatibility, ion selectivity, or kinetic properties [7].

An essential first step in optogenetic stimulation is the delivery of a new gene into a target cell [3, 5]. For future human therapeutic applications this is a potentially controversial issue that could limit adoption [10]. Based on the specific application, different techniques are adapted to deliver the gene to the cell populations of interest. In cultured cells, genes can be delivered by nonviral delivery methods such as calcium phosphate transfection or electroporation. However, viral gene delivery is still the most popular method for in-vivo application and for delivery of genetic constructs smaller than a few kb [11]. Viral gene delivery is often easy, robust, has high infectivity compared to other methods and, so far, has shown no significant iatrogenic effect. By controlling the process of viral injection, the spatial distribution of the expressed protein can be limited almost to one or a few brain sites. Cell-type-specific targeting can be obtained by using appropriate promoters and engineering of the genetic constructs. In recent years, lenti- and adeno-associated viral (AAV) vectors have been exploited to target both neurons and astroglia in rodents and nonhuman primates. The AAV vectors have become more popular for optogenetic gene delivery since they are less immunogenic and offer better transduction efficiency for targeting larger tissue volumes. For applications where more uniform distribution of opsins is required, as for manipulation of large cortical networks, transgenic animals are usually a better choice. Using transgenic animals also simplifies the experimental procedures and reduces the cost and effort needed for each experiment. Different opsin expressing optogenetic animals are produced among which the most famous example is the transgenic mice generated under Thy1 promoter [11]. This promoter is used to target projection neurons in layer V of neocortex.

In this methods manuscript, we will describe a two-part methodology starting with a procedure to inject an optogenetic AAV virus into rat cortex followed by a second procedure to surgically implant an optical cannula for light delivery to the deeper cortical layers.

2 Materials

2.1 Viral Vector

1. Vector: AAV [Serotype: 2 ChR2: AAV-hSyn-hChR2(H134R)-EYFP] was purchased from the University of North Carolina Vector Core (UNC Vector Core, Chapel Hill, North Carolina, USA) (http://www.med.unc.edu/genetherapy/vectorcore). We thaw the 100 µl frozen vector sample upon arrival at our facility and pipette the material into smaller 10 µl aliquots. The small aliquots are then stored in a −80 °C freezer for later use.

2.2 Stereotaxic and Injecting Equipment

1. Stereotaxic: Lab Standard Rat Stereotaxic Instrument with non-rupture ear-bars. We utilize this instrument to make precise distance measurements on the rat skull for injection site location. This device is equipped with a mask to deliver gas anesthesia to the subject rat.

2. Anesthesia Setup: Our laboratory is equipped with a gas anesthesia delivery system that includes a gas-mixing flow-meter, isoflurane Vaporizer, and hospital grade vacuum system to remove excess anesthetic. We use a heated water circulation blanket with a rectal temperature feedback system to maintain constant temperature in the subject rat during surgical procedures.

3. Injection System: We use Quintessential Stereotaxic Injector (QSI) to achieve accurate control of the volume and flow rate of injections.

4. Syringes: Two glass-made Nanofil syringes are needed, one for viral injections into experimental animals and another for vehicle injections into sham animals to prevent cross contamination.

5. Needles: Nanofil 35 gauge blunt needle or 33 gauge beveled needle is required for these experiments.

2.3 Surgical Tools and Micro Drills

1. Surgical Tools: The following tools are needed: scalpel handle, scalpel blades, two ultra wide curved hemostats, Halsey needle holder, small bone curette 0.5 mm, Dumont forceps, two micro spatulas, and a hot bead sterilizer.

2. Electrocautery: Veterinary Electrosurgical Unit.

3. Micro Drills: 0.3 mm carbide twist drill, and Preclinical Drill.

4. Suture: Standard sterile chromic gut suture for wound closure.

5. Drugs: Cefazolin (1 g per vial), Carprofen (50 mg/mL), Lidocaine Hydrochloride (2 % Jelly).

6. General Surgical Supplies: iodine scrub, rubbing alcohol, sterile gauze, sterile needles, sterile plastic syringes, and sterile cotton swabs.

2.4 Skull Fixation Supplies

1. Dental Cement: Orthojet BCA liquid.

2. Glue: Loctite 454.

3. Skull Screws: Nylon Mounting Screws $080 \times 3/32$ Diameter (0–80×3–32 N).

2.5 Optical Cannula

1. Optical Cannula: Fiber Optic Cannula, $\emptyset$1.25 mm Ceramic Ferrule, $\emptyset$200 μm Core, 0.39 NA, length = 2 mm.

3   Methods

All procedures approved by Medical College of Wisconsin Institutional Animal Care and Use Committee (IACUC) and adhere to the "The Guide for the Care and Use of Laboratory Animals." This protocol consists of two surgical procedures separated by 2–3 weeks.

3.1   Intracranial Injection

1. Safety precautions: we perform our surgeries in a Biosafety Level II facility. We use surgical caps, masks, eye protection, and gloves while working with the virus, and load the syringe with the virus inside of a safety cabinet. We use sterile surgical technique for all survival procedures involving animals.

2. Anesthetize the rat in an induction chamber with 4 % isoflurane. Secure the rat in a stereotaxic apparatus with a nose cone attachment and reduce isoflurane to 1.5–3 % as needed for maintenance of the anesthetic plane (Fig. 2).

3. Prepare the rat by shaving the fur off of the scalp and washing with iodine followed by rubbing alcohol. Repeat the washing process twice more.

4. Give preoperative injections of cefazolin (30 mg/kg) antibiotic subcutaneously and carprofen (5 mg/kg) analgesic intramuscularly, and apply lidocaine jelly subcutaneously in the scalp at the injection site. Use a piece of gauze to massage in the lidocaine.

5. Make a small (approximately 2 cm) midline incision in the rat scalp to expose the dorsal surface of the calvaria; be sure to

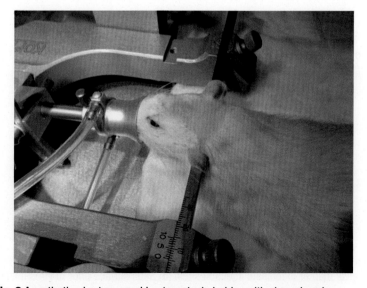

Fig. 2 Anesthetized rat secured in stereotaxic holder with shaved scalp

include bregma and lambda in this area to ensure the accuracy of other coordinate measurements.

6. Bluntly dissect the tissue to expose the bone. Clean the skull with hydrogen peroxide to highlight the cranial sutures. Use electrocautery as needed to achieve hemostasis and eliminate any bleeding from scalp tissue vessels.

7. Using a marker attached to a stereotactic arm, find the dorso-ventral coordinates of bregma and lambda. Adjust the bite bar position until these readings are equal. This ensures the accuracy of further stereotactic depth measurements. Mark bregma and record its coordinates (Fig. 3a, b).

8. Calculate the coordinates of the region of interest and the location of the injection sites. Mark these using the stereotactic system with the marker, and afterward remove the marker and stereotactic arm (Fig. 3c, d) (*see* **Note 1**).

9. Drill a small hole through the calvaria at each injection site using a 0.3 mm micro drill bit by hand using a light touch (Fig. 4a)

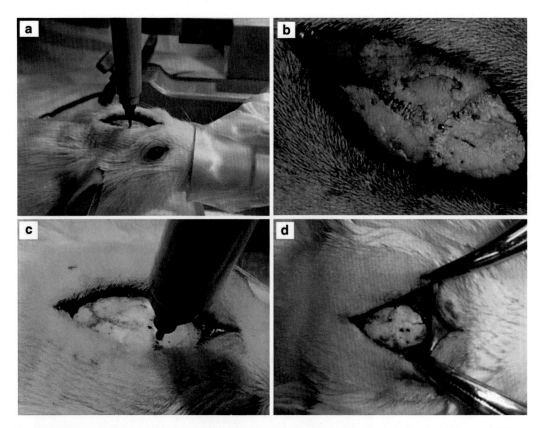

Fig. 3 (**a**) Use a surgical marker to highlight bregma. (**b**) Cleaned skull surface indicating bregma. (**c**) Use a surgical marker to indicate injection sites. (**d**) Skull surface prior to virus injection featuring marks at bregma and three injection sites

(*see* **Note 2**). Nick the dura with the tip of the drill bit if you are using a blunt needle for virus injection to ensure that the needle will penetrate into the brain tissue.

10. Fill the needle with the desired amount of virus, drawing up 0.5–1 µl extra virus. Affix to the stereotactic arm of the injection system. Set the pump to inject at 0.01 µl/min until you acquire your chosen volume. Advance the pump manually (on the touch screen) until a small drop of virus can be seen at the tip of the needle.

11. Advance the needle on the stereotaxic holder until it is at the surface of the skull over the first injection site (Fig. 4b). Record the dorsoventral coordinate. Calculate the coordinate of your desired injection depth. Slowly advance the needle to this depth (*see* **Note 3**). Wait 5 min for the brain to equilibrate to the presence of the needle (*see* **Note 4**).

12. Begin the injection of the virus. When the pump has finished injecting, wait 5 more minutes to allow the brain to equilibrate to the added volume of the injection before slowly withdrawing the needle. At this point you may adjust the setup or touch the rat if needed (*see* **Note 3**).

13. Repeat **steps 9** and **10** for each injection site.

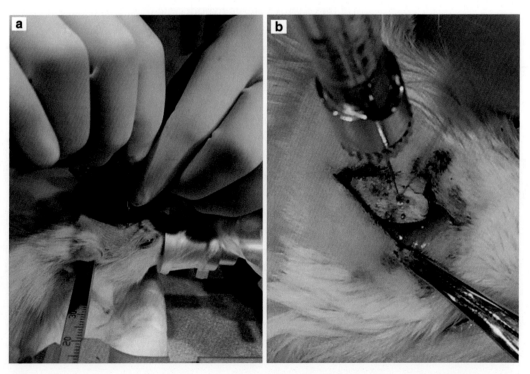

Fig. 4 (a) Drill a small hole through the skull by hand for viral injection using a micro drill bit. **(b)** Viral injection process

14. When the injections are completed and the needle withdrawn, suture the skin closed. It is not necessary to cover the holes in the skull before closing the skin.

15. Provide standard post-surgical care including secondary doses of cefazolin (30 mg/kg) and carprofen (5 mg/kg) subcutaneously.

3.2 Optical Cannula Implantation

1. Two to three weeks after the first protocol prepare the rat as for the injection procedure, **steps 1–3**.

2. Make a midline incision about 3.5 cm long in the rat scalp. Expose the calvaria, bluntly dissecting away the temporalis muscle to expose the sides of the skull as well.

3. Clean the skull with hydrogen peroxide and dry, using electro-cautery to stop any persistent bleeding.

4. Using a marker attached to a stereotaxic arm, mark bregma and record its coordinates (Fig. 3a, b). Realign the dorsoventral coordinates of bregma and lambda as in the injection procedure, **step 5**.

5. Calculate and mark the location for the cannula placement (*see* **Note 5**).

6. Drill a small hole for the cannula by hand using the 0.3 mm micro drill bit (Fig. 5a). Puncture the dura with the tip of the drill bit or with a sterile hypodermic needle to ensure adequate penetration of the optical fiber.

7. Affix the cannula to a stereotactic arm and position above the implantation site (Fig. 5b) (*see* **Note 6**).

8. Lower the cannula into the implantation site, leaving approximately a 0.5 mm gap between the base of the cannula and the surface of the skull (Fig. 5c). Apply a dot of Loctite glue to this area and then lower the cannula until it touches the skull. Allow the glue to dry (Fig. 5d) (*see* **Note 7**).

9. Expose the side of the calvaria. Clean with peroxide and dry; cauterize if necessary to attain hemostasis (Fig. 6a).

10. Drill a hole in the side of the calvaria by hand with the smaller drill bit (Fig. 6b).

11. Enlarge the outer edge of the hole by hand with the larger drill bit. Use the small bone curette to smooth the edges of the hole (Fig. 6b).

12. Trim the nylon screw to 0.5–1 mm in length.

13. Carefully position the nylon screw and screw it fully into the hole. Dry around the screw with a piece of gauze or a cotton swab.

14. Apply a dot of Loctite glue to the skull and the screw. Allow glue to dry (*see* **Note 7**).

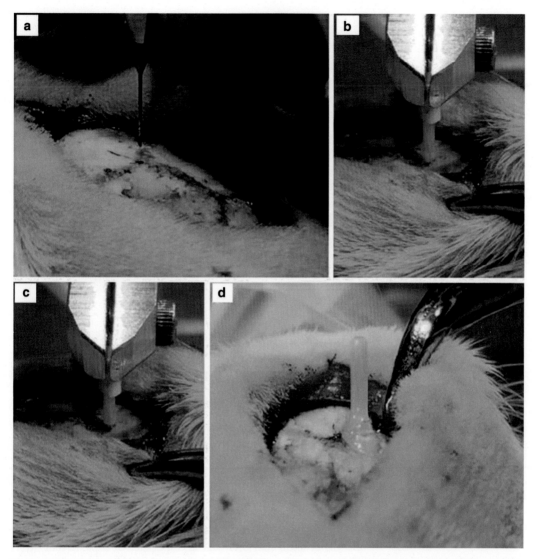

Fig. 5 (a) Drill hole for cannula by hand. **(b)** Align the fiber optic cannula with the hole. **(c)** Advance the implant to 0.5–1 mm above surface of skull. **(d)** Completed cannula dotted with superglue and fully advanced

15. Mix dental cement to a thick consistency and apply around screw using the micro spatulas, covering entirely. Allow to harden. Mix and apply more cement as needed to ensure an even coating (*see* **Note 8**). Also place a trail of cement over the top of the calvaria to the cannula (Fig. 6c).

16. Repeat **steps 9–15** on the other side of the skull.

17. Mix and apply more dental cement to the top of the skull, making a small mound around the cannula. Leave enough space to fit the connector and the fiber optic cable to the laser (*see* **Note 9**). Allow dental cement to harden (Fig. 6d).

18. Suture the skin closed, leaving space for the cannula to poke through (*see* **Note 10**).

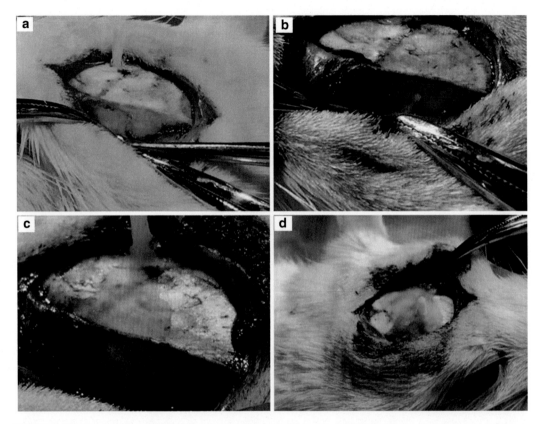

Fig. 6 (**a**) Expose the calvaria. (**b**) Hole in calvaria ready for screw insertion. (**c**) Spread dental cement over screw in calvaria and implant site. (**d**) Final cemented cannula ready for closure

4   Notes

1. All published rat atlases contain brain coordinates from bregma and other skull markers that are specific for a particular strain of rat within a specified size, age, and weight range (for example *see*: [12]). Please refer to the atlas for listed conversion factors when utilizing any other rat strains or ages/weights outside the specified strain used to construct the atlas.

2. Drilling the hole by hand decreases the risk of damaging the brain at the injection site.

3. You can vary the injection process in many ways. Such as injecting small amounts of virus across different cortical layers. The best approach to this method is to start the injections at the desired bottom cortical layer first and then continuing with additional micro injections as you advance the needle towards the brain surface.

4. It is important not to jar the operating surface or the stereotactic setup for the entire time that the needle is in the brain. We recommend not touching the setup or the operating surface

through the entire injection process, except in case of emergency.

5. It may be of benefit not to place the cannula directly over one of the injection sites. This is the case for many imaging applications where the cannula may interfere with imaging the injection sites and the cannula is only needed for light delivery.

6. The optical implant manufacturer includes a small sheath to protect the cannula in the shipping process. We affix this sheath to the stereotactic arm and fit the cannula into the sheath. This holds the cannula firmly enough to be stable for the implantation, but loosely enough that we can easily remove the arm once the cannula is secured in place.

7. A drop of the liquid component of Ortho-Jet BCA dental cement applied directly to the Loctite glue will act as an accelerant, causing it to harden immediately. This greatly reduces the surgical time.

8. Thicker dental cement dries much faster than when mixed to a thinner consistency, so mix it in small batches to reduce waste.

9. In our experience, the connector can be placed mostly onto the fiber optic cable, needing only a millimeter space free on the cannula.

10. If the cannula is more than about 2 mm lateral, you may need to make a small lateral incision to avoid stress on the cannula from the skin tension.

11. Several formulations are developed to model the distribution and penetration depth of light within the brain tissue [13, 14]. These models are used to find a reasonable estimation for the amount of light required to efficiently stimulate target cells within the region of interest. One simple yet practical formulation for optogenetic applications is developed based on the Kubelka-Munk model in which the brain tissue is considered as a highly scattering homogeneous turbid medium in which the absorption is negligible. Based on this model, the relative intensity of 473 nm laser at depth z (mm) along the optical axis of the fiber versus the intensity at the surface of an optical fiber of diameter $\varnothing$ (mm) and numerical aperture NA is given by [13]:

$$\frac{I(z)}{I(z=0)} \cong \frac{\rho^2}{(11z+1)(z+\rho)^2}$$

$$\rho = \frac{\varnothing}{2}\sqrt{(\frac{1.36}{NA})^2 - 1}$$

The $I(z)/I(z=0)$ ratio for a fiber with $\varnothing = 0.2$ mm and NA $= 0.22$ is shown in Fig. 7. Based on this curve, if we inject 10 mW of optical power into the tissue through this fiber, in

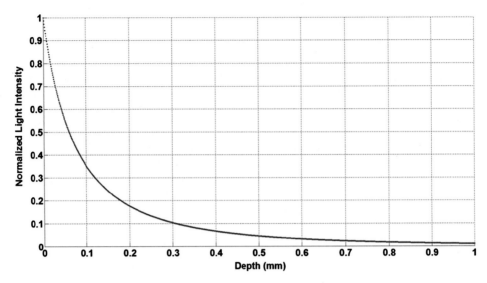

Fig. 7 Normalized intensity of light at different depths along the optical axis of an optical fiber of 0.2 mm diameter and 0.22 numerical aperture

about 0.3 mm distance from the fiber tip, the intensity of light drops to about 1 mW, which is almost the threshold level for the ChR2 protein to be efficiently activated. In case we use the ChR2(H134R) variation of the protein which is about two times more sensitive to light, the penetration depth for effective optogenetic stimulation increases to 0.5 mm or slightly more.

Acknowledgement

This work was supported by National Institutes of Health (NIBIB R01-EB000215NIH) and the Advancing a Healthier Wisconsin Program (5520208). R. Pashaie is supported by National Science Foundation (NSF) Career Award grant # MIL109784.

References

1. Fenno L, Yizhar O, Deisseroth K (2011) The development and application of optogenetics. Annu Rev Neurosci 34:389–412

2. Yizhar O, Fenno L, Davidson T et al (2011) Optogenetics in neural systems. Neuron 71(1):9–34

3. Pashaie R, Anikeeva P, Lee J et al (2014) Optogenetic brain interfaces. IEEE Rev Biomed Eng 7:3–30

4. Zhang F, Aravanis A, Adamantidis A et al (2007) Circuit-breakers: optical technologies for probing neural signals and systems. Nat Rev Neurosci 8:577–581

5. Zhang F, Wang L, Brauner M et al (2007) Multimodal fast optical interrogation of neural circuitry. Nature 446:633–639

6. Deisseroth K (2011) Optogenetics. Nat Methods 8(1):26–29

7. Mattis J, Tye K, Ferenczi E et al (2011) Principles for applying optogenetic tools derived from direct comparative analysis of microbial opsins. Nat Methods 9(2):159–172

8. Gradinaru V, Thompson K, Diesseroth K (2008) eNpHR: a Natronomonas halorhodopsin enhanced for optogenetic applications. Brain Cell Biol 36:129–139

9. Kato H, Zhang F, Yizhar O et al (2012) Crystal structure of the channelrhodopsin light-gated cation channel. Nature 482(7385): 369–374

10. Williams J, Denison T (2013) From optogenetic technologies to neuromodulation therapies. Sci Transl Med 5(177):177ps6

11. Gradinaru V, Thompson K, Zhang F et al (2007) Targeting and readout strategies for fast optical neural control in vitro and in vivo. J Neurosci 27(2):14231–14238

12. Paxinos G, Watson C (2007) The rat brain in stereotaxic coordinates, 6th edn. Academic, New York

13. Aravanis A, Wang L, Feng Z et al (2007) An optical neural interface: in vivo control of rodent motor cortex with integrated fiberoptic and optogenetic technology. J Neural Eng 4:143–156

14. Azimipour M, Baumgartner R, Liu Y et al (2014) Extraction of optical properties and predition of light distribution in rat brain tissue. J Biomed Opt 19(7):075001

Chapter 16

An Optimized Calcium-Phosphate Transfection Method for Characterizing Genetically Encoded Tools in Primary Neurons

Shiyao Wang and Yong Ku Cho

Abstract

In order to characterize genetically encoded tools under the most relevant conditions, the constructs need to be expressed in the cell type in which they will be used. This is a major hurdle in developing optogenetic tools for neuronal cells, due to the difficulty of gene transfer to these cells. Several protocols have been developed for transfecting neurons, focusing on improved transfection efficiency. However, obtaining healthy cells is as important. We monitored transfected cell health by measuring electrophysiological parameters, and used them as a guideline to optimize transfection. Here we describe an optimized transfection protocol that achieves reasonably high efficiency (10–20 %) with no discernable impact on cell health, as characterized by electrophysiology.

Key words Primary neurons, Dissociated culture, Transfection, Electrophysiology, Optogenetics

1 Introduction

When developing genetically encoded tools, it is critical to assess the level of expression, proper folding, and trafficking of the protein construct in the cell type that they will be used. Systematic metagenomic screens that characterized microbial opsin homologues from a wide range of archaea and alga in mammalian neurons found that successful expression and trafficking of these proteins are key to high-performance tools [1, 2]. This was also true for tool optimization; screens based on primary neurons [3] resulted in dramatic further improvement of genetically encoded calcium indicators which had been extensively optimized using bacteria-based screens [4]. Performing screens in primary neurons is especially challenging, due to the difficulty of gene transfer into neuronal cells. To tackle this challenge, several transfection methods, based on electrical, chemical, or viral gene transfer approaches have been developed [5, 6]. Many of these studies focused on

Arash Kianianmomeni (ed.), *Optogenetics: Methods and Protocols*, Methods in Molecular Biology, vol. 1408,
DOI 10.1007/978-1-4939-3512-3_16, © Springer Science+Business Media New York 2016

improved transfection efficiency, but reducing toxicity to obtain healthy cells is also critical. Objective criteria on neuronal cell health may be obtained from electrophysiological characterizations, by measuring properties such as resting potential, membrane resistance, and threshold for generating action potentials. We have found that DNA transfection mediated by calcium-phosphate precipitation results in healthy neurons as characterized by electrophysiological properties, and have successfully used it to identify highly efficient channelrhodopsins [2]. The transfected cells using this approach yielded hippocampal mouse neurons that routinely formed giga-ohm seals in whole-cell patch clamp, have membrane capacitance around 50–60 pF, membrane resistance higher than 250 MΩ, and resting potential at –65 mV [2], which are typical values for untransfected control neurons. The method allows transfection of plasmid DNA regardless of its size, and is labor- and cost-effective compared to viral vectors. The transfection efficiency ranges between 10 and 20 %. In addition, we identify specific parameters to tune for improved efficiency.

The transfection method is based on a previous report [7], but has been optimized with fully defined components, instead of relying on commercial kits. As indicated in the previous study [7], the formation of uniform and small calcium phosphate precipitates, and their removal after transfection were found to be critical. We optimized the amount of DNA, calcium chloride concentration, and phosphate concentration for forming uniform small precipitates. We also identified a slightly lower pH wash buffer (pH 6.9) that effectively dissolves all calcium phosphate precipitates after transfection. The protocol has been optimized using primary mouse hippocampal neurons, and has been tested in cortical neurons.

2   Materials

2.1   Neuronal Culture

1. 24-well plate.

2. Hemocytometer.

3. Round coverglass (25 mm in diameter, 0.15 mm in thickness), autoclaved.

4. Matrigel: Make 250 μL aliquots. Store at –20 °C. Thaw one vial and dilute in 12 mL Dulbecco's modified eagle medium (DMEM) (without phenol red). Diluted solution can be stored at 4 °C.

5. Culture Medium (500 mL): 450 mL minimum essential media (MEM) (without phenol red); 5 g/L Glucose (add 2.5 g), 0.1 g/L Transferrin (Bovine holotransferrin, add 50 mg), 2.38 g/L HEPES (add 1.19 g), 2 mM l-Glutamine, 0.025 g/L

Insulin (Bovine pancreas), 50 mL Heat Inactivated Fetal Bovine Serum, 10 mL B27 Supplement. Adjust pH to 7.3 using 10 M NaOH, sterilize with 0.2 μm filter, make 40 mL aliquots, store at −20 °C (*see* **Note 1**).

6. AraC solution (500 mL): 500 mL MEM, 5 g/L Glucose, 50 mg Transferrin, 1.19 g HEPES, 0.5 mM l-Glutamine, 10 mL B27 Supplement, 4 μM Cytosine β-d-arabinofuranoside hydrochloride (AraC), 25 mL Heat Inactivated Fetal Bovine Serum. Adjust pH to 7.3–7.4, sterilize with 0.2 μm filter, make 40 mL aliquots, store at −20 °C.

2.2 Transfection

1. Transfection buffer (500 mL): 500 mL MEM (without phenol red). pH to 7.15, sterilize with 0.2 μm filter, store at 4 °C (*see* **Note 2**).

2. Calcium chloride solution (50 mL): 50 mL double-deionized H_2O, 2 M $CaCl_2$ (add 14.7 g). Sterilize with 0.2 μm filter, make 1 mL aliquots, store at −20 °C.

3. Sodium phosphate solution (10 mL): Start with about 8 mL double-deionized H_2O, 1.5 M $Na_2HPO_4 \cdot 7H_2O$ (add 4.02 g). Add double-deionized H_2O to final volume of 10 mL. Sterilize with 0.2 μm filter, make 1 mL aliquots, store at −20 °C (*see* **Note 3**).

4. 2× HBS solution (50 mL): 50 mL double-deionized H_2O, 50 mM HEPES (add 0.65 g), 280 mM NaCl (add 0.8 g), 1.5 mM sodium phosphate (add 50 μL of sodium phosphate solution). pH to 7.0, sterilize with 0.2 μm filter, make 1 mL aliquots, store at −20 °C.

5. Sterile water: Filter sterilize double-deionized H_2O, make 1 mL aliquots, store at 4 °C.

6. Wash buffer (500 mL): 500 mL MEM (without phenol red). pH to 6.9, sterilize with 0.2 μm filter, store at 4 °C.

3 Methods

3.1 Neuronal Culture

1. Round coverglasses are placed into each well of a 24-well plate. 75 μL of diluted matrigel is added at the center of coverglass to form a droplet. Transfer the plate carefully to a 37 °C 5 % CO_2 incubator without disturbing the droplets, and incubate for 2 h or longer.

2. Remove the matrigel by aspiration, and dry for 30 min with lid open in a biosafety cabinet.

3. Procedure for preparing dissociated hippocampus follows previously published protocols [8, 9], but uses the culture media described here. 75 μL of dissociated mouse hippocampus (at a density of 50,000 total cells per well) in culture medium are

added at the center of coverglass and incubated for 3 h in a 37 °C 5 % CO_2 incubator.

4. After incubation, the cells are inspected under a microscope to check for adhesion, and 1 mL of prewarmed culture media is added to each well. Cells are placed in a 37 °C 5 % CO_2 incubator.

5. After 1–2 days the glial cells cover 50–70 % of the coverglass, and axonal projections of neurons are visible. At this stage, 1 mL AraC solution is added to each well to inhibit glial proliferation. The cells can be transfected starting 1–2 days after this step.

3.2 Transfection

1. For transfection, total of 1.25 μg of DNA is required for each well (*see* **Note 4**).

2. Prewarm sterile water, calcium chloride, and 2× HBS to room temperature.

3. Remove culture media from wells using a serological pipette, and collect in a 50 mL tube. The collected culture media is placed in a 37 °C bath, and will be added back after the transfection is completed (*see* **Note 5**).

4. Add 0.5 mL of transfection buffer to each well, and place the plate in a 37 °C 5 % CO_2 incubator.

5. For each well, prepare the following transfection mix according to Table 1:

6. For each transfection mix, prepare equal volume of 2× HBS in a 1.5 mL tube. Take the cells out from the incubator.

7. Transfer transfection mix into the 1.5 mL tube containing equal volume of 2× HBS and mix by pipetting 10 times to start calcium phosphate precipitation. Incubate the mixture for 30 s without mixing (*see* **Note 6**).

8. Add 50 μL of the transfection mix plus HBS drop-wise to each well (*see* **Note 7**).

9. Repeat **steps** **7** and **8** until all transfection mix are added. Transfer the plate into a 37 °C 5 % CO_2 incubator, and incubate for 20–30 min (*see* **Note 8**).

Table 1
Preparation of transfection mixture

Component	1 well	2-wells	3-wells
DNA	1.25 μg	2.5 μg	3.75 μg
2 M calcium chloride	3.13 μL	6.25 μL	9.38 μL
ddH$_2$O	Bring to 25 μL	Bring to 50 μL	Bring to 75 μL

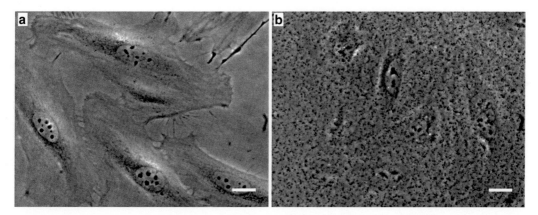

Fig. 1 Bright-field microscope images of small and even calcium phosphate precipitates. Since the precipitates are hard to visualize in neuronal cultures, HeLa cells are shown to give a sense of precipitate size. (**a**) Untransfected HeLa cells. (**b**) HeLa cells being incubated with calcium phosphate precipitates. *Small black dots* covering the cells are the precipitates. Scale bars indicate 20 μm

10. After incubation, view cells under the microscope to check calcium phosphate precipitate size (Fig. 1).

11. Remove the transfection mixture by aspiration, and add 1 mL wash buffer drop-wise to each well. Incubate cells in wash buffer for 10 min in a 37 °C 5 % CO_2 incubator. After this incubation calcium phosphate precipitates should be completely dissolved when visualized under a microscope. If the precipitates remain, repeat this wash step.

12. Remove wash buffer by aspiration, and add 1.5 mL of culture media collected from **step 3** to cells. Add 0.5 mL of fresh culture media to bring the total volume to about 2 mL.

13. The transfected cells remain healthy in a 37 °C 5 % CO_2 incubator for 2 weeks or longer. Typically for high-expressing proteins, expression is detected after 1–2 days. For well-expressing microbial opsins, expression is detected after 1–2 days, and reach optimal levels after a week (Fig. 2).

4   Notes

1. The pH of plating medium is critical for neuronal health. Measure pH of an aliquot again after filtration to make sure pH is at 7.3.

2. The pH of the transfection buffer is important for generating even and small precipitates. If the pH is low, no precipitate will form, and if too high, large aggregates tend to form.

3. This sodium phosphate solution is prepared to accurately control the phosphate concentration in the 2× HBS solution.

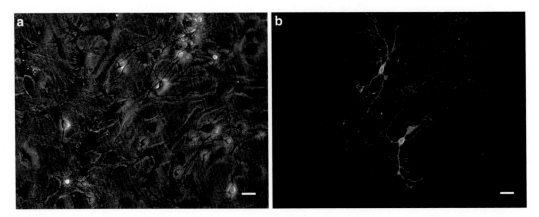

Fig. 2 Primary mouse hippocampal neurons transfected with channelrhodopsin-2 (ChR2) fused with EGFP. (**a**) Bright-field image. (**b**) ChR2 expression visualized by EGFP fluorescence. Images were taken 1 week after transfection. Scale bars indicate 20 μm

The phosphate concentration is critical for obtaining small and even calcium phosphate precipitation [10].

4. The DNA is typically a plasmid, or can be a mixture of multiple plasmids. DNA in double-deionized water works best and the concentration should be higher than 100 ng/μL.

5. The culture media at this point may contain factors released by cells, and therefore is kept for use after transfection.

6. It seems better not to disturb the mixture during this step for obtaining small uniform precipitates. Incubation over 30 s is not necessary.

7. Drop-wise addition of the transfection mix plus HBS is critical for forming uniform precipitates. When doing this, the pipette tip should be about 1 in. above the liquid surface, so that the mixture is added in droplets.

8. This incubation time can be extended to 1 h to improve transfection efficiency, but it tends to increase toxicity to neurons.

Acknowledgements

This work was funded by the University of Connecticut and the Brain and Behavior Research Foundation (NARSAD Young Investigator grant). Shiyao Wang was partially supported by the outstanding scholars program of the UConn graduate school.

References

1. Chow BY, Han X, Dobry AS et al (2010) High-performance genetically targetable optical neural silencing by light-driven proton pumps. Nature 463(7277):98–102

2. Klapoetke NC, Murata Y, Kim SS et al (2014) Independent optical excitation of distinct neural populations. Nat Methods 11(3):338–346

3. Chen TW, Wardill TJ, Sun Y et al (2013) Ultrasensitive fluorescent proteins for imaging neuronal activity. Nature 499(7458):295–300

4. Akerboom J, Chen TW, Wardill TJ et al (2012) Optimization of a GCaMP calcium indicator for neural activity imaging. J Neurosci 32(40):13819–13840

5. Karra D, Dahm R (2010) Transfection techniques for neuronal cells. J Neurosci 30(18):6171–6177

6. Washbourne P, McAllister AK (2002) Techniques for gene transfer into neurons. Curr Opin Neurobiol 12(5):566–573

7. Jiang M, Chen G (2006) High Ca2+-phosphate transfection efficiency in low-density neuronal cultures. Nat Protoc 1(2):695–700

8. Kaech S, Banker G (2006) Culturing hippocampal neurons. Nat Protoc 1(5):2406–2415

9. Beaudoin GM 3rd, Lee SH, Singh D et al (2012) Culturing pyramidal neurons from the early postnatal mouse hippocampus and cortex. Nat Protoc 7(9):1741–1754

10. Jordan M, Schallhorn A, Wurm FM (1996) Transfecting mammalian cells: optimization of critical parameters affecting calcium-phosphate precipitate formation. Nucleic Acids Res 24(4):596–601

Chapter 17

Optogenetic Approaches for Mesoscopic Brain Mapping

Michael Kyweriga and Majid H. Mohajerani

Abstract

Recent advances in identifying genetically unique neuronal proteins has revolutionized the study of brain circuitry. Researchers are now able to insert specific light-sensitive proteins (opsins) into a wide range of specific cell types via viral injections or by breeding transgenic mice. These opsins enable the activation, inhibition, or modulation of neuronal activity with millisecond control within distinct brain regions defined by genetic markers. Here we present a useful guide to implement this technique into any lab. We first review the materials needed and practical considerations and provide in-depth instructions for acute surgeries in mice. We conclude with all-optical mapping techniques for simultaneous recording and manipulation of population activity of many neurons in vivo by combining arbitrary point optogenetic stimulation and regional voltage-sensitive dye imaging. It is our intent to make these methods available to anyone wishing to use them.

Key words Optogenetics, Virus, Transgenic, Mouse, Neuron, Circuit, Brain, Channelrhodopsin (ChR2), Voltage-sensitive dyes (VSD)

1 Introduction

One of the major goals in neuroscience is to map the structural and functional connectivity of the brain. Many neuroscientists believe that this work will lead to important discoveries that will in turn lead to new treatments of neurological disorders and diseases ranging from stroke to autism [1, 2]. The advent of optogenetics has revolutionized neuroscience by enabling the use of light to interrogate genetically identified neurons within neural circuits using opsins, light-activated proteins [3]. Importantly, such control of neurons is reversible and operates on sub-millisecond timescales. With the technique scientists can activate or silence specific classes of neurons to test specific hypotheses. This work can be performed in awake behaving animals or in brain tissue slices to examine short- and long-range functional connections between neurons. These opsin proteins can be inserted into neurons via viral vectors, electrophoresis, or by breeding transgenic animals [1, 2, 4, 5]. These mapping experiments are typically performed in conjunction

Arash Kianianmomeni (ed.), *Optogenetics: Methods and Protocols*, Methods in Molecular Biology, vol. 1408,
DOI 10.1007/978-1-4939-3512-3_17, © Springer Science+Business Media New York 2016

with either electrophysiology or imaging methods depending on the specific research question. In addition to using optogenetics to perturb neural circuits, it can also be used as a reporter. For example, channelrhodopsin can be inserted into small inhibitory neurons and pulses of blue light will invoke spiking in those cells. This method enables the identification of these small, often quiescent, cells for blind in vivo whole-cell or juxtacellular recordings to investigate their role in large-scale mapping experiments [6, 7].

The selectivity of genetically identified cells has led to many new discoveries, such as the role of specific types of inhibitory and excitatory neurons in the cortical minicolumn [8–10]. To further add to the specificity of spatially restricted areas of the circuits of interest, modern lasers, directed by galvanized mirrors can focus beams of light with ~50 µm precision into target tissues, reaching cortical layers 2/3 [11]. For even greater spatial precision, two-photon optogenetic activation of red-shifted opsins can allow for activation and recording of many individual cells simultaneously [12].

The two primary classes of optogenetic opsins are those that cause neurons to depolarize (e.g., Channelrhodopsin-IIa; ChR2) or hyperpolarize (e.g., Archaerhodopsin). ChR2-type proteins can be used to activate neurons that have an excitatory or modulatory role in brain circuits. If the intent is to temporarily ablate specific brain regions, ChR2 can be placed into inhibitory neurons (e.g. Parvalbumin-positive basket cells) or Archaerhodopsin can be inserted into excitatory or modulatory cells. While ChR2 and Archaerhodopsin are two of the most popular opsins, there are many light-activated proteins available for brain mapping experiments (please *see* [13] to review many of the available opsins). These designer proteins are continuously upgraded to enable control with many different wavelengths of light.

Here we describe in vivo optogenetic functional mapping procedures combining optogenetic stimulation with regional voltage-sensitive dye imaging in transgenic mice to assess intrahemispheric and interhemispheric functional relationships. Any transgenic animal that expresses optogenetic opsins within subsets of cortical neurons could be used. Although optogenetic opsins are expressed in axons of passage and some of the transgenic animals might exhibit some variability in expression levels across the brain, these transgenic mice provide advantages over multiple viral injections due to incomplete sampling and potential for tissue damage at each injection site. To monitor and map intracortical population activity, we use organic voltage-sensitive dyes that offer the ability to monitor activity over large spatial scales (up to 50 mm^2) and with millisecond time resolution [14–16].

The combination of optogenetics and imaging to map functional circuits in vivo is paving the way for new insights in neuronal functioning. One of the strongest advantages of the methods described here is the activation of nonprimary sensory areas. This can enable investigations of network activity and functional maps following the stimulation of secondary or even higher level association areas. In the future, development of novel opsins for optogenetic stimulation and genetically encoded activity sensors [17, 18] may allow for longitudinal and simultaneous stimulation and imaging of more than one class of cell, within the same animal, by using opsins activated with different colors light. Such methodological approaches could be used to deduce functional relationships between cortical areas and large-scale circuit organization in various mouse models of human disease, to study the recovery after injury such as stroke or traumatic brain injury.

2 Materials

2.1 Experiment Setup

Once the experimental goals are planned, obtain the required animals with opsins expressed in genetically identified cell classes. Transgenic mice and rats can be purchased from commercial vendors and bred to into standing colonies. For experiments involving other animal species or nontransgenic mice and rats, viral vectors can be purchased for specific experiments. Some researchers elect to use transgenic Cre mice and then inject specific opsins into small spatial regions of the brain. Please note that some transgenic lines of mice are frozen down and require reconstitution of the line which can take up to 2 months.

2.1.1 Illumination Sources

Solid-state blue (473 nm) or yellow (589 nm) laser to activate optogenetic opsins in our transgenic mouse lines, and a 627 nm red LED to excite the VSD.

2.1.2 Light Filtering

1. VSD fluorescence excitation (620–640 nm) and bandpass emission (673–703 nm) filters.

2. Dichroic mirror (590 dcxr). This separates the red LED for VSD and blue or yellow laser for optogenetic stimulation.

3. Neutral density filters to assist with reducing saturation of LEDs.

2.1.3 Voltage-Sensitive Dye

1. RH1691 [19].

2. Syringe filter, 0.2 μm (Pall Acrodisc Supor Membrane; *see* **Note 6**).

2.1.4 VSD Data Collection

1. CCD camera.

2. EPIX E4 or E8 frame grabber with XCAP 3.8 imaging software.

3. 50 and 35 mm front-to-front video lenses.

2.1.5 Photostimulation

1. PCI-6229 M Series Data Acquisition Board (DAQ; National Instruments, Austin, TX, USA).

2. Windows XP computer (*see* **Note 8**).

3. Ephus software [20] (*see* **Note 8**).

4. Galvanometer scan mirrors.

5. Laser power meter (Thorlab, NJ).

2.1.6 Common Lab Supplies

1. Analytic scale.

2. Magnetic stir plate and stir bar.

3. pH meter or strips.

4. Vacuum pump or line.

5. Vortex.

6. Ultrasonic cleaner.

7. Large beaker (>1 L).

8. Graduated cylinder.

9. Vacuum filter.

10. Syringe filter, 0.45 μm (Pall Acrodisc Supor Membrane).

11. Weigh boats/paper.

12. 50 mL screw top tubes.

13. 1.7 mL Eppendorf tubes.

14. Large syringe (>25 mL).

2.2 Surgical Station

2.2.1 Surgical Tools and Equipment

1. Scalpel blade handle.

2. Fine-tipped scissors.

3. Fine tweezers.

4. Tissue grabber forceps.

5. Cauterizer (*see* **Note 3**).

6. Tools for setting up and adjusting hardware (e.g., Screwdrivers, needle-nose pliers).

7. Clippers to shave mouse.

8. Thermodynamic heat blanket with rectal probe.

9. Surgical drill; pneumatic or battery operated.

10. Dissecting scope with at least 3× magnification power.

11. Light source to illuminate surgical site.

2.2.2 Surgical Supplies

1. Kim wipes.

2. Labeling tape (*see* **Note 12**).

3. Cotton swabs.

4. Artificial tears.

5. Chlorhexidine.

6. Ethanol 70 %.

7. Gelfoam.

8. 2–3 cm needle (25–30 G).

9. 16 G needle.

10. 1 cc syringes.

11. #11 scalpel blades.

12. Transfer pipettes.

13. Bone wax (*see* **Note 12**).

14. Sterile saline, ~50 mL (*see* **Note 5**).

15. Sterile brain buffer, ~100 mL (*see* **Note 5**).

16. Lubrication for rectal probe.

17. Drill burs, FG ¼ (Midwest Carbide Burs).

18. Super glue.

19. Agarose (Type-III, Sigma).

20. PE10 tubing, ~8–12 cm.

21. Straight edge razors.

22. Microscope coverslip glass to cover the agar and produce a uniform imaging surface (may need to be cut to size).

2.2.3 Isoflurane Delivery

1. Oxygen tank and regulator.

2. Isoflurane vaporizer.

3. Induction chamber.

4. Tubing to connect isoflurane vaporizer to induction chamber and nose cone.

5. Charcoal canister to collect isoflurane.

6. Exhaust snorkel for escaped isoflurane.

2.2.4 Surgical Stage

1. Isoflurane nose cone assembly.

2. Head plate (we can provide specific details upon request) and supports (*see* **Note 1**).

3. Dental cement powder.

4. Jet accelerant.

5. 12 well porcelain plate and stir sticks.

6. Acetone.

2.2.5 Surgical Drugs

1. Isoflurane.

2. Oxygen.

3. Dexamethasone (to reduce cerebral edema).

4. Lidocaine with Epinephrine (to reduce surgical pain and blood loss).

5. Atropine.

6. Euthansol (to euthanize animal upon completion of experiment).

2.2.6 Reagents	1. Sterile brain buffer: 134 mM sodium chloride, 5.4 mM potassium, 1 mM magnesium chloride hexahydrate, 1.8 mM calcium chloride dihydrate, and 5 mM HEPES sodium, pH balanced with 5 M hydrogen chloride (*see* **Note 5**).
	2. Sterile glucose in brain buffer: 0.5 mM glucose in brain buffer.

3   Methods

3.1   Lab Setup Before Beginning Experiments

Incorporating optogenetic techniques into a research lab requires a fair amount of strategic planning. One of the first steps is to determine which animal species will be used as this will dictate whether transgenic animals or viral injections are needed. Our lab primarily uses transgenic mice, due to readily available mouse lines from commercial sources and the ease of breeding a colony to have mice available for daily use. We also combine our transgenic mouse lines with viral vectors for spatial control of optogenetic expression. This is useful for experiments requiring tight spatial control when illumination areas are larger than desired.

The present protocol focuses specifically on adding optogenetics and VSD imaging to a standing lab. For labs seeking to add other recording techniques in addition to imaging, such as electrophysiology, many reviews and book chapters are already available regarding the setup of these techniques. The three primary light sources of stimulating opsins are lasers, LEDs, and arc lamps. All of these have their own inherent advantages and limitations (for review, *see* [5]). The light source can then be directed onto the cortex either through fiber optic cables or with a network of galvanized mirrors. In our lab we used laser stimulation controlled with scanning galvanized mirrors [1, 14].

Proper setup of imaging equipment will lead to easier data acquisition during experiments. It is far easier to setup and adjust the physical illumination before data acquisition than to try to correct poor images after recording. LEDs are an inexpensive way to illuminate brain tissue. When the LED is first turned on, the intensity may fluctuate and decay for a short time until it is fully warmed up and reaches a stable level. Be sure to test your LEDs. There is also a need to balance sufficient lighting with over-saturation. When the current delivered to the LED is too low, it can cause fluctuations of the light intensity, however currents needed for stable illumination may cause saturation. To solve this problem we use inexpensive neutral density filters. To ensure uniform illumination of the cortex, we use two LEDs on posts positioned approximately 10–15 cm from the brain. Note that the frame rate of the camera and pixel bin size will dictate the amount of light needed to adequately illuminate the brain.

Below we will discuss how to properly secure the mouse skull to a head plate (*see* **Note 1**). This will add to mechanical stability, critical for stable images. To add further stability, many imaging experiments are performed on anti-vibration air tables.

3.2 Surgery Introduction

Successful imaging experiments are highly dependent upon well-done surgeries with undamaged brain tissue. Based on our experience, mice survive surgeries best when they are treated gently. Reduce all unnecessary tissue damage and trauma. This lessens the need for excessive use of anesthetics and lowers the risk of large inflammatory responses. Ensure that the animal's body temperature remains at 37 °C from induction of anesthesia and throughout the entity of the experiment (*see* **Note 10**). Throughout the experiment we keep our saline and buffers warmed to 37 °C to ensure we aren't causing undue shock on the animal.

3.3 Presurgical Procedures

1. Ensure all materials and supplies are on hand (*see* Subheading 2) and that all surgical tools have been thoroughly cleaned and disinfected.

2. Collect the correct animal and place in transport cage. Record in lab notebook: animal ID number, weight, gender, date of birth, genetic strain as well as other information required by your institution, such as its litter, cage, room, and protocol numbers.

3. Ensure water in beaker with brain buffer/saline tubes is warmed to 37 °C.

4. Ensure surgical area is ready and prep yourself for surgery.

5. Place mouse into isoflurane induction chamber and lock lid. Set isoflurane to 3–5 % and slowly turn oxygen flow rate up to 0.5–1 L/min. Wait 2–3 min for mouse to fall asleep and ensure it is areflexic to painful foot pinches.

6. Inject Dexamethasone (8 mg/kg, IP) to reduce the risk of cerebral edema.

7. Inject glucose in brain buffer (10–15 mL/kg, IP) to hydrate the mouse. Inject again after the mouse urinates, about once every 1–2 h.

8. Shave off all hair around surgical site. If the mouse begins to wake up (whisking/voluntary movements) place back into induction chamber. Use labeling tape to remove loose fur/hair from the shaving site and to clean up the shaving station (*see* **Note 12**). This will assist in reducing contaminates from the surgical site.

3.4 Surgical Procedures

1. Place mouse onto surgical table and fit into nose cone. Set isoflurane to 0.5–1.5 % (adjust as needed) and oxygen flow rate to 0.5 L/min (*see* **Note 9**). Check the mouse's respiration rate and ensure it is areflexic throughout entirety of experiment, about every 5 min.

2. Use lubrication to insert rectal thermometer 1 cm and tape it down. Ensure the body temperature is 37 °C (*see* **Note 10**).

3. Moisten eyes with artificial tear lubricant to prevent corneal desiccation.

4. To conveniently deliver meds and hydration fluids we use a catheter inserted into the right intraperitoneal space, which reduces the need for moving the animal when giving injections. Cut a piece of PE10 tube to 8–12 cm in length. Use a marker to make a mark ~1.5 cm from end. Insert the marked side of the tubing into a 16 G needle. Put mouse on its back/left side to expose its right abdomen. Insert the needle into the right intraperitoneal space. You should feel two "pops," the first through the skin and the second through the peritoneum. While holding the needle still, carefully insert the tube into the abdomen (about 1 cm). Pinch the tube through the skin and gently remove the needle. The tube is properly placed when the marked end of the tube is at the surface of the skin. Super glue in place and wait until dry before returning mouse back into prone position (*see* **Note 2**).

5. For pain relief and reducing blood loss, we use 7 mg/kg Lidocaine with Epinephrine SQ, over surgical site. Wait ~ 5 min for drug to diffuse into tissue, then ensure the mouse is areflexic by pinching the skin over the surgical site with forceps, first lightly, then firmly. If the mouse responds, increase the isoflurane until it becomes areflexic (*see* **Note 4**).

6. Scrub surgical site with Chlorhexidine. Begin in the middle of the site and work your way outwards. Repeat three times (*see* **Note 4**).

7. Scrub surgical site with 70 % Ethanol. Begin in the middle of the site and work your way outwards. Repeat three times (*see* **Note 4**).

8. Once the mouse is areflexic to pain, begin removing the skin and muscle tissue over the region of interest for your experiments. Note skull landmarks such as the midsagittal suture, lambda, bregma, or the squamosal bone. Every 3–5 min moisten the skull with brain buffer. This will prevent the dura from adhering to the skull. Stop any noticeable bleeding with gel foam or use the cauterizer for significant hemorrhaging (*see* **Note 3**).

9. In Fig. 1, we show a bilateral hemisphere craniotomy spanning across the midline, from lambda to 3 mm anterior of bregma. Attach head plate to posts and position mouse underneath (Fig. 1a). This takes patience and practice. Use gauze and whatever is needed to position the mouse. The skull is rounded so ensure that the region of interest is centered and level. This creates a large craniotomy window approximately 7 mm^2 [14]. Ensure skull landmarks are present and take a photo (Fig. 1b), which will assist with reconstructing cortical maps with skull landmarks (*see* **Note 11**).

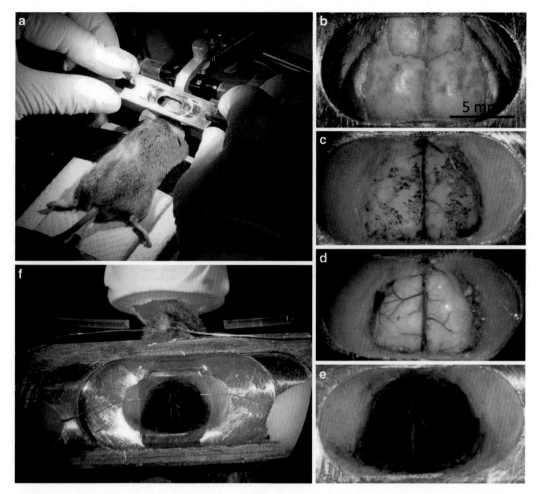

Fig. 1 Surgical set up for a VSD experiment in a ChR2 mouse with a large bilateral craniotomy. (**a**) Following surgical removal of soft tissues, the head plate is positioned over the skull (*see* **Notes 1** and **11**). (**b**) Close up view of the position of the skull underneath the head plate. Note the central suture and bregma are clearly visible. (**c**) Dental cement is added to the edges to secure the head plate to the skull while also serving as a well for the voltage-sensitive dye. Following craniotomy the dura is exposed. (**d**) Completed durotomy, note the brain tissue appears much cleaner than in **c**. (**e**) VSD staining is complete. Note the *deep purple color*. (**f**) Agar and coverslip protect the brain

10. Use super glue around the edges of craniotomy site. The glue should lightly secure the head plate to the skull in the midline anterior and posterior regions (*see* **Note 7**).

11. To reduce cortical pulsations a cisternal drain can help. Once the super glue is dry, remove tissue from top of skull to foramen cisterna. Expose the dura and pierce with a needle or new #11 scalpel. When done correctly, there should be a flash of clear cerebrospinal fluid (CSF). Place a small moistened piece of gauze over the drain to facilitate draining of CSF.

12. Mix dental cement and jet accelerant. Use a razor to cut a cotton swabs at a ~45° angle to make a stir-stick/applicator. Pour the powder into one of the wells in the 12-well porcelain plate, add jet liquid (about 1:1 ratio), and stir well. It will be quite runny at first, but will rapidly harden.

13. Add cement all around the craniotomy site, ensuring that landmarks are fully visible. Keep adding to build the well. This takes patience and many coats. You may need to mix up more cement. Once the well is built, allow to dry for 5–30 min. Ensure it is completely solid before continuing. Before the cement dries use a stir-stick to gently scrape it away from landmarks, such as the central suture or bregma.

14. Figure 1c shows the complete craniotomy and the well made from dental cement. Use the drill to carefully remove excess cement as needed and begin the craniotomy. Ensure the drill burr is sharp and replace as needed during surgery. Since the mouse skull is only about 1 mm thick, carefully remove a few hundred microns of bone at a time. Take extreme care to not punch the drill through the skull, which can cause extensive brain damage or hemorrhaging if large dural vessels are damaged. To remove debris from the well, carefully squirt brain buffer or saline on one side and use another transfer pipette to remove it. For stubborn pieces, carefully remove with fine-tipped forceps. While drilling through the skull, every few minutes stop and gently press on the skull piece that will eventually be removed. Only press down 50–100 μm; when sections are fully drilled the skull will depress into the brain. The drilling is complete when the piece to be removed is no longer attached and is floating on the dura. Be sure to distinguish between free moving skull and skull that has been loosened from poor adhesion to the dental cement. Fill the cranial well with brain buffer, this will assist in removing the skull from the dura (see **Note 5**). With extreme care gently remove the skull from the dura with fine-tipped forceps or a fine-tipped probe. The skull may see-saw, and lifting one end of skull may cause the other end to press into the brain. This is usually a symptom of incomplete drilling. Use a new #11 scalpel blade and gently cut the bone where the drilling is incomplete. Once the skull is removed stop any bleeding with gel foam. Note the condition of the dura. It is possible that it may have been removed with the skull.

15. If a durotomy is desired, use fine-tipped forceps or a microprobe and tiny spring scissors to remove the dura (Fig. 1d). Take extreme care to not damage the brain or blood vessels.

16. Again remove any debris by flushing with brain buffer using extreme care.

17. Fill the well with brain buffer.

3.5 Voltage-Sensitive Dye Preparation

We use the dye RH1691 that can be purchased in 10 mg bottles. We divide this into 20, 0.5 mg aliquots, which is enough for one experiment. Since the dye is light sensitive, protect it from light at all times.

1. The ratio of dye to brain buffer depends on experiment. For preps with a durotomy use 0.6 mg of dye:1 mL brain buffer. For preps with the dura intact use 1 mg of dye::1 mL brain buffer.

2. Mix solution with a vortex for 10 min on high.

3. Place tube in an ultrasonic cleaner for 5 min, ensuring the tube is immersed in the water. This can help to further dissolve the dye.

4. Let rest for 10–20 min, then centrifuge for 1 min at 12,000 rpm (g-force = 13,523).

5. Filter with 0.2 μm Pall Acrodisc filter (*see* **Note 6**).

3.6 Staining the Brain with Voltage-Sensitive Dye

1. Make a dam out of bone wax at the edges of the craniotomy to minimize dye needed. The bone wax may leave a film in the buffer, rinse and replace the brain buffer as needed.

2. Immediately before adding the dye, carefully remove all brain buffer. Twist the end of a Kim wipe to a fine point and under the dissecting scope carefully remove all traces of brain buffer from the edges of the craniotomy. This will greatly assist the staining of the entire brain, otherwise microfluidics will prevent the staining of the edges of the craniotomy.

3. Apply to brain and allow to soak for 60–120 min until the brain is deep purple. Use a 200 μL pipetter to gently mix the dye every 10–20 min. Ensure the brain is inundated at all times and add more as needed since it may evaporate or leak.

4. Rinse off with brain buffer and then fill well with brain buffer (Fig. 1e).

3.7 Protect the Brain and Reduce Movement Artifacts with Agar

1. Mix 200 mg of agar in 15 mL brain buffer (1.3 %) in a small beaker.

2. Bring to boil in the microwave and ensure there is no undissolved agar.

3. Cool while continuously stirring and apply over brain once the agar is <42 °C. Ensure there is no debris, bubbles, or undissolved agar as this can reduce the quality of the images by scattering light.

4. Immediately after applying agar, gently place glass coverslip. Place one end down first and slowly lower to force out air bubbles. When done correctly, the glass should rest on the bottom layer of the head plate (Fig. 1f).

5. Wait a few min for agar to set, then place a couple of pieces of gel foam at the corners of the agar and drop brain buffer every ~10 min to prevent the agar from desiccating.

3.8 Mapping

1. Figure 2a depicts our imaging set up. We illuminate the brain with two red LEDs (620–640 nm excitation filter) positioned ~10 cm from the brain. On laser stimulation trials, a 473 nm laser is positioned over cortical regions of interest via galvanometer scan mirrors before passing through a dichroic mirror to the brain. Light emitted from the brain surface passes through a bandpass filter (673–703 nm) and is recorded with a CCD camera.

2. To image VSD activity, focus the camera into the brain until the blood vessels are blurry, a depth of approximately 500 μm.

3. Each trial should begin with 300–500 ms of illumination with the red LED, followed by laser stimulation of the region of interest or physical stimulation of the sensory system. The pre-trial recording period allows for a baseline to compare the VSD fluorescence before and after stimulation. The red LED illumination should continue for 1–2 s after stimulation, with 10 s between trials to ensure the brain does not adapt or habituate to the stimulation (this may vary depending on the specific experiment). In Fig. 2b we show cortical activity following visual stimuli and laser stimulation of the primary visual cortex. We find that 20 trials are usually sufficient to obtain robust stimulus-evoked responses.

4. To arbitrary optogenetically stimulate neurons, we use a 473 or 589 nm diode pumper solid-state lasers. We position the laser beam on the cortex using Ephus software, which control galvanometer scan mirrors, via analog output voltage from the DAQ. The Ephus program controls the overall timing of individual stimulation trials with TTL triggers to XCAP.

5. Set the laser output to 5 mW using a laser power meter.

6. Position laser over region of interest. This can be accomplished by controlling galvanized scanning mirrors [1]. Figure 2c shows an example of an optical stimulation grid. Each blue circle represents a site of optogenetic laser stimulation. Five examples of cortical responses to optogenetic stimulation are shown on the right.

7. Previous work in our lab has demonstrated that 1 ms, 5 mW laser pulses are sufficient to evoke robust firing from ChR2 expressing neurons [1].

4 Notes

1. While there are many ways to immobilize the animal's head, we find that using a head plate attached/sealed to skull with dental cement is the best option. This is because ear bars or customized bite bar/nose cone assemblies require the building of a well for the VSD staining procedure. The head plate performs both functions simultaneously.

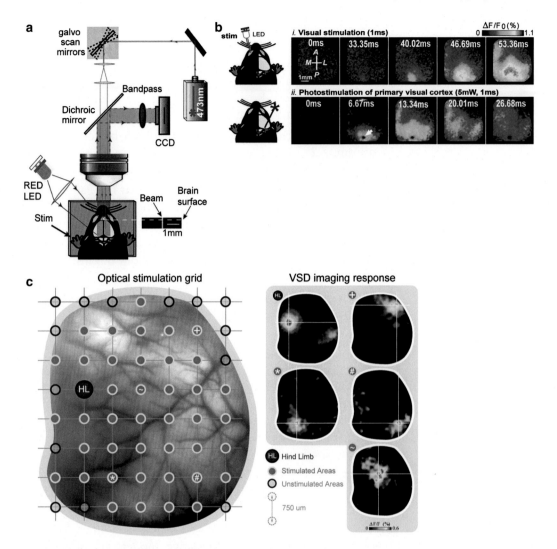

Fig. 2 Mapping interhemispheric and intrahemispheric connectivity using ChR2 stimulation and VSD imaging. (**a**) Experimental set up for simultaneous ChR2-photoactivation and VSD imaging. Galvanometer scan mirrors position a 473 nm laser at specific cortical locations while VSD fluorescence is monitored in epi-fluorescence mode. An image of laser beam demonstrates that it is relatively collimated. (**b**) Example of VSD responses in a unilateral craniotomy preparation during visual stimulation of the contralateral (*left*) eye (i), or direct photo-stimulation (*white arrow*) of the right V1 (ii). (**c**) Optical stimulation grid (*left*) and VSD imaging responses (*right*). Stimulated areas are denoted by *blue circles* spaced 750 μm. *Special symbols* superimposed over *blue circles* highlight responses shown on the *right*. Figure 2a and b reproduced from Lim et al., 2012, Figs. 1a and 2d, respectively, with permission from Frontiers in Neural Circuits

2. When placing the catheter, we find that some combinations of marker ink and glue cause an exothermic reaction that can melt the tubing and prevent injections. Be sure to practice this procedure before performing it on an animal.

3. We use a cauterizer to assist with stopping bleeding during surgery. Since surgical cauterizers are expensive, we instead

use a fly-tying cauterizer, typically used for making fly fishing lures. It is an excellent alternative for a fraction of the price.

4. Shaving the surgical site can cause small cuts in the skin that may cause a painful response when applying alcohol during the surgical prep. Be sure to wait for the lidocaine to fully infuse before proceeding.

5. Always use normal saline or isotonic brain buffer (290–300 mOsm, 7.2–7.4 pH) on exposed tissue; never use pure water as it is extremely hypotonic and destructive to cells.

6. Through much trial and error we found that the Pall Acrodisc 0.2 μm filter must be used to filter the dye, as other filter brands somehow prevent the dye from functioning properly.

7. For easy and controlled application of super glue, we remove the plunger from a 1 cc syringe to add about 0.5 mL. Replace the plunger to apply very small quantities.

8. The free open-source Ephus software platform requires Windows XP, although future updates are scheduled to work with Windows 7.

9. If the mouse makes raspy/gurgling/clicking sounds, we find that turning down the isoflurane usually solves the problem. Atropine can also help (1–2 mg/kg SQ) by reducing tracheal secretions. Note that atropine is fast acting and is quickly metabolized. Therefore, preventative doses at the beginning of each procedure are unlikely to prevent mucus buildup later in the experiment.

10. If maintaining the mouse at 37 °C becomes difficult, ensure the tail is on the heat blanket and drape an insulated blanket over their body.

11. We strongly advise taking photos before and after the craniotomy. As an inexpensive alternative, modern cell phone cameras often have excellent resolution. With practice the camera can be positioned by hand over the eyepiece of the dissecting scope.

12. Bone wax and tape are extremely useful disposable tools during surgery as well as setting up and during the experiment. We use small pea-sized blobs of bone wax and/or labeling tape to hold things down. Both of these can be used for temporary or long-term solutions.

Acknowledgments

This work was supported from the Natural Sciences and Engineering Research Council of Canada Discovery Grant, the Alberta Alzheimer Research Program, and the Alberta Innovates: Health

Solutions to M.H.M. and NSERC CREATE BIP Postdoctoral Trainee Grant to M.K. M.H.M. is the holder of the CAIP chair in Brain Function in Health and Dementia.

References

1. Lim DH, Mohajerani MH, LeDue J et al (2012) In vivo large-scale cortical mapping using channelrhodopsin-2 stimulation in transgenic mice reveals asymmetric and reciprocal relationships between cortical areas. Front Neural Circuits 6:1–19. doi:10.3389/fncir.2012.00011
2. Lim DH, LeDue JM, Mohajerani MH, Murphy TH (2014) Optogenetic mapping after stroke reveals network-wide scaling of functional connections and heterogeneous recovery of the peri-infarct. J Neurosci 34:16455–16466. doi:10.1523/JNEUROSCI.3384-14.2014
3. Boyden ES, Zhang F, Bamberg E et al (2005) Millisecond-timescale, genetically targeted optical control of neural activity. Nat Neurosci 8:1263–1268
4. Packer AM, Roska B, Häusser M (2013) Targeting neurons and photons for optogenetics. Nat Neurosci 16:805–815. doi:10.1038/nn.3427
5. Lin JY (2012) Optogenetic excitation of neurons with channelrhodopsins: Light instrumentation, expression systems, and channelrhodopsin variants. In: Knopfel T, Boyden ES (eds) Prog Brain Res. pp 29–47
6. Lima SQ, Hromádka T, Znamenskiy P, Zador AM (2009) PINP: a new method of tagging neuronal populations for identification during in vivo electrophysiological recording. PLoS One. doi:10.1371/journal.pone.0006099
7. Moore AK, Wehr M (2013) Parvalbumin-expressing inhibitory interneurons in auditory cortex are well-tuned for frequency. J Neurosci 33:13713–13723. doi:10.1523/JNEUROSCI.0663-13.2013
8. Gentet LJ, Kremer Y, Taniguchi H et al (2012) Unique functional properties of somatostatin-expressing GABAergic neurons in mouse barrel cortex. Nat Neurosci 15:607–612. doi:10.1038/nn.3051
9. Pfeffer CK, Xue M, He M et al (2013) Inhibition of inhibition in visual cortex: the logic of connections between molecularly distinct interneurons. Nat Neurosci 16:1068–1076. doi:10.1038/nn.3446
10. Shepherd GMG, Stepanyants A, Bureau I et al (2005) Geometric and functional organization of cortical circuits. Nat Neurosci 8:782–790. doi:10.1038/nn1447
11. Chen S, Mohajerani MH, Xie Y, Murphy TH (2012) Optogenetic analysis of neuronal excitability during global ischemia reveals selective deficits in sensory processing following reperfusion in mouse cortex. J Neurosci 32:13510–13519. doi:10.1523/JNEUROSCI.1439-12.2012
12. Packer AM, Russell LE, Dalgleish HWP, Häusser M (2015) Simultaneous all-optical manipulation and recording of neural circuit activity with cellular resolution in vivo. Nat Methods 12:140–146. doi:10.1038/nmeth.3217
13. Chow BY, Han X, Boyden ES (2012) Genetically encoded molecular tools for light-driven silencing of targeted neurons, 1st ed. Prog Brain Res 196:49–61. doi:10.1016/B978-0-444-59426-6.00003-3
14. Mohajerani MH, Chan AW, Mohsenvand M et al (2013) Spontaneous cortical activity alternates between motifs defined by regional axonal projections. Nat Neurosci 16:1426–35. doi:10.1038/nn.3499
15. Chan AW, Mohajerani MH, Ledue JM et al (2015) Mesoscale infraslow spontaneous membrane potential fluctuations recapitulate high-frequency activity cortical motifs. Nat Commun. doi:10.1038/ncomms8738
16. Grinvald A, Hildesheim R (2004) VSDI: a new era in functional imaging of cortical dynamics. Nat Rev Neurosci 5:874–885. doi:10.1038/nrn1536
17. Akemann W, Mutoh H, Perron A et al (2012) Imaging neural circuit dynamics with a voltage-sensitive fluorescent protein. J Neurophysiol 108:2323–2337. doi:10.1152/jn.00452.2012
18. Chen T-W, Wardill TJ, Sun Y et al (2013) Ultrasensitive fluorescent proteins for imaging neuronal activity. Nature 499:295–300. doi:10.1038/nature12354
19. Shoham D, Glaser DE, Arieli A et al (1999) Imaging cortical dynamics at high spatial and temporal resolution with novel blue voltage-sensitive dyes. Neuron 24:791–802, doi: S0896-6273(00)81027-2 [pii]
20. Suter BA, O'Connor T, Iyer V et al (2010) Ephus: multipurpose data acquisition software for neuroscience experiments. Front Neurosci 4:1–12. doi:10.3389/fnins.2010.00053

Chapter 18

Optogenetic Tools for Confined Stimulation in Deep Brain Structures

Alexandre Castonguay, Sébastien Thomas, Frédéric Lesage, and Christian Casanova

Abstract

Optogenetics has emerged in the past decade as a technique to modulate brain activity with cell-type specificity and with high temporal resolution. Among the challenges associated with this technique is the difficulty to target a spatially restricted neuron population. Indeed, light absorption and scattering in biological tissues make it difficult to illuminate a minute volume, especially in the deep brain, without the use of optical fibers to guide light. This work describes the design and the in vivo application of a side-firing optical fiber adequate for delivering light to specific regions within a brain subcortical structure.

Key words Deep brain stimulation, Optogenetics, Side-firing optical fiber

1 Introduction

The understanding of brain function ineluctably passes through the investigation of the role played by distinct brain structures. To reveal the function of a specific neuronal structure, one can modulate its activity and measure engendered changes [1]. Classical techniques for neural activity modulation include microelectrical stimulation, drugs delivery, genetic manipulations, or lesions making. While these methods led to clear advances in our understanding of the brain, they are associated with poor spatial resolution, slow kinetics, irreversibility, or collateral effects. Recently, optogenetics emerged as a technique allowing high temporal and spatial resolution control over neural activity [2]. Early investigations using optogenetics have mostly been limited to the neocortex [3–5]. The main reason for this is the limited light penetration in biological tissue [6]. However, to fully understand information processing in the brain, it is also necessary to study the function of deep brain structures. Many groups have shown interest in developing tools for deep brain optogenetic stimulation [7, 8], which has proven to be quite challenging. Indeed, due to the absorption

Arash Kianianmomeni (ed.), *Optogenetics: Methods and Protocols*, Methods in Molecular Biology, vol. 1408,
DOI 10.1007/978-1-4939-3512-3_18, © Springer Science+Business Media New York 2016

and scattering of light in the wavelength range used in optogenetics [9], a spatially restricted stimulation of deep brain structures remains difficult without the use of optical fibers to guide light to specific regions.

In this protocol, we present an optogenetic experimental setup that enables the modulation of activity of distinct neural populations in subcortical nuclei of mice expressing ChannelRhodopsin-2. A side-firing optical fiber was developed to shine light in a cylindrical pattern by rotating and translating the fiber around its axis, thus allowing the stimulation of restricted populations of neurons within a thalamic region. Such an approach minimizes the number of fiber penetration needed to probe a given structure, and potentially reduces tissue damage.

Aiming for specific thalamic regions in small animal models such as mice can be quite challenging because of their small size. To assess the spatial extent of optogenetic stimulation, tests were conducted in the visual system, taking advantage of its highly structured network. Indeed, in the visual system of most studied mammals, the organization of the visual field is topographically preserved along the visual pathways through cell distributions and synaptic connections, from the retina to the visual cortex [10]. Here, we used the designed fiber to sequentially stimulate subpopulations of neurons in the lateral geniculate nucleus (LGN), a thalamic structure that receives topographically organized inputs from the retina and projects in an orderly manner to the visual cortex (V1). Effective optogenetic stimulation of subpopulations of neurons in the LGN was validated using intrinsic optical imaging (IOI) of the visual cortex [11].

This protocol is comprised of two main methodological challenges: the development of a precisely rotating side-firing optical fiber and the in vivo optogenetic stimulation of a small thalamic region.

2 Materials

2.1 Optical Setup

2.1.1 Side-Firing Optical Fiber

1. One 1 m Optical fiber (SMA connector suggested) with small NA and a core diameter of 200 μm or less to minimize tissue damage (Suggested: M92L01—ø200 μm, 0.22 NA, SMA-SMA, 1 m, Thorlabs).

2. Fiber stripping tool matching optical fiber size (T01S13, Thorlabs).

3. Glass micropipette.

4. Micropipette beveller (World Precision Instrument, 48000) with sandpapers of various grit sizes (from 10 to 0.3 μm suggested).

5. Electron Beam Physical Vapor Deposition system.

2.1.2 Mechanical System

1. Stepper motor (Matsushita, 55SI-DAYA, precision: 7.5°) and driver (UCN5804B).
2. Two gears with timing belt.
3. Mono-coil tube (HAGITECH, FS-6).
4. Two metal guide tubes (suggested diameter/length: 15 mm/250 mm and 5 mm/300 mm).
5. Two bearings with outer diameter equal to large metal guide inner diameter and an inner diameter equal to small guide tube outer diameter (ABEC-7).

2.1.3 Laser Coupling

1. Optical fiber with same connector, NA and core diameter as in Subheading 2.1.1, step 1.
2. Fiber optic rotary joint (FRJ-v4, Doric).
3. Microscope objective (MV-20x, Newport).
4. Neutral density variable filter (NDC-25C-4M, Thorlabs).
5. Mirror with kinematic mirror mount (KCB1, Thorlabs).
6. X and Y micromanipulators (PT1, Thorlabs).
7. Laser source (DHOM-L-473-50 mW).
8. Optical breadboard (MB1218, Thorlabs).

3 In Vivo Optogenetic Stimulation of Deep Brain Structures

3.1 Reagents

1. Anesthetic (Urethane, 2 g/kg).
2. Local anesthetic (Xylocaine®, 2 %).
3. Skin disinfectant (Povidone-Iodine).
4. Agarose (1 %).
5. Oxygen.

3.2 Equipment

1. Standard surgery tools.
2. Micro drill.
3. Tracheal tube.
4. Needle (24G).
5. Tungsten microelectrode (1–2 MΩ).
6. Electric shaver.

3.3 Setup

1. Side-firing optical fiber setup.
2. Stereotaxic system with stereotaxic arm and electrode holder (Kopf).
3. Electrocardiogram.
4. Feedback-controlled heating pad.

5. Microelectrode amplifier.

6. Imaging data acquisition hardware (Imager 3000, Optical Imaging).

7. Light source (halogen bulb) and fiber optic bundle.

8. Bandpass spectral filters ($\lambda_0 = 545$ nm and 630 ± 15 nm).

9. CCD camera (1M60, Dalsa, Colorado Springs, USA).

10. Lens (Nikon, AF Micro Nikkor, 60 mm, 1:2.8D).

4 Methods

4.1 Optical Setup

4.1.1 Side-Firing Optical Fiber

1. Commercial optical fiber patch cables are usually composed of a core where light propagates, a cladding to maintain light guided in the core and a coating to protect the integrity of the fiber. Cut off the connector from one end of the fiber and remove protective coating to keep only the core and cladding. Using the appropriate fiber-stripping tool, remove cladding 2 cm from the cut optical fiber tip.

2. Insert the bare tip of the optical fiber in a glass micropipette and fix it on the micropipette beveller at an angle of 45°. The glass micropipette will serve as support for the fragile exposed core.

3. Start polishing the silica fiber tip with coarse polishing paper (suggested grit size: 10 µm) and progressively reduce grit size to have a fine polished surface. If final grit size is smaller than the illumination wavelength, specular reflection will occur on the tip. A 0.3 µm grit will lead to such reflection in optogenetic applications.

4. A thin layer (approximately 100 nm) of highly reflective metal, such as chrome, at the fiber tip will ensure an almost 100 % reflection at the fiber tip (Fig. 1a). Electron Beam Physical Vapor Deposition necessitates a fully trained technician to use; therefore the protocol for E-beam usage is skipped here.

4.1.2 Mechanical System

1. The metal guide tube assures that the optical fiber turns on its axis to avoid unbalanced rotation that would damage tissue. Fix the two bearings at each end of the larger metal guide tube, and then insert the smaller guide tube into the bearings (troubleshooting, *see* **Note 1**). At this point, a small guide tube can freely rotate around its axis, in the center of the large guide tube.

2. To fix the optical fiber in the center of the guide tube, use two cylindrical adaptors with a diameter that fits tightly in the small guide tube and drill a small hole in the center to fit the optical fiber with the cladding. By inserting the fiber in both cylindrical

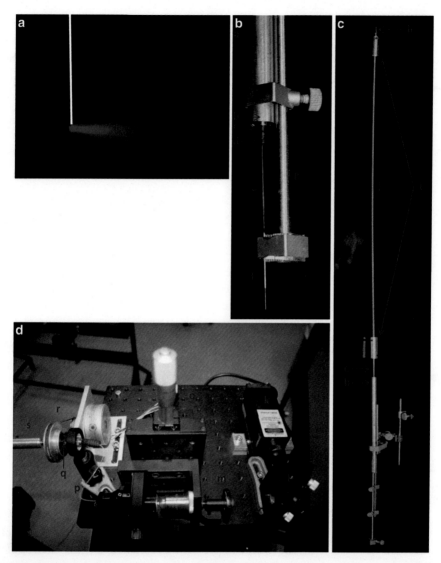

Fig. 1 Opto-mechanical setup. (**a**) Illumination pattern of the designed side-firing optical fiber. White bar indicates the fiber's position. (**b**) Tip of the freely rotating optical fiber. Brass guide tube is fixed to an electrode holder. Black plastic adaptor holds optical fiber in the center of the copper guide tube, which freely rotates in the brass guide tube with high-precision bearings (not visible). A 24G needle fixed on the electrode holder stabilizes rotation of the fiber. (**c**) Inner copper guide tube is linked to mono-coil tube, in which passes the optical fiber, terminated with SMA connector. (**d**) Laser coupling. Free laser beam is attenuated with an adjustable filter wheel and coupled in the optical fiber with a microscope objective. Kinematic mirror at 45° and X and Y micromanipulators allow precise adjustment of beam alignment. Stepper motor puts the rotary joint into motion via timing belt, to which the SMA connector in (**c**) is connected. *Component id.* a: large guide tube (brass), b: bearing (hidden), c: small guide tube (copper), d: cylindrical adaptors (plastic), e: optical fiber, f: 24G needle, g: microelectrode holder, h: SMA connector, i: homemade adaptors, j: mono-coil, k: laser source, l: mirror, m: attenuator, n: collimator, o: micromanipulator, p: optical fiber, q: rotary joint (hidden), r: gears and timing belt, s: top end of side-firing fiber. Figure adapted from [15]

adaptors and placing them at each end of the small guide tube, the optical fiber should rotate smoothly around its axis (Fig. 1b).

3. Fix a microelectrode holder to the end part of the outer metal guide tube and attach a 24G needle. Pass the exposed core of the optical fiber through the needle, which serves to stabilize the optical fiber while rotating (Fig. 1b, c).

4. The rotary joint allows for free rotation of the optical fiber. One side of the rotary joint (stator) should be fixed in placed using a post. The other part of the rotary joint (rotor) is placed in a gearing.

5. The stepper motor and driver are used to rotate the rotary joint. Place the second gear on the shaft of the stepper motor and connect the two gears with the timing belt (Fig. 1d).

6. The mono-coil is a hollow flexible shaft that transmits rotary motion from the rotary joint to the guide tube. Fix one end of the mono-coil to the polished optical fiber connector and the other end to the metal guide tube with homemade adaptors (Fig. 1c).

7. Connect the optical fiber to the rotor of the rotary joint.

4.1.3 Laser Coupling

1. Fix laser source on breadboard (*see* Fig. 1d).

2. Place mirror with kinematic mirror mount at 45° with incident laser beam to deflect beam at right angle.

3. Place neutral density filter in the beam path. This will allow modulating output power of the fiber.

4. Fix collimator on an X and Y micromanipulator and direct laser beam in collimator.

5. Connect one end of the optical fiber to the collimator.

6. Using the mirror kinematic mount to make fine adjustments on the beam angle and the X and Y micromanipulators to precisely center the collimator on the beam, maximize light coming out of the connected optical fiber.

7. Connect the free end of the optical fiber to the stator of the rotary joint.

4.2 In Vivo Optogenetic Stimulation of Deep Brain Structures

In this study, mice expressing channelrhodopsin-2 (ChR2) fused to Yellow Fluorescent Protein under the control of the mouse thymus cell antigen 1 promoter were used (strain name: B6.Cg-Tg (Thy-COP4/EYFP), Jackson Laboratory). All procedures were carried out in accordance with the guidelines of the Canadian Council for the Protection of Animals and the experimental protocol was approved by the Ethics Committee of the Université de Montréal.

Mice were anesthetized with an intra peritoneal injection of urethane (2 g/kg). After testing reflexes on hind paw, 2 % Xylocaine® was injected subcutaneously in the region covering the trachea. The neck and the head of the mice was shaved and disinfected with iodine solution. Tracheotomy was performed in order to ease breathing.

The animal was then placed in stereotaxic frame. A flow of pure oxygen was directed toward the tracheal tube. Body temperature was maintained through the experiment at 38 °C with a heating pad and the electrocardiogram was continuously monitored (Fig. 2c).

Finally, the head of the mouse was stereotaxically aligned. Targeted brain structure, LGN, in mice is of very small size (<1 mm³). It is therefore of prime importance to have a precise stereotaxic alignment if we are to reach the LGN of the mice.

To reach a deep brain structure with an optical fiber, one must first determine its exact stereotaxic position and find the best possible way to reach it. In this particular case, the LGN is situated directly under the primary visual cortex V1 (Fig. 2b) where we wish to image intrinsic optical signals. To avoid blocking access to V1, the optical fiber was inserted in the cortex with a 57° angle relative to the vertical. A small craniotomy (1 mm diameter) was performed with a micro drill (0.5 mm in diameter) at −2.3 mm relative to bregma in the sagittal axis and at 4 mm on the lateral axis.

To insure that the right location was found, a tungsten microelectrode was first used to record neural activity in the LGN (*see* **Note 2**). The microelectrode was fixed to the optical fiber stereotaxic holder (*see* Subheading 4.1.2, step 3). The microelectrode was then lowered through the bone opening and towards the LGN until robust visual multiunit responses were evoked by light flashes (approximately 2.7 mm from the surface) (troubleshooting, *see* **Note 3**).

Once the localization of the LGN was confirmed, the microelectrode was removed and replaced by a 24G needle. The exposed core of the optical fiber was then passed through the needle, which served two purposes: first to maintain the same coordinates as the microelectrode and second to stabilize the optical fiber during its rotation. The optical fiber was then lowered in the brain at the same depth where visual responses were obtained, in order to optogenetically induce neuronal activity in the LGN.

1. Intrinsic optical signals

The recording of neural activation was achieved with intrinsic optical imaging (IOI), which represents a powerful technique to visualize the global functional architecture of cortical areas in vivo. One approach is to monitor the slow

a

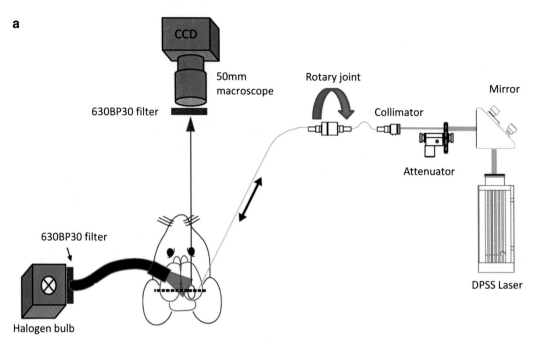

b

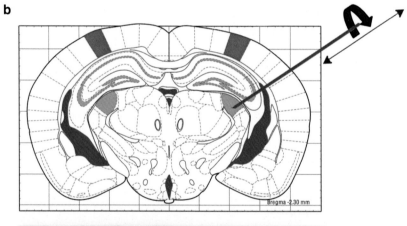

c

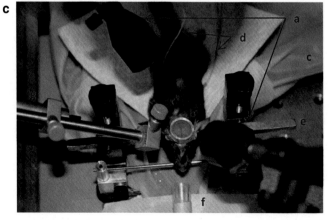

intrinsic changes in the optical properties of the active cortex over time. The main source for these activity-dependent intrinsic signals come from local variations in the concentration of deoxy- and oxy-hemoglobin within capillaries in response to the increased metabolic demand of active neurons [12]. The cortex was imaged directly through the skull by placing a circular ring above the visual cortex as an imaging chamber. The mouse skull is very thin and translucent, making intrinsic optical signals visible without the need for a craniotomy. The chamber was filled with agarose (1 %) and sealed with a coverslip to keep the skull hydrated and transparent.

A 12 bit CCD camera fitted with a macroscopic lens was used to record cortical activity. First, the brain was illuminated at 545 nm to image the cortical vasculature at high contrast in order to adjust the focus of the camera on the cortex. Light was guided from a filtered halogen bulb to the cortex using a fiber optic bundle (Fig. 2a). Intrinsic optical signals were acquired by illuminating the cortex with a 630 ± 15 nm light, a spectral band where absorption is sensitive to variations in deoxyhemoglobin concentration [13].

2. Stimulation and data acquisition

Precise control of fiber illumination conditions allowed us to define a finite volume of optogenetic activation. Single illumination pulses of 250 ms were used, a paradigm shown to be appropriate for the stimulation of channelrhodopsin 2 [14]. To limit the volumetric extent of excited neural tissue by light, the output power from the optical fiber tip was varied from 1 to 10 mW/mm^2 [15] (*see* **Note 4**).

Stimulus synchronization between optogenetic light pulse and IOI system was achieved using VDAQ software and Imager 3001 data acquisition hardware (Optical Imaging Ltd, Israel). Trials lasted a total of 20 s and camera frames were acquired at a rate of 4 Hz. A pre-stimulation period of 2 s preceded the continuous light pulse stimulation of 250 ms, followed by a post-stimulation period of 18 s, during which the slow hemodynamic response occurred. Each trial was repeated 20 times for any given radial or axial position of the fiber.

Fig. 2 Experimental setup. (**a**) Simplified experimental design. A halogen bulb with a 630BP60 filter is used for illumination in order to measure intrinsic signal from the cortex with a CCD camera placed over the head. Side-firing optical fiber is lowered in the LGN by triangulation, where it is free to rotate and slide along its axis. (**b**) Coronal slice of the *dashed line* in (**a**). Adapted from Franklin and Paxinos, 2013 [17]. Fiber is inserted at an angle of 57° through a small craniotomy into the lateral geniculate nucleus (*green* region) using a stereotaxic holder (not shown). Primary visual cortex is presented in *red*. (**c**) Animal preparation. Fiber bundles guide red light to shine the cortex, through an agarose-filled chamber. Optical fiber is inserted lateral to the chamber. Mouse head is fixed with ear bars. *Component id.* a: fiber optic bundle, b: imaging chamber, c: heating pad, d: ECG electrodes, e: ear bars, f: oxygen flow, g: side-firing optical fiber

5 Typical Results

1. Electrophysiological recording in the LGN

   A tungsten microelectrode is used to confirm the position of the lateral geniculate nucleus. Figure 3a shows a typical recording of multiunit responses in the LGN evoked by visual stimuli (flash).

2. IOI following optogenetic stimulation of LGN

   Hemodynamic responses are slow compared to neural activity, lasting 4–10 s, with maximal signal variation occurring approximately 2–3 s after neural stimulation. Figure 3b shows signal variation between camera frame acquired 2.5 s after optogenetic stimulation and pre-stimulation condition (0–2 s) over primary visual cortex ipsilateral to stimulated LGN, over 20 trials. A neural activation is characterized by a drop of reflectance signal, as shown in blue. By averaging the signal in a region of the primary visual cortex (square region in Fig. 3b) and subtracting control trials, we can visualize signal variation over time to obtain the hemodynamic response time course over 20 trials (Fig. 3c). We see a rapid loss of signal at time $t=0$ s where optogenetic stimulation occurs, followed by a progressive rise of reflectance, characteristic of a hemodynamic response.

   To have an insight on the spatial extent of optogenetic stimulation in the LGN, the visual cortex area consequently activated was measured. A Student t test indicated areas having a significant reflectance difference between peak activation (2.5 s after stimulation) and baseline (pre-stimulation average), thus delimiting a precise activation zone, as presented in Fig. 3d.

   The activation map presented in Fig. 3d is for a precise position of the optical fiber in the LGN of a mouse. By rotating the fiber around its axis, varying its depth and by modulating the output power of the fiber, it is possible to activate different neuron populations, giving rise to different activation maps in primary visual cortex. *For further investigation using the present setup for LGN optogenetic activation, please consult* [15].

6 Notes

1. Bearing and guide tube fitting

   Fitting the bearings firmly in the metal guide tubes may reveal to be challenging. If the inner metal guide is too big to fit in the bearing, try sanding or cooling the metal guide tube to reduce its diameter to fit in the bearing. Also, for fitting the bearing in the larger guide tube, heating the rod will temporarily expand the tube to give a little wiggle room. Teflon can be used to tighten any loose fits.

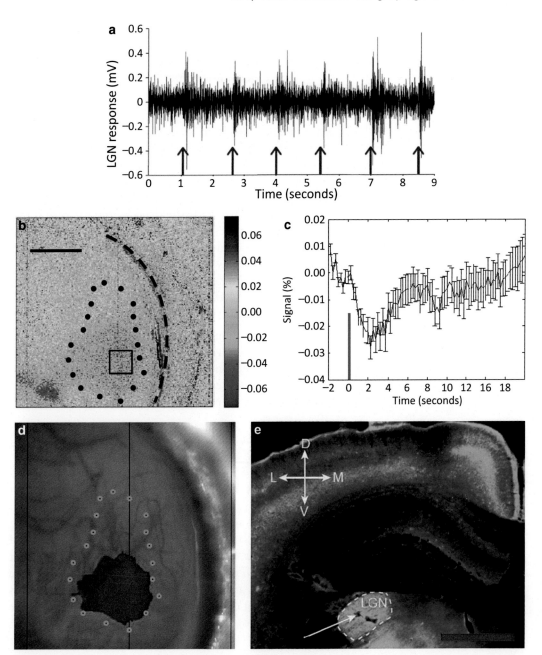

Fig. 3 In vivo optogenetic stimulation and IOI recordings. (**a**) Extracellular multiunit recordings in the LGN in response to visual stimuli (*red arrows*) (adapted from [15]). (**b**) Reflectance signal variation on cortical surface, 2.5 s following optogenetic stimulation of the LGN. Boundaries of primary visual cortex are delimited with *dots*, whereas *dashed line* indicates edge of imaging chamber. Color code represents signal variations in percentage. Optical power at fiber tip: 3,2 mW/mm². Scale bar: 1 mm. (**c**) Time course of relative reflectance signal variations (in %) for the region identified by a *square* in (**b**), averaged over 20 trials. Light pulse duration is represented in *blue*. Error bars show the standard error. (**d**) Activation map in (**b**) after delimitation with a Student *t* test ($\alpha = 0.01$) superimposed on the anatomical image of mouse cortex. Boundaries of primary visual cortex are delimited with *dots*. (**e**) Immunofluorescence image of ChR2 expression from coronal brain section taken at the level of the LGN, delimited by *dashed line* (adapted from [15]). The optical fiber insertion path is indicated by the *arrow*. Dorsoventral (D–V) and mediolateral (M–L) axes are presented. Scale bar: 1 mm. Figure adapted from [15]

2. Optrode

In this protocol, an electrode is used to first confirm the right stereotaxic alignment for the optical fiber. Optrodes are now being developed and commercialized [7, 16], which could be used in this protocol for both recording electrical activity and optogenetic stimulation.

3. Difficulty reaching deep brain structures

Reaching the LGN can be quite difficult due to its small size (<1 mm3). Precise stereotaxic alignment of the brain is of prime importance. Make sure that the Lambda and Bregma intersection points [17] are at the same height and in a straight line on the mediolateral axis. If the deep brain structure is not reached from the first descent of the electrode, remove and insert it at a distance equivalent to the diameter of the structure from the initial entry point. Continue insertions around the initial entry coordinates in a grid pattern until recordings from the desired structure are obtained.

4. Monte Carlo simulation

To evaluate the volumetric extent of optogenetic stimulation in deep brain structures, our group has used Monte Carlo simulation [18]. Based on estimated tissue optical properties, such simulations allow users to estimate the volume of excited tissues by various output power of the fiber [15].

Acknowledgment

Supported by a Fonds de Recherche du Québec-Nature et Technologies (FRQ-NT) grant #165075 (Projet de recherche en équipe) to F.L. and C.C. and by Natural Sciences and Engineering Research Council of Canada (NSERC) grants #194670 and 239876 to C.C. and F.L., respectively. A.C. was supported in part by scholarships from the Fonds de Recherche en Santé du Québec (FRSQ) vision network and the Faculté des Études Supérieures et Postdoctorales-Institut de Génie Biomedical (FESP-IGB) of the Université de Montréal.

References

1. Berman RA, Wurtz RH (2008) Exploring the pulvinar path to visual cortex. Prog Brain Res 171:467–473, Elsevier

2. Boyden ES, Zhang F, Bamberg E et al (2005) Millisecond-timescale, genetically targeted optical control of neural activity. Nat Neurosci 8:1263–1268

3. Scott NA, Murphy TH (2012) Hemodynamic responses evoked by neuronal stimulation via channelrhodopsin-2 can be independent of intracortical glutamatergic synaptic transmission. PLoS One 7:e29859

4. Mateo C, Avermann M, Gentet LJ et al (2011) In vivo optogenetic stimulation of neocortical excitatory neurons drives brain-state-dependent inhibition. Curr Biol 21:1593–1602

5. Yizhar O, Fenno LE, Prigge M et al (2011) Neocortical excitation/inhibition balance in information processing and social dysfunction. Nature 477:171–178

6. Fodor L, Ullmann Y, Elman M (2011) Light tissue interactions. In: Aesthetic applications of intense pulsed light. Springer, London

7. LeChasseur Y, Dufour S, Lavertu G et al (2011) A microprobe for parallel optical and electrical recordings from single neurons in vivo. Nat Methods 8:319–325

8. Zorzos AN, Scholvin J, Boyden ES, Fonstad CG (2012) Three-dimensional multiwaveguide probe array for light delivery to distributed brain circuits. Opt Lett 37:4841

9. Fenno L, Yizhar O, Deisseroth K (2011) The development and application of optogenetics. Annu Rev Neurosci 34:389–412

10. Chalupa L, Williams R (2008) Eye, retina and visual system of the mouse. The MIT Press, Cambridge

11. Grinvald A, Lieke E, Frostig RD et al (1986) Functional architecture of cortex revealed by optical imaging of intrinsic signals. Nature 324:361–364

12. Hillman EMC (2007) Optical brain imaging in vivo: techniques and applications from animal to man. J Biomed Opt 12:051402

13. Frostig RD, Masino SA, Kwon MC, Chen CH (1995) Using light to probe the brain: intrinsic signal optical imaging. Int J Imaging Syst Technol 6:216–224

14. Wang J, Wagner F, Borton DA et al (2012) Integrated device for combined optical neuromodulation and electrical recording for chronic *in vivo* applications. J Neural Eng 9:016001

15. Castonguay A, Thomas S, Lesage F, Casanova C (2014) Repetitive and retinotopically restricted activation of the dorsal lateral geniculate nucleus with optogenetics. PLoS One 9:e94633

16. Lin S-T, Gheewala M, Wolfe JC, et al (2011) A flexible optrode for deep brain neurophotonics. Paper presented at the 5th International IEEE EMBS Conference on Neural Engineering, Cancun, April 27–May 1 2011

17. Franklin KBJ (2013) Paxinos and Franklin's: the mouse brain in stereotaxic coordinates, 4th edn. Academic, Amsterdam, An imprint of Elsevier

18. Boas D, Culver J, Stott J, Dunn A (2002) Three dimensional Monte Carlo code for photon migration through complex heterogeneous media including the adult human head. Opt Express 10:159–170

Chapter 19

Remote Patterning of Transgene Expression Using Near Infrared-Responsive Plasmonic Hydrogels

Francisco Martín-Saavedra and Nuria Vilaboa

Abstract

The development of noninvasive technologies for remote control of gene expression has received increased attention for their therapeutic potential in clinical scenarios, including cancer, neurological disorders, immunology, tissue engineering, as well as developmental biology research. Near-infrared (NIR) light is a suitable source of energy that can be employed to pattern transgene expression in plasmonic cell constructs. Gold nanoparticles tailored to exhibit a plasmon surface band absorption peaking at NIR wavelengths within the so called tissue optical window (TOW) can be used as fillers in fibrin-based hydrogels. These biocompatible composites can be loaded with cells harboring heat-inducible gene switches. NIR laser irradiation of the resulting plasmonic cell constructs causes the local conversion of NIR photon energy into heat, achieving spatially restricted patterns of transgene expression that faithfully match the illuminated areas of the hydrogels. In combination with cells genetically engineered to harbor gene switches activated by heat and dependent on a small-molecule regulator (SMR), NIR-responsive hydrogels allow reliable and safe control of the spatiotemporal availability of therapeutic biomolecules in target tissues.

Key words Hydrogel, Gene therapy, Gold, Infrared, Nanoparticle, Biomaterial, Scaffold, Transgene, Spatiotemporal, Plasmon

1 Introduction

Heat-shock protein genes encode a small group of proteins (HSPs) found in virtually all living organisms from bacteria to higher vertebrates. HSPs accumulate transiently to elevated levels in cells exposed to certain physical and chemical stressors, heat being the most powerful inducer of these genes [1–4]. Induced expression of *hsp* genes is mediated by heat-shock factors (HSFs) that interact with heat-shock elements located in their promoters [5]. The human HSP70B gene is one of the most highly inducible HSP genes [6–8]. Its promoter has a very low basal activity, which can be induced several thousandfold upon thermal treatment. Interestingly, the magnitude of HSP70B promoter activation is a function of the intensity of the heat treatment in terms of both

Arash Kianianmomeni (ed.), *Optogenetics: Methods and Protocols*, Methods in Molecular Biology, vol. 1408,
DOI 10.1007/978-1-4939-3512-3_19, © Springer Science+Business Media New York 2016

temperature and length of exposure, an attribute that makes this promoter extremely attractive for the deliberate control of transgene expression [9]. The main problems associated with uses of HSP promoters in gene therapy are the insufficiently long periods of transgene expression as well as the possibility of inadvertent expression of transgenes [10–13]. These problems are avoided by gene switches that combine an HSP70B promoter and an SMR-activated transactivator. Gene switches employing transactivators activated by different SMRs successfully provide spatiotemporal control of transgene activity [14, 15].

To administer heat in a focused fashion, technologies based in ultraviolet (UV), short-wavelength visible (vis), or infrared (IR) laser-light irradiation have been developed. However, the main limitation of these approaches arises from the poor penetration of UV, vis, or IR light in biological tissues, which are less prone to absorb or scatter light in the wavelength range of the TOW (i.e., 650–1100 nm). As a consequence, NIR laser light may penetrate at least 10 cm through deep tissue [16, 17]. On account of the phenomenon of localized surface plasmon resonance, gold nanoparticles (GNPs) tailored to strongly absorb NIR light can be used as highly efficient nanotransducers for converting this kind of photon energy into heat. This nanotechnology resource, known as plasmonic photothermia, has been successfully employed for inducing the expression of transgenes driven by the human HSP70B promoter [18, 19].

This chapter describes the detailed procedure for preparing fibrin-based hydrogels that integrate in their structure GNPs with a plasmon surface band absorption peaking at NIR wavelengths within the TOW. The protocol describes the steps for encapsulating cell populations within the NIR-responsive biomaterial to obtain three-dimensional assemblies or injectable cell constructs. The chapter also includes the procedures for in vitro and in vivo NIR irradiation of plasmonic scaffolds populated by genetically modified cells, which enable a tight spatiotemporal control of transgene expression patterns.

2   Materials

Prepare all reagents inside a tissue culture hood, at room temperature. Diligently follow waste disposal regulations when disposing waste materials.

2.1   Fibrin-Based Plasmonic Hydrogels

1. Lab scale.
2. Ultrasonic bath.
3. Refrigerated benchtop centrifuge.
4. Sterile metallic bent spatula.

5. Sterile tweezers.

6. Disposable biopsy punches.

7. Conical centrifuge tubes, 15 and 50 mL.

8. Hemocytometer.

9. Eppendorf tubes, 1.5 mL.

10. Syringes.

11. Sterile syringe filter with a 0.2 μm polyethersulfone (PES) membrane.

12. Sterile multiwall plates.

13. Double-distilled water.

14. Penicillin-streptomycin (P/S) solution 10,000 U/mL.

15. Fetal bovine serum heated for 30 min at 56 °C with mixing to inactivate complement (FBSi).

16. Dulbecco's phosphate-buffered saline (DPBS).

17. Dulbecco's modified Eagle's medium incorporating L-glutamine, high glucose (4.5 g/L) and phenol red (DMEM).

18. Trypsin-EDTA solution: 0.25 % Trypsin/0.91 mM EDTA with phenol red. Prewarm to 37 °C before use.

19. GNP stock solution. Weigh lyophilized nanomaterial in a glass vial and add sterile double-distilled water to a final concentration of 1 mg/mL. In order to break particle agglomerates and facilitate nanomaterial dispersion in water, the vial is kept immersed in ice water for 30 min in an ultrasonic bath operating at 40 kHz and a power of 80 W. Sterilize GNP solution by ultraviolet germicidal irradiation for 12 h. Store GNP stock solution at 4 °C.

20. P/S-DMEM solution: Add penicillin-streptomycin solution to DMEM at a final concentration of 1,000 U/mL. Store at 4 °C.

21. Cell culture media: Add 0.1 volumes of FBSi to P/S-DMEM. Store at 4 °C.

22. Fibrinogen suspension: Weigh bovine fibrinogen (Sigma) in a 50 mL centrifuge tube and add P/S-DMEM to a final concentration of 20 mg/mL of clottable protein. Gently stir the solution with a vortex for 20 s and incubate at 37 °C in a water bath for 10 min. Repeat this step twice and then centrifuge at $3,000 \times g$ for 5 min in a benchtop centrifuge refrigerated at 4 °C. Check that fibrinogen is completely solubilized and there are no traces of foam or insolubilized fibrinogen. Otherwise, gently stir the solution with vortex for 20 s, incubate for an additional 10 min at 37 °C, and then repeat centrifugation step at 4 °C. Filter solution using a sterile syringe filter with a 0.2 μm PES membrane.

23. Fibrinogen conjugate suspension: Labeled fibrinogen from human plasma is commercially available in four fluorescent colors (Alexa Fluor -488, -546, -594, -647 or Oregon-488; Molecular Probes). Reconstitute 5 mg of lyophilized product in 3.33 mL of 0.1 M sodium bicarbonate (pH 8.3) to obtain a 1.5 mg/mL stock solution. Complete solubilization may take 1 h or more with occasional gentle mixing. Divide the solution into aliquots and freeze at −20 °C protected from light.

24. Thrombin solution: Add double-distilled water to lyophilized thrombin from bovine plasma (Sigma) to a final concentration of 500 U/mL. Store at −20 °C. Prepare the working solution by diluting stock solution at 40 U/mL in chilled P/S--DMEM. Filter solution using a sterile syringe filter with a 0.2 μm PES membrane.

2.2 Animal Experimentation

1. Table-top anesthesia system: Isoflurane vaporizer with O_2 H-Tank regulator, fluosorber charcoal canister, oxygen hose, and rodent nose cone (Harvard Apparatus).

2. Bead sterilizer.

3. Heat lamp for use during small animal surgery (75 W infrared bulb).

4. Warming blanket.

5. Electric clipper for small animals.

6. Surgical instruments.

7. 7 mm wound clips, wound clips applier, and wound clip remover.

8. Sterile gloves.

9. Sterile syringes.

10. 20 and 27 G needles.

11. High-density foam.

12. Tegaderm™ absorbent clear acrylic dressing (3 M).

13. Aluminum foil.

14. Ophthalmic ointment.

15. Topical depilatory cream.

16. Povidone-iodine 10 % solution.

17. Sterile gauze pads.

18. Sterile surgical drapes.

2.3 Patterning of Transgene Expression in Fibrin-Based Plasmonic Hydrogels

1. Rapamycin (Invivogen): Dissolve rapamycin solid in *N*-dimethylacetamide (DMA) to prepare a stock solution of 3 mg/mL and store at −20 °C. For in vitro assays use rapamycin at a final concentration of 10 nM. For in vivo assays, dilute stock in a mixture of 50 % N,N-dimethylacetamide, 45 %

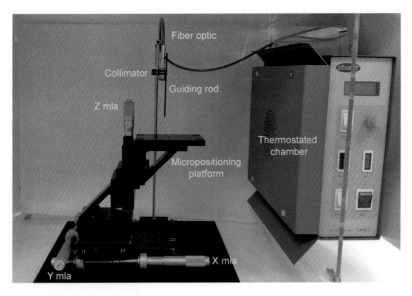

Fig. 1 Experimental setup for NIR irradiation. *mla* micro-linear actuator

polyethylene glycol (average molecular weight of 400 Da), and 5 % polyoxyethylene sorbitan monooleate.

2. NIR irradiation setup (Fig. 1) consisting of a laser diode emitting at a wavelength of 808 nm coupled to a fiber optic of 400 μm of diameter (MXL-III(FC) model, Changchun New Industries Optoelectronics Technology Co., Ltd) connected to a fixed focus collimator (Thorlabs). The collimator is mounted on a micropositioning platform to ensure that the laser beam illuminates the sample orthogonally. For in vitro irradiation assays, the fiber-optic and the positioning system are placed inside a thermostatically controlled cabinet to establish the environmental temperature at 37 °C. For in vivo irradiation assays, the micropositioning platform is covered with a thermal blanket to maintain the corporal temperature of the anesthetized animal. Control measures of NIR laser hazard include laser warning signs, laser safety barriers, and safety laser goggles (all from Thorlabs).

3   Methods

To illustrate the methodology, the procedures detailed below use cells, derived from C3H/10T1/2 cell line (ATCC CCL-226), which stably harbor a firefly luciferase gene under the control of a heat-activated and rapamycin-dependent gene switch (Fig. 2). In rapamycin-treated cells, levels of transgenic luciferase expression increase as function of the intensity of an activating heat treatment.

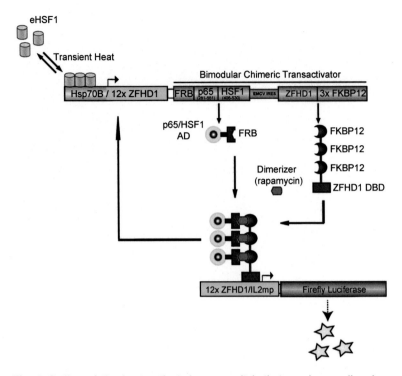

Fig. 2 Outline of the heat-activated gene switch that employs a dimerizer-regulated transactivator to control firefly luciferase expression. A bicistronic gene encoding the two component proteins of a transactivator regulated by a dimerizer (rapamycin) is expressed under the control of promoter cassette HSP70B/12xZFHD12 that responds to activated endogenous heat-shock factor 1 (eHSF1), the transcription factor that mediates heat-induced expression of HSP genes, as well as the dimerizer-activated transactivator. Transactivator-responsive promoter 12xZFHD12/IL2mp drives the expression of the linked transgene. *AD* activation domain. *DBD* DNA-binding domain. *See* ref. 15 for a detailed description of the gene-switch components

Note that if heating conditions are too harsh, the level of induced luciferase activity may decrease substantially due to cell damage caused by thermal stress [20].

3.1 Fibrin-Based Plasmonic Hydrogels Encapsulating Genetically Modified Cells

3.1.1 Preparation of Cellular Component of Fibrin-Based Plasmonic Hydrogels

Carry out all procedures under aseptic conditions in a tissue culture hood.

1. Remove culture medium and eliminate residual serum by rinsing cell monolayers with 5 mL of DPBS per 25 cm^2 of surface area of cell culture. Let the DPBS sit on cells for at least 30 s to remove as much extracellular proteins as possible.

2. Aspirate DPBS and add 1 mL of trypsin-EDTA solution per 25 cm^2 of surface area of cell culture (*see* **Note 1**).

3. Incubate at the tissue culture incubator (37 °C/5 % CO$_2$/100 % humidity) for ~5 min (*see* **Note 2**).

4. Neutralize the trypsin by adding 10 mL of culture medium per mL of trypsin-EDTA solution and distribute cell suspension in centrifuge tubes of appropriate size.

5. Centrifuge cells at $200 \times g$ for 5 min in a benchtop centrifuge.

6. Following centrifugation, aspirate the media and resuspend the cell pellet in 0.25 mL of P/S-DMEM per 25 cm² of initial surface area of cell culture.

7. Count cells using a hemocytometer.

8. Dilute cells in P/S-DMEM to a final concentration of 5×10^6 cells/mL and keep cell suspension on ice.

3.1.2 Preparation of the Polymerizable Mixture for Fabricating Fibrin-Based Plasmonic Cell Constructs

Carry out all procedures under aseptic conditions in a tissue culture hood. All samples must be kept on ice unless otherwise specified.

1. To break up agglomerates of GNPs formed during storage of the sample, immerse the vial containing the GNP stock solution for 10 min in an ultrasonic bath filled with ice water that operates at 40 kHz and a power of 80 W.

2. Add 0.01–0.1 volumes of sonicated GNP stock solution to 0.5 volumes of fibrinogen suspension and pipette briefly to ensure uniform dispersion of the nanomaterial (*see* **Note 3**).

3. According to the volume used of GNP stock solution, add prechilled P/S-DMEM to make up to 0.7 volumes of plasmonic hydrogel mix. Pipette briefly to homogenize the solution (*see* **Note 4**).

4. Add 0.2 volumes of cell suspension. Pipette gently to distribute the cells among the other components of the plasmonic hydrogel mix.

5. Add 0.1 volumes of chilled thrombin working solution to the plasmonic hydrogel mix that includes the cellular component. Pipette gently to homogenize the sample. The solution is ready for the fabrication of three-dimensional plasmonic assemblies described in Subheading 3.1.3, or for subcutaneous polymerization of plasmonic cell constructs as indicated in Subheading 3.2.2 (*see* **Note 5**).

3.1.3 Fabrication of Three-Dimensional Plasmonic Assemblies

1. Add 1 volume of FBSi to the well selected as mold for the gelation of the plasmonic hydrogel-incorporating cells (*see* **Note 6**).

2. Incubate the well-plate at the tissue culture incubator for 5 min. Aspirate the FBSi.

3. Dispense 1 volume of the mixture obtained in Subheading 3.1.2, **step 5**, into the well pretreated with FBSi. Leave for 5 min at room temperature. Monitor visually the sol-gel transition and incubate at the tissue culture incubator for 1 h to allow consolidation of the plasmonic cell construct.

4. Add 0.9 volumes of cell culture medium and 0.1 volumes of FBSi to the top of the hydrogel. Incubate at the tissue culture incubator for 1 h to balance the serum content of the hydrogel.

5. Use a bent spatula for unmolding the plasmonic cell construct and transfer the assembly to a larger well containing 4 hydrogel volumes of cell culture medium.

6. Incubate in the tissue culture incubator for 24 h.

3.2 In Vivo Implantation of Fibrin-Based Plasmonic Cell Constructs

3.2.1 Surgical Preparation of the Animal

1. Administer preemptive analgesics according to the institutional procedure for Animal Care and Use.

2. Anesthetize the animal with isoflurane. Administer the anesthetic at 5 % for induction and at 2 % for maintenance in O_2 flowing at 1 L/min. Protect animal eyes from desiccation using an ophthalmic ointment.

3. Remove hair from implantation site using an electric clipper.

4. Treat the skin in the region of implantation with depilatory cream for 5 min.

5. Remove loose hair and debris from the animal and clean the surgical area with water and gauzes.

6. Place the animal in the surgical area.

7. Scrub surgical site with povidone-iodine solution.

8. Rinse with 70 % ethanol.

9. Repeat soap scrub and rinse process three times.

10. Drape the animal with sterile, impermeable covering to isolate the disinfected area.

3.2.2 Subcutaneous Polymerization of Fibrin-Based Plasmonic Cell Constructs

1. Pre-chill a 1 mL syringe on ice.

2. Load the syringe with 1 volume of the mixture obtained in Subheading 3.1.2, **step 5**.

3. Attach a 20-gauge needle to the syringe.

4. Insert the needle, bevel up and slightly angled, in the animal skin to access to the subcutaneous space.

5. Lift the skin slightly with the needle and inject slowly the contents of the syringe into the subcutaneous space.

6. Remove carefully the needle from the area of implantation (*see* **Note 7**).

7. Keep the anesthetized animal stationary for 5 min under the infrared lamp to complete the polymerization of the hydrogel.

3.2.3 Subcutaneous Implantation of Three-Dimensional Plasmonic Assemblies

1. Use a scalpel blade to make the smallest possible incision.

2. Use sterile tweezers to open a subcutaneous pocket of sufficient size to fit the hydrogel assembly obtained in Subheading 3.1.3.

3. Introduce the plasmonic assembly in the subcutaneous pocket using sterile forceps.

4. Close the skin incision with 7 mm clips stapled with a wound clip applier (*see* **Note 8**).

3.3 Patterning Transgene Expression in Fibrin-Based Plasmonic Cell Constructs by NIR Irradiation

The coherence and low divergence angle of a NIR laser, aided by focusing from the lens of an eye, can concentrate the radiation into a spot on the retina leading to irreversible damage. When operating with the NIR laser diode, refer to institutional laser safety procedures and always wear certified laser safety glasses (ANSI Z136 and CE).

3.3.1 Patterning Transgene Expression In Vitro

1. To immobilize the hydrogel, with the help of a bent spatula transfer the three-dimensional plasmonic assembly obtained in Subheading 3.1.3 to a well-plate of similar size to that used for gelation. Add 2 volumes of cell culture medium containing rapamycin at a final concentration of 10 nM.

2. Incubate the well-plate in the tissue culture incubator.

3. Place the well-plate containing the three-dimensional hydrogel assembly on the micropositioning platform placed inside the thermostated chamber set at 37 °C.

4. Use the Z micro-linear actuator to position the collimator 1 mm above the top of the lid of the well-plate.

5. Use the X-Y micro-linear actuators to align concentrically the hydrogel-well with the collimator. This position is established as the center of coordinates of the system (*see* **Note 9**).

6. Use the scale graduation of the micro-linear actuators to set the coordinates for the first irradiation spot of the projected pattern.

7. Select continuous-wave (CW) or pulsed mode of NIR irradiation in the driver unit. For the pulsed regime of irradiation, select duty cycle and pulse repetition rate. Switch on the laser driver unit and adjust to the desired output power.

8. After irradiation at the selected illumination spot, turn off the laser driver unit and use the micro-linear actuators to position the collimator at the next irradiation spot.

9. Repeat **steps 5** and **6** to complete the desired pattern of transgene expression.

10. Incubate the well-plate in the tissue culture incubator for the required period of time and measure transgene activity by bioluminescence assay using a coupled charged device (CCD) camera and D-luciferin substrate.

3.3.2 Patterning Transgene Expression In Vivo

The use of a mask for referencing NIR-irradiation spots on the body of the animal is highly recommended to accurately induce patterns of transgene expression in vivo.

1. Use a sheet of paper to draw the desired irradiation pattern based on circles of 2.5 mm of diameter.

2. Place the template on the top of a sheet of high-density foam.

3. Place a piece of Tegaderm™ transparent film dressing on the template.

4. Use a biopsy punch of 2.5 mm of diameter to trim the Tegaderm™ membrane at points that shape the projected pattern.

5. One hour before irradiation, inject rapamycin intraperitoneally at a dose of 1 mg/kg in a final volume of 50 μL using a 27 G needle.

6. To proceed with NIR irradiation, anesthetize the animal with isoflurane and cover its head with aluminum foil to protect eyes from laser light scattering.

7. Place the irradiation mask at the implantation site.

8. Place the animal on the micropositioning platform covered with the thermal blanket.

9. Operate on Z micro-linear actuator to position the collimator ~5 cm above the region of implantation.

10. Use X-Y micro-linear actuators to align the guiding rod with the first irradiation spot masked in the Tegaderm™ membrane (*see* **Note 10**).

11. Use X micro-linear actuator to position the collimator at the known distance between the center of the collimator lens and the guiding rod, thereby focusing the laser beam at the center of targeted irradiation spot.

12. Select CW or pulsed mode of NIR irradiation in the driver unit. For the pulsed regime of irradiation, select duty cycle and pulse repetition rate. Switch on the laser driver unit and adjust to the desired output power.

13. After irradiation, turn off the laser driver unit and use the micro-linear actuators and the guiding rod to position the collimator at the next irradiation spot.

14. Repeat **steps 6–9** to complete the masked irradiation pattern.

15. Remove irradiation mask and transfer the animal to the recovery cage.

16. After the desired period of time, measure induced transgene activity by bioluminescence assay using a CCD camera and D-luciferin substrate.

4   Notes

1. Trypsin-EDTA solution is suitable for most but not for all adherent cells. Use cell dissociation solution and cell culture medium recommended for subculturing and harvesting the cells of choice.

2. Examine cells under an inverted microscope. Fully trypsinized cells should appear rounded up and no longer attached to the surface of the culture plasticware. Otherwise, increase the time of incubation in the tissue culture incubator for an additional 5 min.

3. Upon NIR irradiation, GNPs embedded in fibrin-based plasmonic hydrogels promote high induction of transgenic activity with negligible cell damage. Optimal GNP concentration in plasmonic cell constructs must be determined empirically by performing procedure 3.3.1 followed by quantification of induced transgene activity and cell viability in the plasmonic cell construct.

4. To monitor dynamic degradation of plasmonic cell constructs, add 0.02 volumes of a human fibrinogen conjugate solution that must be substracted to the amount of P/S-DMEM to complete 0.7 volumes of plasmonic hydrogel mix. Hydrogel degradation over time can be estimated by fluorometric quantification techniques or fluorescence imaging.

5. After thrombin addition, slow fibrin polymerization may occur in chilled samples. Work fast and minimize handling time between adding the thrombin solution and dispensing/injecting the plasmonic hydrogel mix containing cells.

6. Multiwell-plates of 6-, 12-, 24-, and 48-wells are convenient formats to generate three-dimensional assemblies. If a smaller size is desired, glass, polystyrene, or Permanox® plastic slides with a removable silicone chamber for cell culture (Nunc) can be used as mold. Alternatively, biopsy punches of different diameters can be used to obtain small samples from gelated three-dimensional assemblies of higher size.

7. Needle should be removed gradually from the injection area to minimize leaking of the polymerizable solution. At the time of withdrawing the needle tip, wait for 20 s to allow clogging of the punctured skin by fibrin polymerization.

8. After the skin incision has healed, surgical clips must be withdrawn using the wound clip remover.

9. To facilitate the positioning of the collimator, a circle with same diameter as the collimator housing and concentric to the center of the target well can be plotted on the lid of the multiwellplate. Using X, Y, and Z micro-linear actuators, the collimator can be positioned according to the marked reference.

10. To position the collimator for in vivo irradiation, a plastic rod of 0.5 mm of diameter can be attached to one side of the collimator. The known distance between the center of the collimator lens and the guiding rod will be used to align the collimator over the masked irradiation spot.

Acknowledgment

This work was supported by grants PI12/01698 from Fondo de Investigaciones Sanitarias (FIS, Spanish Ministry of Economy and Competitiveness, MINECO, Spain) and SAF2013-50364-EXP (MINECO, Spain) to N.V.

References

1. Parsell DA, Lindquist S (1993) The function of heat-shock proteins in stress tolerance: degradation and reactivation of damaged proteins. Annu Rev Genet 27:437–496

2. Welch WJ (1993) How cells respond to stress. Sci Am 268:56–64

3. Cotto JJ, Morimoto RI (1999) Stress-induced activation of the heat-shock response: cell and molecular biology of heat-shock factors. Biochem Soc Symp 64:105–118

4. Voellmy R (2004) Transcriptional regulation of the metazoan stress protein response. Prog Nucleic Acid Res Mol Biol 78:143–185

5. Christians ES, Benjamin IJ (2006) Heat shock response: lessons from mouse knockouts. Handb Exp Pharmacol: (172) 139–152

6. Voellmy R, Ahmed A, Schiller P, Bromley P et al (1985) Isolation and functional analysis of a human 70,000-dalton heat shock protein gene segment. Proc Natl Acad Sci U S A 82:4949–4953

7. Schiller P, Amin J, Ananthan J, Brown ME et al (1988) Cis-acting elements involved in the regulated expression of a human HSP70 gene. J Mol Biol 203:97–105

8. Dreano M, Brochot J, Myers A, Cheng-Meyer C et al (1986) High-level, heat-regulated synthesis of proteins in eukaryotic cells. Gene 49:1–8

9. Vilaboa N, Voellmy R (2006) Regulatable gene expression systems for gene therapy. Curr Gene Ther 6:421–438

10. Huang Q, Hu JK, Lohr F, Zhang L et al (2000) Heat-induced gene expression as a novel targeted cancer gene therapy strategy. Cancer Res 60:3435–3439

11. Vekris A, Maurange C, Moonen C, Mazurier F et al (2000) Control of transgene expression using local hyperthermia in combination with a heat-sensitive promoter. J Gene Med 2:89–96

12. Locke M, Noble EG, Tanguay RM, Feild MR et al (1995) Activation of heat-shock transcription factor in rat heart after heat shock and exercise. Am J Physiol 268:C1387–C1394

13. Shastry S, Toft DO, Joyner MJ (2002) HSP70 and HSP90 expression in leucocytes after exercise in moderately trained humans. Acta Physiol Scand 175:139–146

14. Vilaboa N, Fenna M, Munson J, Roberts SM et al (2005) Novel gene switches for targeted and timed expression of proteins of interest. Mol Ther 12:290–298

15. Martin-Saavedra FM, Wilson CG, Voellmy R, Vilaboa N et al (2013) Spatiotemporal control of vascular endothelial growth factor expression using a heat-shock-activated, rapamycin-dependent gene switch. Hum Gene Ther Methods 24:160–170

16. Konig K (2000) Multiphoton microscopy in life sciences. J Microsc 200:83–104

17. Weissleder R (2001) A clearer vision for in vivo imaging. Nat Biotechnol 19:316–317

18. Miyako E, Deguchi T, Nakajima Y, Yudasaka M et al (2012) Photothermic regulation of gene expression triggered by laser-induced carbon nanohorns. Proc Natl Acad Sci U S A 109:7523–7528

19. Cebrian V, Martin-Saavedra F, Gomez L, Arruebo M et al (2013) Enhancing of plasmonic photothermal therapy through heat-inducible transgene activity. Nanomedicine 9:646–656

20. Martin-Saavedra FM, Cebrian V, Gomez L, Lopez D et al (2014) Temporal and spatial patterning of transgene expression by near-infrared irradiation. Biomaterials 35:8134–8143

Chapter 20

Optogenetic Light Crafting Tools for the Control of Cardiac Arrhythmias

Claudia Richter, Jan Christoph, Stephan E. Lehnart, and Stefan Luther

Abstract

The control of spatiotemporal dynamics in biological systems is a fundamental problem in nonlinear sciences and has important applications in engineering and medicine. Optogenetic tools combined with advanced optical technologies provide unique opportunities to develop and validate novel approaches to control spatiotemporal complexity in neuronal and cardiac systems. Understanding of the mechanisms and instabilities underlying the onset, perpetuation, and control of cardiac arrhythmias will enable the development and translation of novel therapeutic approaches. Here we describe in detail the preparation and optical mapping of transgenic channelrhodopsin-2 (ChR2) mouse hearts, cardiac cell cultures, and the optical setup for photostimulation using digital light processing.

Key words Cardiac dynamics, Optogenetics, Optical mapping, Photostimulation, Arrhythmias

1 Introduction

Optogenetic tools are widely used in neurobiological applications to control cell activity using light [1, 2]. This approach is based on microbial opsins or bacteriorhodopsins, which are light-responsive membrane proteins developed in microorganisms for a variety of physiological processes [2, 3]. Activated by light, opsins may induce cellular depolarization or hyperpolarization. The application of opsins to cardiomyocytes has been demonstrated recently [3].

In the heart, the normal sinus rhythm is triggered by regular, anisotropic waves of electric excitation in tissue. Complex spatiotemporal excitation patterns have been shown to be associated with cardiac ventricular fibrillation [5–7]. The control of the spatiotemporal complexity underlying atrial and ventricular arrhythmias is extremely challenging due to the interaction of vortex-like rotating excitation waves with the heterogeneous anatomical tissue substrate.

Arash Kianianmomeni (ed.), *Optogenetics: Methods and Protocols*, Methods in Molecular Biology, vol. 1408,
DOI 10.1007/978-1-4939-3512-3_20, © Springer Science+Business Media New York 2016

For lack of alternative strategies, high-energy electric shocks are used to terminate life-threatening cardiac arrhythmias (up to 360 J for transthoracic external cardiac defibrillation) [8]. However, such high-energy shocks can have severe side effects, including intolerable pain and electrocution of cells, indicating a significant medical need for better strategies [9–11].

This medical need may now be addressed by novel therapeutic approaches including low-energy anti-fibrillation pacing (LEAP) [10, 11], which aims at simultaneously controlling the dynamics of multiple vortices to improve therapeutic efficacy and limit side effects.

In addition, novel optogenetic tools and structured light illumination allow for the implementation of optical multi-site pacing devices to identify arrhythmia triggers and to potentially control arrhythmogenic mechanisms experimentally [13].

2 Materials

2.1 Isolation and Langendorff Perfusion

1. The Langendorff perfusion system consists of bubble trap, tissue bath, water-jacketed reservoirs, waste bottles, tubes (Tygon tubing), and other consumables (three-way cocks, Luer locks, etc.).

2. The heart is perfused with Tyrode solution: 130 mM NaCl, 4 mM KCl, 1 mM $MgCl_2$, 24 mM $NaHCO_3$, 1.8 mM $CaCl_2$, 1.2 mM KH_2PO_4, 5.6 mM glucose, 1 % albumin/BSA; prepare solution with Aqua dest. on a magnetic stirrer at room temperature. For an experiment lasting 3 h usually 2–3 l is sufficient.

3. Carbogen (5 % CO_2 and 95 % O_2) to aerate the Tyrode solution.

4. For anesthesia we use Isoflurane (Forane® 100 % (V/W)).

5. Heparin to avoid coagulation of blood during procedure (Heparin-Rotexmedica 25,000 I.E./5 ml).

6. The heart extraction is done using a stereomicroscope, surgical instruments (e.g., fine forceps and microsurgical scissors), surgical consumables (suture material, small-volume syringes, and blunt needles).

2.2 Optical Mapping of Membrane Voltage and Intracellular Calcium

1. Fluorescence dyes for membrane voltage RH1691, for intracellular calcium Rhod-2 or Quest®Rhod-4, prepare a 1 mM stock solution with dimethyl sulfoxide (DMSO), which can be split into smaller portions for easy freeze and thaw.

2. Blebbistatin may be used for electromechanical uncoupling to eliminate motion artifacts; prepare a 1 mM stock solution using DMSO. For the experiment mix 2.5 µM in 500 ml Tyrode solution.

3. The optical setup consists of dichroic mirrors ($DM_1 = 555$ nm; $DM_2 = 490$ nm), excitation filter ($Filter_{Exc} = 531 \pm 22$ nm), emission filter ($Filter_{Em} > 610$ nm), excitation light source (e.g., tungsten halogen lamp), several optomechanical components, a high-speed complimentary metal-oxide-semiconductor (CMOS) camera at 1 kHz and an ECG recording system (monophasic action potential recording), DLP Lightcrafter 4500 (or similar projection device), optic fibers (Multimode Fibers, with SMA connector on one end and the other end to be polished by customer), and a high-power LED in-housed with SMA connection (emitting wavelength in agreement with the opsin excitation wavelength).

2.3 Preparation of Optogenetic Collagen Matrices

1. To prepare a collagen matrix: collagen I from rat tail, 1 N NaOH, Matrigel® Basement Membrane Matrix, cell culture medium, phosphate buffer (DPBS).

2. For transfection: Lipofectamine® 2000, Opti-MEM®, pcDNA3.1/hChR2(H134R)-mCherry (provided by Karl Deisseroth (Addgene plasmid # 20938), Zhang et al. 2007 [19]).

3 Methods

3.1 Optical Mapping of Optogenetic Stimulation

For the Langendorff perfusion experiments we used a constitutive transgenic mouse model, alpha-MHC-ChR2, which restricts expression of ChR2 specifically to the heart through a cardiac-specific promotor system. Compared to wild-type (WT) animals, alpha-MHC-ChR2 transgenic mice aged 14–24 weeks showed normal electrical action potential propagation throughout cardiac tissues as well as light capture not present in WT hearts. ChR2 expression was further verified via PCR.

The macroscopic optical mapping system described in this section can be easily adapted to other experimental setups including a charge-coupled device (CCD) camera, a photodiode array (PDA), or even microscope-based protocols (*see*, e.g., Stirman et al. [14], Herron et al. [15]). The crucial parts of optogenetic optical mapping in general are the choice of fluorescence sensors (membrane voltage and/or intracellular calcium) and the control of photostimulation patterns. Therefore, this section focuses on description of tissue preparation and perfusion protocols, as well as in-coupling of photostimulation sources. All experiments were done under dark or low-light conditions, respectively, regarding photobleaching of fluorescence dyes and uncontrolled activation.

3.1.1 Experimental Setup and Photostimulation In-Coupling

1. First, build up the optical setup containing the camera, the excitation light source, and filter components regarding your specifications. Here, in regard to the imaging axis tangential excitation illumination was implemented by applying a

dichroic mirror between the excitation filter and emission filter (*see* Fig. 1). Thereby the working distance is restricted by the used camera zoom lens and the desired magnification of the sample (*see* **Note 1**).

2. To couple photostimulation sources into the optical setup, like a digital light-processing (DLP) projector, it is necessary to intervene the fluorescence light path. Therefore, place a second dichroic mirror between sample and first dichroic. The dichroic specification should correspond to the excitation wavelength of the opsins (*see* **Note 2**). For the in-coupling of an optic fiber this step is obsolete (*see* Fig. 1).

3. Since projectors are designed to project an image on a comparatively large screen, it becomes necessary to change the focusing of the DLP projector. Remove the magnifying optics in front of the DLP sensor by loosening the corresponding screws next to the sensor chamber. Now it is possible to use separate lenses to focus the projector onto the sample without losing the spatial resolution of the DLP sensor chip. In order to operate the DLP device use the manufacturer-delivered software or commonly used presentation software, which allows changing and display stimulation patterns in online time. Here we implemented the DLP control into our custom-made camera software.

4. To couple an optic fiber connect the SMA connection sides of LED and fiber, and place the abraded end of the fiber next to the sample with a micromanipulator. Observe the fiber bending capability by doing so. Connect the in-housed LED with your trigger device (e.g., an arbitrary waveform and function generator) (*see* **Note 3**).

3.1.2 Optical Mapping of Whole Hearts In Vitro

1. Aerate the prepared Tyrode solution with Carbogen for 30 min at room temperature, and then adjust the pH to 7.4 with NaOH. Fill 150 ml of Tyrode in a separate beaker, and place on ice (will be used for heart isolation).

2. Inject each mouse with 0.1 ml heparin (eq: 1000 Units), 30 min before isolation procedure.

3. Split Tyrode into two equal portions: Fill the first portion into one of the reservoirs of the perfusion system and deaerate the tubing and bubble trap carefully (*see* **Note 4**); add 2.5 μM blebbistatin to the second Tyrode portion and fill the solution in another reservoir of the perfusion system. Make arrangements for light sensitivity of blebbistatin. Heat the whole perfusion system (water-jacketed glass parts) up to 37 °C with a water heat pump.

4. Fill a 6 cm petri dish as well as a 2 ml syringe with ice-cold Tyrode and place both under the stereomicroscope. Anesthetize

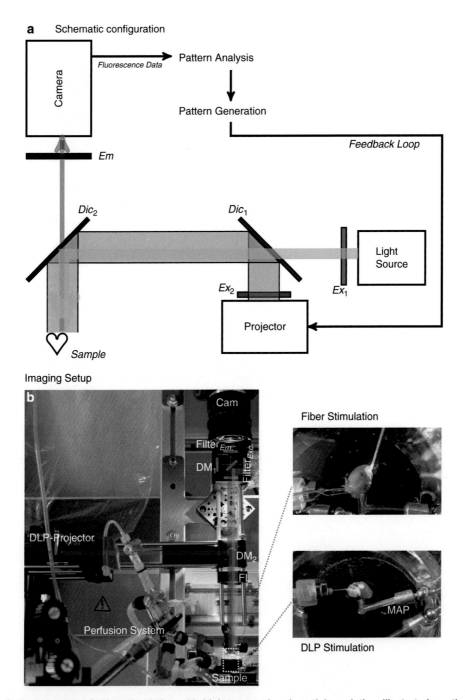

Fig. 1 Optical imaging of photostimulation with high temporal and spatial resolution. Illustrated are the fluorescence light paths (*green* excitation and *red* emission) and the in-coupling of optical stimulation pulses (*blue*). The sample details show the difference between photostimulation via optical fiber and DLP. Cam = high-speed CMOS camera with zoom lens, Filter$_{Em}$ = emission filter (Em), Filter$_{Exc}$ = excitation filter (Ex), DM/Dic = dichroic mirror, FL = focusing lens, MAP = monophasic action potential electrode

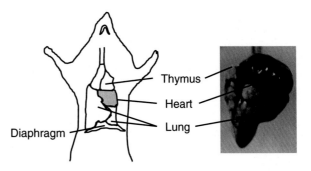

Fig. 2 In situ and extracted mouse heart

with isoflurane and sacrifice one mouse according to your animal welfare guidelines (*see* **Note 5**). Open the chest by cutting firstly the skin of the abdominal region and secondly the costal arches on both sides. Be careful not to injure the subjacent heart with scissor tips. The sternum could now be opened and the heart is accessible (*see* Fig. 2). Cut the aorta approx. 0.5 cm above the aortic arch and put the whole heart/thymus/fatty tissue package into the prepared 6 cm petri dish (the heart stops beating because of the temperature). The fine preparation should be done under the stereomicroscope. Remove the thymus and fatty tissue parts with microsurgical instruments; afterwards wind up the aorta on the blunt needle (mounted on the Tyrode-containing syringe) and fix the vessel with suture material. To control if everything is mounted and tightened well inject ice-cold Tyrode into the heart (*see* **Note 4**). This step has also the effect to rinse the remaining blood out of the heart. Then transfer the mounted heart to the perfusion system and start perfusion flow. As control: the heart should start beating after a short time of accumulation with normally 6–8 Hz. The heart should be covered with warm Tyrode (*see* **Note 5**).

5. After about 10–30 min it is possible to switch to the blebbistatin-containing perfusion circle, to eliminate mechanical motion. After this point the vital parameters are checked with the help of the ECG system (frequency changes, changes in potential shape, etc.) (*see* **Note 6**).

6. Fluorescence staining was done applying the dyes RH-1691 as indicator for membrane voltage and Rhod-2/-4 for calcium visualization [16, 17]. Therefore, stock solutions of 1 mM in concentration were prepared and diluted in perfusate to an end concentration of 40 μM. The dye dilution was directly injected into the bubble trap to ensure an even staining throughout the entire heart (*see* **Note 7**).

7. For analysis of fluorescence signals we use a custom-made python program (PythonAnalyzer), which allows filtering the

Table 1
Ingredients for matrix preparation

Ingredients	Volume (µl/ml)
Collagen stock solution	418
1 N NaOH	0.6
Matrigel®	100
Culture medium	482

Volume information in µl refers to 1 ml collagen gel

camera data in time and space. Potential cross talk of photo-stimulation light and fluorescence signal can possibly be eliminated by averaging over several light pulses and afterwards subtraction of background.

3.2 Photostimulation of Transfected Collagen Matrices

1. For matrix preparation we used a mixture displayed in Table 1. The collagen stock solution contained of 850 µl collagen from rat tail and 150 µl PBS. Consolidation of gels was performed for 5 min at 37 °C. Gels were made in 6 cm petri dishes and 2–3 mm thick.

2. The hearts of 1–2-day-old Wistar rats were extracted and after rinsing with phosphate buffer freed of the atria and other tissue beside the ventricles. Afterwards they were minced into 1–3 mm^3 pieces and transferred to the digestion with collagenase II (300 U/ml, solved in phosphate buffer). There were four digestive steps in total, each for 15 min at 37 °C, whereas the first step was removed to keep cell garbage out of the cultures. After counting with a hemocytometer (after Neubauer) cells were used for chemical transfection with Lipofectamine®. Therefore prepare one solution of 300 µl Opti-MEM® with 12 µl Lipofectamine® and one solution with 300 µl Opti-MEM® and 7 µg rhodopsin DNA; afterwards mix both solutions under constant stirring. Incubate the Opti-MEM-Lipofectamine-DNA solution for 5 min at room temperature and then add the mixture to the cell suspension. Transfected cells were plated onto the prepared collagen gels in DMEM/F12 (containing 10 % FCS, 1 % penicillin/streptomycin with 100 U/ml). These matrices were cultivated under physiological conditions (37 °C, 5 % CO$_2$, humidified atmosphere) and the change of medium took place every 48 h.

3. Photostimulation of at least 14-day-old matrices was realized by fiber optical light pulses (*see* Fig. 3b). Motion of the collagen matrix was captured using spatial correlation and motion tracking algorithms.

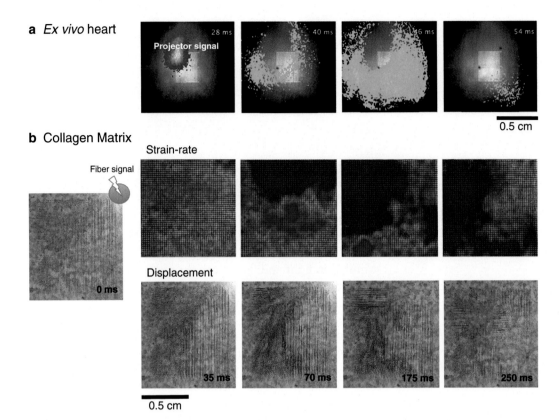

Fig. 3 Experimental results of photostimulation. (**a**) Integration of a noninvasive photostimulation applying video projector signals on a Langendorff-perfused mouse heart. Shown are the optical mapping data of photo-induced calcium wave propagation (*green*) spreading from a projector signal (*blue*)*. (**b**) Planar contractile wave activity in a 3 mm thick collagen matrix seeded with transfected neonatal cardiomyocytes after 14 days of cultivation. The image series shows time steps of propagation visualized by strain-rate tracking and displacements (*red* vectors) of the tissue computed along a regular lattice during the application of a fiber-coupled light pulse. *Raw images contain a visible camera artifact, which occurs as *white shadow* in the *image center*. This does not influence the image-analyzing process

4   Notes

1. Distances between optical components are among several component specifications dependent for example on applied focus length. In principle, keep the working distance short, since long distances between sample and camera can also mean loss of fluorescence signal. Furthermore, the distance between excitation light source output and excitation filter should be observed, since the tip of a light guide may get hot during operation and could irreversibly damage the filter coating.

2. For the usage of different opsins simultaneously we use multiple band-pass filters and corresponding dichroic mirrors. This also applies if the photostimulation should be coupled into commercial microscopes. Thus, the photostimulation source is

inserted between the excitation light source and the fluorescence filter cube.

3. In order to evoke cardiac action potentials it is necessary to achieve the minimal required irradiance [mW/mm²] sufficient for the opsins used. The required irradiance depends on the illuminated size of region [4, 16]. According to Bruegmann et al. [4] applied on a region of 0.05 mm² (left ventricle), with pulsed photostimulation, 7.2 mW/mm² is sufficient to initialize new activity (ChR2). In general, higher irradiance in terms of higher LED current corresponds to shorter possible pulse durations.

4. Do the deaeration step carefully, since remaining bubbles getting into the perfused heart could cause vascular obliteration.

5. All our experiments were done in accordance with the German animal welfare law and were reported to our animal welfare representatives; humane welfare-oriented procedures required at least a laboratory animal introductory course after recommendations of the Federation of Laboratory Animal Science Associations (FELASA B). Mice for our measurements were euthanized by cervical dislocation. During the isolation procedure it is absolutely necessary not to injure the heart or the aorta with the microsurgical scissor tips, since every tissue lesion leads to unpredictable circumstances in the following perfusion and measuring methods.

6. The shape and frequency of cardiac action potentials offer valuable clues to putative cardiac misbehavior (e.g., arrhythmia, AV-block; *see*, e.g., K. Josephson [18] or en.ecgpedia. org/wiki). To early recognize, e.g., prolongation of action potentials it is necessary that the monophasic action potential electrode never loses contact to the heart surface, but without interference with the muscle contraction or impression of individual tissue parts.

7. Choosing fluorescence dyes one has to consider the compatibility of the excitation and emission spectra with the used opsins [16, 17]. In our experiments the calcium fluorescence sensors Rhod-2/-4 as well as the membrane voltage dye RH-1691 are used, because of their red-shifted emission spectra. This also allows for potential fluorescent reporter genes of the opsin [20].

Acknowledgements

The pcDNA3.1/hChR2(H134R)-mCherry was provided by Karl Deisseroth (Addgene plasmid # 20938). We also want to thank M. Kunze and T. Althaus for their technical assistance. The research leading to the results has received funding from the European

Community's Seventh Framework Programme FP7/2007-2013 under Grant Agreement No. HEALTH-F2-2009-241526, EUTrigTreat. The authors acknowledge support from the German Federal Ministry of Education and Research (BMBF) (project FKZ 031A147, GO-Bio), the German Research Foundation (DFG) (Collaborative Research Centers SFB 1002, Projects B05 and C03 and SFB 937 Project A18), and the German Center for Cardiovascular Research (DZHK e.V.).

References

1. Knollmann BC (2010) Pacing lightly: optogenetics gets to the heart. Nat Methods 7:889–891

2. Deisseroth K (2011) Optogenetics. Nat Methods 8:26–29

3. Rein ML, Deussing JM (2012) The optogenetic (r)evolution. Mol Genet Genomics 287:95–109

4. Bruegmann T, Malan D, Hesse M, Beiert T, Fuegemann CJ, Fleischmann BK, Sasse P (2010) Optogenetic control of heart muscle in vitro and in vivo. Nat Methods 7:897–900

5. Davidenko JM, Pertsov AV, Salamonsz R, Baxter W, Jalife J (1992) Stationary and drifting spiral waves of excitation in isolated cardiac muscle. Nature 355:349–351

6. Panfilov AV, Holden AV (1990) Self-generation of turbulent vortices in a two dimensional model of cardiac tissue. Phys Lett A 151:23–26

7. Jalife J, Gray RA, Morley GE, Davidenko JM (1998) Self-organization and the dynamical nature of ventricular fibrillation. Chaos 8:79–93

8. Walcott GP, Killingsworth CR, Ideker RE (2003) Do clinically relevant transthoracic defibrillation energies cause myocardial damage and dysfunction? Resuscitation 59:59–70

9. Zipes DP, Jalife J (eds) (2009) Cardiac electrophysiology: from cell to bedside. Saunders/Elsevier, Philadelphia. ISBN 9781416059738

10. Fenton FH, Luther S, Cherry EM, Otani NF, Krinsky VI, Pumir A, Bodenschatz E, Gilmour RF (2009) Termination of atrial fibrillation using pulsed low-energy far-field stimulation. Circulation 120:467–476

11. Luther S, Fenton FH, Kornreich BG, Squires A, Bittihn P, Hornung D, Zabel M, Flanders J, Gladuli A, Campoy L, Cherry EM, Luther G, Hasenfuss G, Krinsky VI, Pumir A, Gilmour RF Jr, Bodenschatz E (2011) Low-energy control of electrical turbulence in the heart. Nature 475:235–239

12. Bruegmann T, Malan D, Hesse M, Beiert T, Fuegemann CJ, Fleischmann BK, Sasse P (2011) Channelrhodopsin2 expression in cardiomyocytes: a new tool for light-induced depolarization with high spatiotemporal resolution in vitro and in vivo. Thorac cardiovasc Surg 59-S01:MO19. doi:10.1055/s-0030-1269109

13. Sasse P (2011) Optical pacing of the heart: the long way to enlightenment. Circ Arrhythm Electrophysiol 4:598–600

14. Stirman JN, Crane MM, Husson SJ, Gottschalk A, Lu H (2012) A multispectral optical illumination system with precise spatiotemporal control for the manipulation of optogenetic reagents. Nat Protoc 7:207–220

15. Herron TJ, Lee P, Jalife J (2012) Optical imaging of voltage and calcium in cardiac cells & tissues. Circ Res 110:609–623

16. Ambrosi CM, Entcheva E (2014) Chapter 19—Optogenetic control of cardiomyocytes via viral delivery. In: Milica Radisic, Lauren D. Black III (eds) Methods in molecular biology, vol 1181, chapter 19. Springer Science+Business Media. pp 215–228

17. Loew LM (2010) Design and use of organic voltage sensitive dyes. In: Canepari M, Zecevic D (eds) Membrane potential imaging in the nervous system: methods and applications, chapter 2. Springer Science+Business Media pp 13–23

18. Josephson ME (2008) Clinical cardiac electrophysiology: techniques and interpretations. Solution (Lippincott Williams & Wilkins). Wolters Kluwer Health/Lippincott Williams & Wilkins. ISBN 9780781777391

19. Zhang F, Wang LP, Brauner M, Liewald JF, Kay K, Watzke N, Wood PG, Bamberg E, Nagel G, Gottschalk A, Deisseroth K (2007) Multimodal fast optical interrogation of neural circuitry. Nature 446:633–639

20. Jia Z, Valiunas V, Lu Z, Bien H, Liu H, Wang H-Z, Rosati B, Brink PR, Cohen IS, Entcheva E (2011) Stimulating cardiac muscle by light: cardiac optogenetics by cell delivery. Circ Arrhythm Electrophysiol 4:753–760

Chapter 21

Inscribing Optical Excitability to Non-Excitable Cardiac Cells: Viral Delivery of Optogenetic Tools in Primary Cardiac Fibroblasts

Jinzhu Yu and Emilia Entcheva

Abstract

We describe in detail a method to introduce optogenetic actuation tools, a mutant version of channelrhodopsin-2, ChR2(H134R), and archaerhodopsin (ArchT), into primary cardiac fibroblasts (cFB) in vitro by adenoviral infection to yield quick, robust, and consistent expression. Instructions on adjusting infection parameters such as the multiplicity of infection and virus incubation duration are provided to generalize the method for different lab settings or cell types. Specific conditions are discussed to create hybrid co-cultures of the optogenetically modified cFB and non-transformed cardiomyocytes to obtain light-sensitive excitable cardiac syncytium, including stencil-patterned cell growth. We also describe an all-optical framework for the functional testing of responsiveness of these opsins in cFB. The presented methodology provides cell-specific tools for the mechanistic investigation of the functional bioelectric contribution of different non-excitable cells in the heart and their electrical coupling to cardiomyocytes under different conditions.

Key words Optogenetics, Cardiac, Non-excitable cells, Fibroblasts, ChR2, ArchT

1 Introduction

Non-excitable cells in the heart (fibroblasts, endothelial cells, smooth muscle cells, pericytes, and macrophages) are an under-investigated mystery compared to the excitable muscle cells. This is particularly true for the largest non-myocyte cell population, the cardiac fibroblasts, cFB [1]. There are no unique markers to identify them [2, 3], and hence no robust cell-specific promoters to genetically target them [4–6] for sensing or manipulation purposes.

Recent evidence has accumulated pointing to quite important and "exciting" roles of non-excitable cells in the heart [7–10], including their key contributions to cardiac electrical activity. For instance, fluctuations in the cFB membrane potential can modulate

Arash Kianianmomeni (ed.), *Optogenetics: Methods and Protocols*, Methods in Molecular Biology, vol. 1408,
DOI 10.1007/978-1-4939-3512-3_21, © Springer Science+Business Media New York 2016

the operation of the sinoatrial node cells and hence can influence pacing frequency [11]. Through paracrine signaling, cFB can also influence important ion channel expression levels, such as SCN5A and KCNJ2, and hence can actively remodel myocyte electrophysiology [12]. Following cardiac injury, cFB undergo electrical phenotypic transformation to myofibroblasts [11]. This can lead to altered action potential characteristics of co-cultured cardiomyocytes through both paracrine signaling [12] and increased direct electrical coupling via gap junctional (Cx43) channels between the cFB and the cardiomyocytes [13]. Pathological responses of cFB, triggered by cardiac injury, affect their interactions with cardiomyocytes and can become the culprit of a range of pro-arrhythmic consequences, including conduction slowing or block, and enhanced automaticity in the myocytes bordering the injury zone [14–17]. Overall, the lack of investigational research tools to specifically manipulate cFB has hampered the study and understanding of their functional contributions to cardiac activity.

Optogenetic actuation tools (light-sensitive microbial ion channels), when genetically expressed in mammalian cells, can mediate fast and cell-specific optically induced changes in their membrane potential, i.e., can cause depolarization or hyperpolarization [18–20]. Along with optogenetic sensors [21], they can facilitate the mechanistic study not only of excitable (action potential-generating) cells but also of non-excitable cells in the complex multicellular environment of the heart [22, 23]. Pending the identification and development of better cFB-specific promoters or other ways to target cFB [6], optogenetic actuators can allow for unique interrogation of these cells to quantify their contribution to cardiac electrical activity during normal function and during injury response and remodeling in vivo [2]. Separately, in vitro high-throughput coupling assays can be developed [24] in conjunction with micro-patterning or other ways to produce controlled heterotypic cell pairs or hybrid tissue equivalents [25, 26] that can be useful in evaluating stem cell-derived biologics for cardiac tissue repair.

So far, optogenetic actuation tools have found limited use in manipulating excitable cardiac myocytes [27–32]. Furthermore, we have shown that optical manipulation of cardiac syncytium can also be achieved through dedicated donor cells as light-responsive units that are not excitable themselves, i.e., "tandem-cell-unit" approach, demonstrated using different generic somatic cells [33, 34]. This idea can be extended to primary cardiac fibroblasts, as demonstrated in a preliminary study [24] and described here in detail for in vitro application.

2   Materials

2.1   Culturing Non-excitable Cardiac Cells

1. Neonatal rat ventricular tissue, dissected from 2–3-day-old Sprague-Dawley rats, temporarily submerged in Hanks' Balanced Salt Solution (HBSS).

2. Tissue digestion enzymes: Trypsin dissolved in HBSS at 1 mg/mL, and collagenase also dissolved in HBSS at 1 mg/mL. Both enzyme solutions are prepared right before usage, short-term storage at 4 °C.

3. Cell culture media M199 (500 mL bottle), supplemented with 12 μM l-glutamine, 3.5 μg/mL glucose, 0.05 μg/mL penicillin-streptomycin, 200 μg/mL vitamin B12, 10 mM HEPES, and 2 or 10 % fetal bovine serum (FBS).

4. 175 cm² Flasks.

2.2   Optogenetic Transgene Delivery by Adenovirus Infection and Optimization

1. For opsin delivery, we used custom-generated adenovirus vectors (Ad–): Ad-CAG-ChR2-eYFP from the plasmid pcDNA3.1/hChR2(H134R)-EYFP [32] and Ad-CAG-ArchT-eGFP from the plasmid pAAV-CAG-ArchT-GFP. The plasmids are purchased from Addgene (Cambridge, MA) and deposited by K. Deisseroth's lab (plasmid 20940) and Ed Boyden's lab (plasmid 29777). Our stock viruses with titer concentration at 10^{12} units/mL are stored at –20 °C.

2. 1× Phosphate-buffered saline (PBS).

3. 0.05 % Trypsin-EDTA in HBSS.

4. 10 % Bleach.

2.3   Quantification of Gene Expression and Cell Viability

1. Prepare a 5× Tyrode's solution for long-term storage at 4 °C by dissolving the following compounds in 1 L of diH₂O: 1.02 g MgCl₂, 2.01 g KCl, 39.45 g NaCl, 0.2 g NaH₂PO₄, and 5.96 g HEPES. Adjust the pH to 7.4 with 1 M HCl or 0.1 N NaOH at room temperature. For usage, dilute 100 mL 5× stock Tyrode's solution with 400 mL diH₂O to make a total of 500 mL working solution. Add 0.46 g d-glucose, 0.75 mL of 1 M CaCl₂ (for 1.5 mM Ca²⁺ in final volume), and pH the solution to 7.4 at the temperature at which experiments will be done.

2. Prepare propidium iodide (PI) solution at 2 μg/mL in 1× Tyrode's solution. Dye solution should be prepared immediately before usage/imaging, and wrapped in aluminum foil.

2.4   Examining Opsin Functionality

1. Glass-bottom dishes with glass number 1.

2. Human fibronectin is prepared at a stock concentration of 5 mg/mL in filtered diH₂O, which could be stored at –20 °C for long term. We recommend making aliquots of the stock fibronectin. For usage, dilute stock aliquots with 1× PBS to

final concentration of 50 μg/mL. Diluted fibronectin could be stored at 4 °C and should be used within 2 weeks.

3. Sylgard® 184 Silicone Elastomer Kit.

4. Voltage-sensitive fluorescent dyes: FluoVolt™ Membrane Potential Kit; Di-4ANBDQBS (obtained from Dr. L. Loew, University of Connecticut), dissolved in ethanol at stock concentration of 10 mg/mL. Stock dye is stored at –20 °C. For usage, dilute 2 μL stock dye in 1 mL of Tyrode's solution to make 35 μM to stain one glass-bottom sample in the dark.

5. Calcium-sensitive fluorescent dye Quest Rhod-4 AM is dissolved in DMSO (with 20 % Pluronic F17) to make 0.5 mM stock dye concentration. Typically, aliquots of dye are made at 20 μL tubes (each is enough for a 35 mm glass-bottom dish) and stored at –20 °C for long term. For usage, dilute stock aliquots with 1 mL Tyrode's solution to final concentration of 10 μM immediately before staining.

3   Methods

The overall process from extracting primary cFB through their transduction with opsins to their functional testing requires proper time management (Fig. 1).

3.1 Culturing Non-excitable Cardiac Cells

Dissection of cardiac tissue and cell isolation steps have been reported in detail previously [34, 35] and are briefly summarized in this section.

1. Remove neonatal rat ventricular tissue from HBSS and put into 30 mL trypsin solution (1 mg/mL) to digest away extracellular matrix for 12–14 h at 4 °C on a shaker at 75 rpm.

2. Remove trypsin solution using pipette and avoid picking up tissue. Wash tissue with 20 mL M199 with 10 % FBS at 37 °C for 4 min with swirling motion. Remove M199 and start isolating cardiac cells with collagenase solutions at 37 °C five

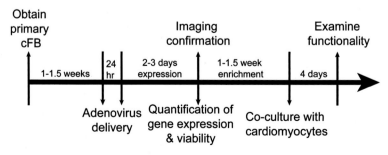

Fig. 1 Summary time line. Each study point listed in Subheading 3 is labeled at the *arrows*, and optimized time is listed

times, each time adding 10 mL and swirling at 75 rpm for 2 min. Throw away the first 10 mL collagenase solution and collect the later four repetitions.

3. Spin down cells at 300 × g for 8 min, remove collagenase, and resuspend cells in HBSS. Filter out remaining tissue chunks and change HBSS to M199 with 10 % FBS.

   Separation of non-myocyte cells (*see* **Note 1**) from cardiomyocytes is done using their ability to rapidly adhere to culturing substrates.

4. Plate the heterogeneous cell mixture from the previous step in a 175 cm² flask; the typical time frame to avoid cardiomyocyte adhesion but ensure non-myocyte cell attachment is between 45 min and 1 h.

5. Repeat once by gently transferring floating cells into a second flask for a total of 90 min to 2 h. The plated non-myocyte cells can be maintained in M199 with 2 % FBS until confluent or ready for adenovirus infection; ensure media change every 3–4 days; re-plating should be considered when stored for over 2 weeks.

3.2 Optogenetic Transgene Delivery and Optimization

The following steps specifically focus on transducing cFB with ChR2 and ArchT using adenoviral vectors starting with cell suspension to increase efficiency of infection [32]. This protocol can be scaled up or down by cell number based on experimental design. All solutions should be pre-warmed to 37 °C.

1. Wash cells with 1× PBS twice.

2. Remove the second PBS wash, and add 5 mL of 0.05 % trypsin in HBSS to the flasks or enough to make sure that the entire monolayer is submerged. Immediately put the cell and trypsin mixture in a 37 °C incubator and wait for 5 min. At the end of 5 min, agitate the cell culture container (swirl and tap on the sides) and observe under a light microscope for floating cells in a quick manner. Stop trypsin digestion by adding equal volume of M199 with 2 % FBS and collect the cells in a conical. Wash the culture substrate once with M199 with 2 % FBS.

3. Perform cell counting. *Optional*: Add trypan blue to the cell sample to label nonviable cells and subtract from the total count.

4. Spin cells down at 180 × g for 3 min, remove the supernatant mixture of trypsin, and add fresh M199 with 2 % FBS (*see* **Note 2**) to keep concentration between 7.2 and $12 × 10^4$ for plating on glass-bottom dishes or enough volume to keep density between 1 and $2 × 10^4$ cells/cm² for enrichment (*see* **Note 3**).

5. Remove stock virus from –20 °C freezer and put on ice. Invert or tap virus stock, but avoid vortexing.

6. Add virus to cFB solution at multiplicity of infection (MOI) of 2000 viral particles per cell. Primary cFB are much more resistant to adenoviral infection compared to cardiomyocytes and require higher MOIs [23]; for both ChR2 and ArchT infection, we recommend a starting MOI 2000. An optimized MOI is determined based on expression efficiency and cell viability post-infection. For first-time infection experiments, we recommend to test a large range of MOI covering 3 orders of magnitude from several viral particles per cell to several 10^3 viral particles per cell (*see* **Note 4**) using the appropriate virus solution volume:

$$virus volume = \frac{cell \# \times MOI}{Virus titer}$$

7. Mix cells and virus by pipetting up and down (*see* **Note 5**) before plating the mixture. If samples are to be used for imaging, plate cFB on glass-bottom dishes (pre-coated with fibronectin for 2 h at 37 °C) at $3–5 \times 10^4$ cells/cm². Let samples sit in the laminar flow hood for 15 min to allow cells to settle to the bottom. Then move the dishes into 37 °C incubator for 2 h before filling up the entire dish with additional 2 mL M199 with 2 % FBS. If samples are to be used for flow cytometry/ sorting, plate cells in flasks or tissue culture-treated dishes at similar density. Add more media if necessary and swirl to distribute cells evenly.

8. Keep virally infected cFB in a 37 °C incubator for 24 h. Incubation duration is strongly dependent on cell type; we recommend testing a wide range from brief exposure (2 h) up to 2 days.

9. At the end of the incubation period, remove virus by simply aspirating off the media from cFB monolayer and adding fresh M199 with 2 % FBS (10 % FBS could be used to stimulate growth). Wait for 2 days before examining expression.

3.3 Quantification of Gene Expression and Cell Viability

1. Initial confirmation of ChR2 or ArchT expression in cFB can come from reporter gene fluorescence, eYFP or eGFP, respectively, under fluorescence microscopy (*see* **Note 6**).

   To quantify viability by flow cytometry, we recommend applying PI stain before harvesting the cells with trypsin to avoid unnecessary excessive uptake of the dye and raised background from partially compromised cell membranes (*see* **Note 7**).

2. Aspirate cell culture media from the 35 mm dish sample, add 1 mL of 2 µg/mL PI stain, and wait for 2 min.

3. Remove PI stain, add 0.05 % trypsin-EDTA in HBSS, and repeat the same harvesting procedure described in **step 2** of Subheading 3.2.

4. *Optional*: Count cells and adjust concentration in the next step by replacing trypsin solution with cell culture media. The optimal cell concentration should be based on the specific flow cytometer. Typically, 0.5×10^6 cells in 0.5 mL media are needed for BD Calibur flow cytometer.

5. Spin cells down, aspirate trypsin mixture, and add M199 with 2 % FBS to dispense the cell pellet. Make sure to pipette the cells up and down multiple times to break any clusters that can clog the flow cytometer tube. Run the cells through a filter with appropriate pore size. For cFBs, a filter with 40 μm diameter is recommended.

6. Transfer cFBs into flow cytometry tubes and perform analysis (*see* **Note 8**).

7. If ChR2 or ArchT expression is low, especially in difficult-to-transduce cells, enrichment by flow-cytometry-assisted cell sorting (FACS) can be applied. However, this step adds an additional cycle of lifting and plating, which may be undesirable for primary cells.

This infection protocol, based on high virus dose and longer virus incubation (24 h), has provided consistent and improved expression efficiency for both ChR2 and ArchT in cFB, without detrimental effects on cell viability (Fig. 2).

3.4 Opsin Functionality Testing	The most direct way to ensure opsin functionality is to measure light-evoked ChR2- or ArchT-photocurrents in single cells using patch clamp methods [30]. Alternatively, opsin functionality in cFBs can be tested within multicellular preparations. We use a co-culture of opsin-transformed primary cFBs and cardiomyocytes, and probe cFB responsiveness to light by measuring the cardiomyocyte activity, based on the "tandem-cell-unit" concept [34].
3.4.1 Co-culture with Cardiomyocytes	Different patterns of co-culture of ChR2-cFB and cardiomyocytes can be created, e.g., diffuse uniform co-culture or spatially localized ChR2-cFB cluster surrounded by cardiomyocytes. Typically, the clustered pattern of opsin-expressing non-myocytes yields better optical excitability. Cell patterning can be done using polydimethylsiloxane (PDMS) stencils. The thickness of the stencil determines the volume of cells that can be deposited, and it can be easily adjusted by the amount of elastomer cured in a fixed area.

1. The stencil stiffness can be varied by the ratio of elastomer to curing agent; here 10:1 ratio is used. A combined weight of 9.5 g makes approximately 1 mm thick stencils in a 100 mm wide Petri dish. In a plastic cup, weigh out elastomer and curing agent at the desired volume and ratio. Mix thoroughly.

2. Pour the elastomer mixture into 100 mm Petri dish and swirl the dish to cover the entire bottom surface.

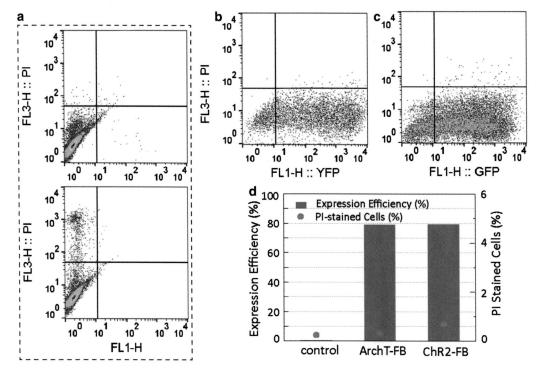

Fig. 2 Cell viability and expression efficiency. Flow cytometry analysis indicates that cells subjected to 24-h infection incubation show high expression efficiency and low toxicity. (**a**) Negative (*top*) and positive (*bottom*) control for PI stain on non-transduced cFB (positive control of PI stain was done with prolonged trypsin digestion to break cell membrane). (**b**) ChR2-cFB analysis (5000 events) and (**c**) ArchT-FB analysis (20,000 events) were acquired based on gates determined in control condition. (**d**) Both ChR2 and ArchT delivery at MOI2000 and 24-h incubation achieved over 70 % efficiency

3. Put the Petri dish in a desiccator and turn on vacuum to remove bubbles for 60 min. Occasional de-pressure helps draw bubbles out. However, when opening up the chamber, slowly turn the air valve to open to avoid disturbing the sample by strong flow of air.

4. Put de-bubbled elastomer mixture on a leveled surface in oven, and bake at 50–60 °C for 2 h. Paper towel can be put below the petri dish to ensure even heating.

5. The stencils will be applied to the glass area of glass-bottom dishes (14 or 20 mm). Cut out 1 cm×1 cm squares (or desired size) using a pattern printout placed below the petri dish (Fig. 3a). Puncture a circle Ø=0.4 cm in the middle of the square with a glass puncher.

6. Sterilize the PDMS stencils by submerging them in ethanol.
     Depending on the specific cell type and their ability to couple with cardiomyocytes and to proliferate, the ratio for co-culturing optically sensitized cells with cardiomyocytes needs to be calculated. Overloading the co-culture system with cFB can

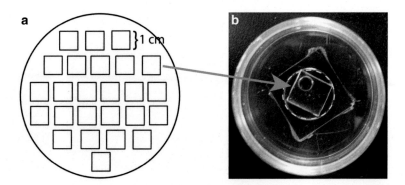

Fig. 3 PDMS stencil for cell patterning. (**a**) Template printout for trace-cut cured PDMS in a 100 mm petri dish. (**b**) Each PDMS square dimension is 1 cm by 1 cm to fit into the glass well, indicated by *yellow dash line*, of a 14 mm glass-bottom dish. A circle with Ø=0.4 cm is punctured through the PDMS square to create the stencil space for cFB cluster

cause overcrowding and tissue culture peeling, since their proliferation potentials are reinforced by high serum (10 %) media required by cardiomyocytes during the first days of culturing. Vice versa, a low ratio of cFB to cardiomyocytes may make it difficult to optically excite (despite the presence of functional opsins) due to source-load mismatch. A ratio of 1 opsin-expressing cFB to 16 cardiomyocytes in co-cultures has been reliable in our experiences in testing optical excitability of ChR2-cFB with relatively low light levels.

When ChR2-cFB or ArchT-cFB come to confluence and are ready to be experimented on (*see* **Note 9**), they should be collected in parallel with the cardiomyocyte dissection/isolation procedure (Fig. 1).

7. Coat 14 mm glass-bottom dishes with 0.25 mL fibronectin, and incubate dishes at 37 °C for at least 2 h. If PDMS pattern stencil is used, rinse the ethanol off them by submerging them in PBS twice. Put one pattern on the fibronectin coating.

8. Isolate neonatal cardiac tissue using previously described methods (Subheading 3.1). After obtaining purified cardiomyocytes, determine cell count.

9. During the step of separating cardiomyocytes from non-myocytes (usually 90 min to 2 h), harvest ChR2-cFB or ArchT-cFB using previously described trypsin digestion steps (Subheading 3.2, **steps 1** and **2**) and count cell number.

For a diffused pattern (Fig. 4a): we typically use plating density of 0.35×10^6 cells/cm² on a 14 mm glass-bottom dish (area = 1.43 cm²) that holds 0.6 mL in the well, leading to mixed cell concentration of 0.83×10^6 cells/mL.

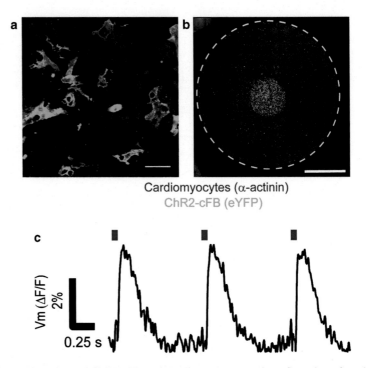

Cardiomyocytes (α-actinin)
ChR2-cFB (eYFP)

Fig. 4 Co-culture of ChR2-cFB and cardiomyocytes and confirmation of opsin functionality. (**a**) Fluorescent images of diffuse plating. Scale bar is 50 μm. (**b**) Panoramic macroscopic fluorescent image of patterned ChR-cFB cluster surrounded and covered by cardiomyocytes. Scale bar is 4 mm. (**c**) Optical activation of ChR2-cFB leads to excitation in cardiomyocytes. *Blue bars* indicate time of 470 nm light pulse. Optical stimulation was 10-ms pulse at 0.31 mW/mm². Action potentials were acquired via Di4-ANBDQBS optically at 390 frames/s

10. Combine the two cell types into one conical tube at a ratio of 1 opsin-expressing cFB:16 CM. Adjust the concentration of the cell mixture to 0.83×10^6 cells/mL by spinning the cells down at 1000 rpm for 4 min and remove/add M199 with 10 % FBS.

11. Remove fibronectin coating from the glass-bottom dishes and deposit 0.6 mL of well-mixed cells in the glass well.

    For a localized/confined cluster pattern (Fig. 4b): We typically pipette 30 μL of opsin-expressing cFB in the circular cutout space of the PDMS stencil. At the 1 opsin-expressing cFB:16 CM ratio and plating density of 0.35×10^6 cells/cm², each glass-bottom dish will have 0.03×10^6 opsin-expressing cFB and 0.471×10^6 cardiomyocytes, necessitating concentration of the opsin-expressing cFBs at 0.98×10^6/mL and cardiomyocyte concentration at 0.785×10^6/mL.

12. Remove the fibronectin coating from the glass-bottom dishes. Leave the dishes uncovered to allow the glass to become completely dry so that the PDMS stencil seals to the glass. Keep the side of the stencil touching the fibronetin on the glass (Fig. 3b).

13. Deposit the 30 μL of opsin-expressing cFB at 0.98×10^6 cells/ mL into the PDMS stencil and let it sit in the humidified 37 °C incubator for at least 1 h for cell attachment.

14. After incubation, in a quick manner, gently aspirate off the cell media from the surface of the PDMS stencil, remove the PDMS stencil, and deposit 0.6 mL cardiomyocytes at 0.785×10^6/mL to fill up the glass-bottom well.

15. Leave the samples in the hood still for 15 min to avoid spillage of cells from the glass to the plastic region. Then move the samples to the humidified 37 °C incubator and wait for 2 h before filling up the entire glass-bottom dish with 2 mL of M199 with 10 % FBS.

16. On the next day approximately 24 h after plating, remove cell culture media from the samples and wash in PBS with shaking for 5 min to remove loosely attached dead cells. Remove PBS, fill glass-bottom dishes with M199 with 10 % FBS, and keep samples incubated in humidified 37 °C incubator for another 48 h. At this point, cardiomyocytes no longer need high-serum supplement and samples will be switched to M199 with 2 % serum for the rest of their maintenance.

3.4.2 Optical Mapping

1. Using the 1 opsin-expressing cFB:16 CM conditions, experiments can be conducted on the 4th day after plating. Remove sample from the incubator and aspirate culture media. Stain ChR2-cFB with 1 mL of 10 μM of the calcium-sensitive dye Quest Rhod4-AM for 20 min and wash the sample once with 20-min incubation of Tyrode's solution. Stain ArchT-cFB samples with the voltage-sensitive dye FluoVolt™ following the manufacturer's manual. Incubate sample in dye (and PowerLoad™) for 15 min and wash sample twice with Tyrode's solution. Alternatively, both opsin's activation spectra are compatible with that of the voltage-sensitive dye Di4-ANBDQBS. Incubate samples in 35 μM Di4-ANBDQBS in 1 mL Tyrode's for 5 min. Wash samples with 5-min incubation in Tyrode's solution. Remove the wash and add fresh Tyrode's solution.

2. Place stained sample on optical mapping setup, which has been described in detail previously [34] (Fig. 5).

3. Set up appropriate filters for the fluorescent dye used. We used 525 nm excitation light sheet and 605/70 nm emission filter behind a macro-lens (50 mm, $f/1.0$) to collect Quest Rhod4-AM signals. Similarly, 475 nm excitation light sheet and 535/50 nm emission filter to collect FluoVolt™ signals. For Di 4-ANBDQBS, the excitation and emission filters are 655/40 nm and 700 nm long pass, respectively.

4. Apply optical stimulation of ChR2-cFB from underneath the glass-bottom dish using a pulsed 470 nm laser or LED

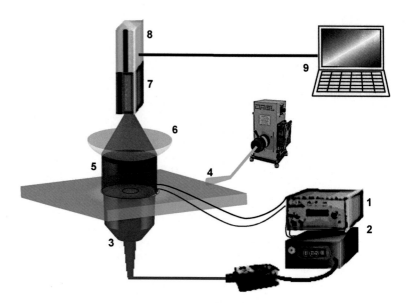

Fig. 5 Ultrahigh-resolution high-speed optical imaging system was used to capture electrically and optically induced excitation waves, reported by calcium-sensitive fluorescent dye Rhod-4 or voltage-sensitive fluorescent dye FluoVolt™: (1) Pulse generator for electrical pacing (analog modulation) and optical pacing (TTL modulation). (2) Fiber optic-coupled 470 nm laser as activation light source for ChR2 or 590 nm LED as activation light source for ArchT. (3) Global illumination of co-cultured ChR2-cFB/cardiomyocytes or ArchT-cFB/cardiomyocytes was delivered from below the sample. (4) Excitation for Rhod-4 or FluoVolt was delivered to the sample as a light sheet from a filtered arc lamp. (5 and 6) Rhod-4 and FluoVolt emission fluorescence was collected through appropriate filters for high-resolution macroscopic lens. (7) Gen III MCP intensifier. (8) Pco 1200hs CMOS camera. (9) Computer system and software for image acquisition

(Thorlabs) coupled to an optical fiber (core $\emptyset = 1$ mm). The pulse duration and the needed irradiance are inversely related; hence more light is needed if pulse duration is shortened. Typically a spot size covering 75–80 % of the plated glass bottom with irradiance of 0.5 mW/mm^2 is enough for pulses longer than 10 ms. Successful excitation of cardiomyocytes by light via optically stimulated ChR2-cFB indicates functional opsins.

5. ArchT is an inhibitory opsin that can be used to suppress spontaneous or electrode-triggered activity. Optical stimulation of ArchT-cFB may require co-stimulation from electrodes in quiescent cells to check opsin functionality. Apply electrical stimulation (10 V and 5 ms pulses) to tissue. When the co-cultured tissue follows pacing reliably, introduce a 10-s pulse of 590 nm light to activate ArchT. Successful suppression of electrically evoked excitation or shortening of the action potentials indicates functional opsins.

Optical excitation or suppression of electrical activity in cardiomyocytes can be applied to extract different electrophysiology-relevant parameters, such as action potential duration or conduction velocity (Fig. 4c).

4 Notes

1. The majority of non-myocytes are cardiac fibroblasts and small percentage of endothelial cells, pericytes, smooth muscle cells, and macrophages [7–9]. Hence we treat the non-myocyte population as a clean fibroblast culture. Small amount of residual cardiomyocytes from the separation step eventually lose functionality and get washed away during media change in the duration of subsequent maintenance and infection process.

2. Virus entry can be less effective in high-serum condition.

3. This is a good opportunity to set some cells aside as a control group without infection for viability testing. The control group should go through same infection process except without the actual virus.

4. For low MOI using high-virus titers, a dilution in PBS could be done.

5. Waste disposal of virus contaminated tools, such as pipette tips and pipettes, should be put in 10 % bleach.

6. Pattern of expression varies among cell types [23]. cFB tend to show more cytosolic expression of ChR2-eYFP or ArchT-eGFP without strong membrane-bound fluorescence.

7. PI gets incorporated into the nuclei of cells with compromised membrane in a short time. However, prolonged staining incubation will eventually stain most cells' nuclei and bias signal-to-background ratio.

8. Depending on filter settings, eYFP fluorescence can appear stronger than eGFP fluorescence. If both ChR2- and ArchT-infected cells are examined in the same flow cytometry experiment, extra attention should be paid in setting the compensation for bleed thru. In addition, if PI stain for viability is done in dual channel with reporter genes, a separate sample with no PI stain is needed to establish baseline.

9. Cardiac fibroblasts have limited proliferation cycles. Functional examination of opsins via the TCU approach needs coordination between the age of the ChR-cFB and extracting fresh cardiomyocytes. Our experience has been using 2- to 4-week-old fibroblasts since their isolation.

Acknowledgements

This work was supported by NIH-NHLBI grant R01-HL-111649 and NSF-Biophotonics grant 1511353 (to E.E.), and partially by a NYSTEM grant C026716 to the Stony Brook Stem Cell Center. We thank Christina Ambrosi and Aleks Klimas for helpful discussions.

References

1. Lajiness JD, Conway SJ (2013) Origin, development, and differentiation of cardiac fibroblasts. J Mol Cell Cardiol. doi:10.1016/j.yjmcc.2013.11.003

2. Vasquez C, Benamer N, Morley GE (2011) The cardiac fibroblast: functional and electrophysiological considerations in healthy and diseased hearts. J Cardiovasc Pharmacol 57(4):380–388. doi:10.1097/FJC.0b013e31820cda19

3. Bursac N (2014) Cardiac fibroblasts in pressure overload hypertrophy: the enemy within? J Clin Invest 124(7):2850–2853. doi:10.1172/JCI76628

4. Doetschman T, Azhar M (2012) Cardiac-specific inducible and conditional gene targeting in mice. Circ Res 110(11):1498–1512. doi:10.1161/CIRCRESAHA.112.265066

5. Zeisberg EM, Kalluri R (2010) Origins of cardiac fibroblasts. Circ Res 107(11):1304–1312. doi:10.1161/CIRCRESAHA.110.231910

6. Krenning G, Zeisberg EM, Kalluri R (2010) The origin of fibroblasts and mechanism of cardiac fibrosis. J Cell Physiol 225(3):631–637. doi:10.1002/jcp.22322

7. Baudino TA, Carver W, Giles W, Borg TK (2006) Cardiac fibroblasts: friend or foe? Am J Physiol 291(3):H1015–H1026. doi:10.1152/ajpheart.00023.2006

8. Nag AC (1980) Study of non-muscle cells of the adult mammalian heart: a fine structural analysis and distribution. Cytobios 28(109):41–61

9. Camelliti P, Borg TK, Kohl P (2005) Structural and functional characterisation of cardiac fibroblasts. Cardiovasc Res 65(1):40–51. doi:10.1016/j.cardiores.2004.08.020

10. Kohl P, Gourdie RG (2014) Fibroblast-myocyte electrotonic coupling: does it occur in native cardiac tissue? J Mol Cell Cardiol. doi:10.1016/j.yjmcc.2013.12.024

11. Rohr S (2012) Arrhythmogenic implications of fibroblast-myocyte interactions. Circ Arrhythm Electrophysiol 5(2):442–452. doi:10.1161/circep.110.957647

12. Pedrotty DM, Klinger RY, Kirkton RD, Bursac N (2009) Cardiac fibroblast paracrine factors alter impulse conduction and ion channel expression of neonatal rat cardiomyocytes. Cardiovasc Res 83(4):688–697. doi:10.1093/cvr/cvp164, cvp164 [pii]

13. Vasquez C, Mohandas P, Louie KL, Benamer N, Bapat AC, Morley GE (2010) Enhanced fibroblast-myocyte interactions in response to cardiac injury. Circ Res 107(8):1011–1020. doi:10.1161/CIRCRESAHA.110.227421

14. Miragoli M, Salvarani N, Rohr S (2007) Myofibroblasts induce ectopic activity in cardiac tissue. Circ Res 101(8):755–758. doi:10.1161/CIRCRESAHA.107.160549

15. Miragoli M, Gaudesius G, Rohr S (2006) Electrotonic modulation of cardiac impulse conduction by myofibroblasts. Circ Res 98(6):801–810. doi:10.1161/01.RES.0000214537.44195.a3

16. Vasquez C, Morley GE (2012) The origin and arrhythmogenic potential of fibroblasts in cardiac disease. J Cardiovasc Transl Res 5(6):760–767. doi:10.1007/s12265-012-9408-1

17. Chen W, Frangogiannis NG (2013) Fibroblasts in post-infarction inflammation and cardiac repair. Biochim Biophys Acta 1833(4):945–953. doi:10.1016/j.bbamcr.2012.08.023

18. Nagel G, Brauner M, Liewald JF, Adeishvili N, Bamberg E, Gottschalk A (2005) Light activation of channelrhodopsin-2 in excitable cells of Caenorhabditis elegans triggers rapid behavioral responses. Curr Biol 15(24):2279–2284. doi:10.1016/j.cub.2005.11.032, S0960-9822(05)01407-7 [pii]

19. Han X, Chow BY, Zhou H, Klapoetke NC, Chuong A, Rajimehr R, Yang A, Baratta MV, Winkle J, Desimone R, Boyden ES (2011) A high-light sensitivity optical neural silencer: development and application to optogenetic control of non-human primate cortex. Front Syst Neurosci 5:18. doi:10.3389/fnsys.2011.00018

20. Chow BY, Han X, Dobry AS, Qian X, Chuong AS, Li M, Henninger MA, Belfort GM, Lin Y, Monahan PE, Boyden ES (2010) High-performance genetically targetable optical neural silencing by light-driven proton pumps.

Nature 463(7277):98–102. doi:10.1038/nature08652

21. Dugue GP, Akemann W, Knopfel T (2012) A comprehensive concept of optogenetics. Prog Brain Res 196:1–28. doi:10.1016/B978-0-444-59426-6.00001-X

22. Entcheva E (2013) Cardiac optogenetics. Am J Physiol Heart Circ Physiol 304(9):H1179–H1191. doi:10.1152/ajpheart.00432.2012

23. Ambrosi CM, Klimas A, Yu J, Entcheva E (2014) Cardiac applications of optogenetics. Prog Biophys Mol Biol 115(2-3):294–304. doi:10.1016/j.pbiomolbio.2014.07.001

24. Yu JZ, Boyle PM, Ambrosi CM, Trayanova NA, Entcheva E (2013) High-throughput contactless optogenetic assay for cellular coupling: illustration by Chr2-light-sensitized cardiac fibroblasts and cardiomyocytes. Circulation 128(22)

25. Pedrotty DM, Klinger RY, Badie N, Hinds S, Kardashian A, Bursac N (2008) Structural coupling of cardiomyocytes and noncardiomyocytes: quantitative comparisons using a novel micropatterned cell pair assay. Am J Physiol Heart Circ Physiol 295(1):H390–H400. doi:10.1152/ajpheart.91531.2007

26. Nguyen H, Badie N, McSpadden L, Pedrotty D, Bursac N (2014) Quantifying electrical interactions between cardiomyocytes and other cells in micropatterned cell pairs. Methods Mol Biol 1181:249–262. doi:10.1007/978-1-4939-1047-2_21

27. Arrenberg AB, Stainier DY, Baier H, Huisken J (2010) Optogenetic control of cardiac function. Science 330(6006):971–974. doi:10.1126/science.1195929

28. Bruegmann T, Malan D, Hesse M, Beiert T, Fuegemann CJ, Fleischmann BK, Sasse P (2010) Optogenetic control of heart muscle in vitro and in vivo. Nat Methods 7(11):897–900. doi:10.1038/nmeth.1512

29. Abilez OJ, Wong J, Prakash R, Deisseroth K, Zarins CK, Kuhl E (2011) Multiscale computational models for optogenetic control of cardiac function. Biophys J 101(6):1326–1334. doi:10.1016/j.bpj.2011.08.004

30. Williams JC, Xu J, Lu Z, Klimas A, Chen X, Ambrosi CM, Cohen IS, Entcheva E (2013) Computational optogenetics: empirically-derived voltage- and light-sensitive channelrhodopsin-2 model. PLoS Comput Biol 9(9):e1003220. doi:10.1371/journal.pcbi.1003220

31. Vogt CC, Bruegmann T, Malan D, Ottersbach A, Roell W, Fleischmann BK, Sasse P (2015) Systemic gene transfer enables optogenetic pacing of mouse hearts. Cardiovasc Res. doi:10.1093/cvr/cvv004

32. Ambrosi C, Entcheva E (2014) Optogenetic control of cardiomyocytes via viral delivery. In: Radisic M, Black Iii LD (eds) Cardiac tissue engineering, vol 1181, Methods in molecular biology. Springer, New York, pp 215–228. doi:10.1007/978-1-4939-1047-2_19

33. Nussinovitch U, Shinnawi R, Gepstein L (2014) Modulation of cardiac tissue electrophysiological properties with light-sensitive proteins. Cardiovasc Res 102(1):176–187. doi:10.1093/cvr/cvu037

34. Jia Z, Valiunas V, Lu Z, Bien H, Liu H, Wang HZ, Rosati B, Brink PR, Cohen IS, Entcheva E (2011) Stimulating cardiac muscle by light: cardiac optogenetics by cell delivery. Circ Arrhythm Electrophysiol 4(5):753–760. doi:10.1161/CIRCEP.111.964247

35. Entcheva E, Bien H (2009) Mechanical and spatial determinants of cytoskeletal geodesic dome formation in cardiac fibroblasts. Integr Biol 1(2):212–219. doi:10.1039/B818874b

Chapter 22

Optogenetic Engineering of Atrial Cardiomyocytes

Iolanda Feola, Alexander Teplenin, Antoine A.F. de Vries,
and Daniël A. Pijnappels

Abstract

Optogenetics is emerging in the cardiology field as a new strategy to explore biological functions through the use of light-sensitive proteins and dedicated light sources. For example, this technology allows modification of the electrophysiological properties of cardiac muscle cells with superb spatiotemporal resolution and quantitative control. In this chapter, the optogenetic modification of atrial cardiomyocytes (aCMCs) from 2-day-old Wistar rats using lentiviral vector (LV) technology and the subsequent activation of the light-sensitive proteins (i.e., ion channels) through light-emitting diodes (LEDs) are described.

Key words Optogenetics, Atrial cardiomyocytes, Lentiviral vectors, Optical mapping, Light-emitting diode (LED)

1 Introduction

Optogenetics is a new technology to control cellular function by a combination of genetic engineering and light application. Optogenetics has revolutionized neuroscience by offering the possibility to control at high spatial and temporal resolution various neurological processes both in vitro and in vivo [1]. Despite these unique features, optogenetics has not yet been extensively explored in cardiac research. Nevertheless, several scientific contributions have shown how light-gated ion channels can modulate the transmembrane potential of cardiomyocytes giving rise to excitatory or inhibitory responses, controlled in time, space, and magnitude [2, 3]. Channelrhodopsins [4] are excitatory proteins that upon illumination can evoke an action potential (AP) in excitable cells due to the passive inward flow of positively charged ions. When instead light induces protons or potassium ions to leave the cell or chloride ions to enter the cell, cellular excitation is inhibited [5–9]. Due to their specific properties, light-activated ionophoric proteins may be very helpful to gain additional insight into the mechanisms underlying cardiac arrhythmias, which may inspire the

Arash Kianianmomeni (ed.), *Optogenetics: Methods and Protocols*, Methods in Molecular Biology, vol. 1408,
DOI 10.1007/978-1-4939-3512-3_22, © Springer Science+Business Media New York 2016

exploration of new strategies for treating electrical disturbances in the heart. Besides via light-gated ionophoric proteins, optogenetics offers many other possibilities to control cellular behavior, including transcription as well as intracellular and receptor signaling [10–12], and could thereby further improve our understanding of cardiomyocyte biology and heart function.

The following section describes how aCMCs, isolated from the hearts of 2-day-old Wistar rats, can be successfully optogenetically engineered by forced expression of a depolarizing optogenetic tool, i.e., Ca^{2+}-permeable channelrhodopsin (CatCh) [13] following LV-mediated transgene delivery. In addition, details are provided on how to study the functional consequences of LED-mediated CatCh activation by optical voltage mapping. For example, exposure of CatCh-expressing aCMCs to 10-ms blue light pulses induces a photocurrent strong enough to evoke an AP in these cells [14].

2 Materials

2.1 Reagents for Isolating aCMCs

1. Isoflurane.

2. Fibronectin working solution (100 μg/ml): Dilute the fibronectin stock solution ten times with phosphate-buffered saline (PBS) and store at 4 °C.

3. Solution A: 0.02 g/l Phenol red, 136 mM NaCl, 5 mM KCl, 0.2 mM $Na_2HPO_4\cdot2H_2O$, 0.44 mM KH_2PO_4, 5.5 mM D-glucose, and 20 mM acid-free HEPES. Adjust the pH to 7.4–7.5 by adding 5 M NaOH at 21 °C. Sterilize by filtration through a 0.22 μm pore size cellulose acetate bottle top filter and store at 4 °C.

4. Solution B: 1 mM $CaCl_2\cdot2H_2O$, 60 mM $MgCl_2\cdot6H_2O$ in Solution A. Store at –20 °C.

5. Solution C: DNase I solution. Mix 500 units of deoxyribonuclease with 800 μl Solution A. Store at –20 °C.

6. Dissociation medium: Mix 90,000 units of collagenase type I (Worthington Biochemical, Lakewood, NJ, USA) with 200 ml of Solution A, 2 ml of solution B, and 200 μl of solution C. Sterilize by filtration through a 0.22 μm pore size cellulose acetate bottle top filter and store at –20 °C.

7. Growth medium: Mix 450 ml of 1× Ham's F10 nutrient mix with 10 ml of heat-inactivated fetal bovine serum (HI-FBS), 10 ml of heat-inactivated horse serum (HI-HS), and 10 ml of penicillin-streptomycin stock solution (5000 U/ml). Store at 4 °C.

8. Solution D: 2.8 mM sodium ascorbate, 107 mM pyruvic acid, 500 mM D-glucose, and 5.7 % (w/v) bovine serum albumin (BSA).

9. CMC medium: Add 15 ml of solution D to 250 ml of Dulbecco's modified Eagle's medium (DMEM)-low glucose. Next, add 25 ml of 10× Ham's F10, 192.5 ml of sterile water, 4 ml of 7.5 % sodium bicarbonate, 0.305 ml of 0.2 M L-glutamine, and 10 ml of penicillin-streptomycin solution. Store at 4 °C.

10. Sterile glass beats, 6.7–7.3 mm ∅.

11. Mitomycin-C solution: Dissolve 2 mg mitomycin-C powder (from *Streptomyces caespitosus*) in 4 ml PBS. Sterilize by filtration through a 0.22 µm pore size cellulose acetate syringe filter and store at 4 °C.

12. Round glass cover slips (15 mm ∅).

13. Primaria cell culture dishes (60 mm ∅).

14. Falcon 70 µm mesh size cell strainers.

15. Costar 24-well clear TC-treated multiple well plates.

2.2 Reagents for Producing Self-Inactivating LVs (SIN-LVs)

1. 293T cells [15].

2. SIN-LV shuttle plasmid pLV.MHCK7.CatCh~eYFP. WHVPRE [14] and derivatives thereof.

3. SIN-LV packaging/helper plasmids psPAX2 and pLP/VSV-G.

4. DMEM-high glucose (HG) + 10 % FBS.

5. DMEM-HG + 5 % FBS + 25 mM HEPES-NaOH (pH 7.4).

6. TrypLE Express.

7. 150 mM Sterile NaCl solution.

8. 1 mg/ml PEI solution (pH 7.4): Weigh 45 mg of linear polyethylenimine (Mw 25,000), and add 100 µl of 1 M HCl and 40 ml of sterile water of 80 °C. Shake vigorously to facilitate dissolution. Adjust the pH to 7.4 with 37 % HCl. Increase the total volume to 45 ml with sterile water. Sterilize by filtration through a 0.22 µm pore size cellulose acetate syringe filter and store in 1.8 ml aliquots at −80 °C.

9. 20 % sucrose solution: Weigh 100 g of sucrose and add 50 ml of 10× PBS. Increase the volume to 500 ml with sterile water. Use a magnetic stirrer to facilitate dissolution. Sterilize by filtration through a 0.22 µm pore size cellulose acetate bottle top filter and store in 50 ml aliquots at 4 °C.

10. Polyallomer ultracentrifuge tubes for SW28 or SW32 rotor.

11. Millex sterile 0.45 µm pore size syringe filters (33 mm ∅).

12. BD Plastipak sterile 50 ml syringes with Luer Lock system.

13. PBS-1 % BSA: Dissolve 1 g of BSA in 100 ml of PBS by gentle stirring with a magnetic stirrer. Sterilize by filtration through a 0.22 µm pore size cellulose acetate syringe filter and store in 1.6 ml aliquots at 4 °C.

2.3 Combination of LED and Optical Mapping System

1. HEPES-buffered, phenol red-free DMEM/F12.

2. Di-4-ANEPPS solution (1 mg/ml): Dissolve 5 mg of di-4-ANEPPS powder in 5 ml sterile dimethyl sulfoxide and store in 500 μl aliquots at 4 °C and protected from light.

3. MiCAM ULTIMA-L imaging system (SciMedia, Costa Mesa, CA, USA).

4. Excitation filter (530 nm pass, Semrock, Rochester, NY, USA).

5. Dichroic mirror (520–560 nm reflect >90 %, >600 nm pass >85 %, Semrock).

6. Emission filter (>590 nm pass, Semrock).

7. 470 nm Rebel LED mounted on a 25 mm ∅ round CoolBase (Luxeonstar, Brantford, Ontario, Canada).

8. Plano-convex lens (1 in. ∅, 25.4 mm focal length) (Thorlabs, Munich, Germany).

9. Stimulus generator STG2004 (Multichannel System, Reutlingen, Germany).

10. Custom-made platinum electrode pair for bipolar point stimulation.

11. Power meter PM100D (Thorlabs).

12. Brain Vision Analyzer 1208 software (Brainvision, Tokyo, Japan).

3 Methods

3.1 aCMC Isolation and Culture

3.1.1 Isolation of aCMCs

1. Anesthetize 2-day-old Wistar rats ($n \approx 50$) via 4–5 % isoflurane inhalation. Ensure adequate anesthesia by checking the absence of reflexes.

2. Rapidly excise the hearts and collect them in a plastic Petri dish containing ±10 ml ice-cold Solution A.

3. Separate the atria from the ventricles (*see* **Note 1**).

4. Remove Solution A until the bottom of the Petri dish just stays covered with fluid. Chop the atrial tissues in small pieces and rinse the tissue with 4 ml of Solution A to get rid of erythrocytes (*see* **Note 2**).

5. Prepare 12 ml of an ice-cold 1:1 mixture of dissociation medium and Solution A (Solution E). Remove solution A as much as possible with a 1000 μl pipetman and add 7 ml of Solution E to the Petri dish with atrial tissue pieces.

6. Transfer the atrial tissue pieces in Solution E to a sterile 50 ml Erlenmeyer flask with screw-cap containing sterile glass beats.

7. Incubate under gentle agitation at 37 °C for 35 min.

8. Transfer the fully digested material (5 ml) to a sterile 15 ml polypropylene screw-cap tube on ice and leave the undigested material (2 ml) in the Erlenmeyer flask.

9. Add 4 ml of fresh Solution E to the Erlenmeyer flask and incubate once again under agitation at 37 °C for 35 min.

10. Transfer the content of the Erlenmeyer flask to the 15 ml tube with the remainder of the atrial digest and pellet the cells by centrifugation for 10 min at $150 \times g$. Suspend the cell pellet in 12 ml of prewarmed growth medium. Equally divide the cell suspension over four Primaria cell culture dishes and incubate at 37 °C in a humidified 95 % air/5 % CO_2 atmosphere (culture conditions) for 120 min to allow preferential attachment of non-cardiomyocytes (mainly cardiac fibroblasts).

11. Pass the growth medium with the non-adhered cells (mainly cardiomyocytes) through a nylon cell strainer with a mesh size of 70 μm to remove undigested tissue fragments and cell aggregates.

12. Count and seed the cells on fibronectin-coated glass cover slips in 24-well cell culture plates. To obtain confluent aCMC monolayers seed 8×10^5 cells/well.

3.1.2 Preparation of Fibronectin-Coated Glass Cover Slips

1. Place a sterile glass cover slips on the bottom of each well of a 24-well cell culture plate.

2. Add 300 μl of fibronectin solution to each well.

3. Incubate for 1 h under culture conditions.

4. Collect the fibronectin solution (can be used two more times) and leave the cover slips to dry in the flow hood for about 60 min.

3.1.3 Antiproliferative Treatment with Mitomycin-C [16]

1. The day after aCMC isolation dilute the mitomycin-C solution 50 times in growth medium and add 300 μl/well of a 24-well cell culture plate.

2. Incubate no longer than 2 h under culture conditions. Of note, as mitomycin-C is highly mutagenic handle with extreme care.

3. Rinse the cells twice with PBS.

4. Add 1 ml of CMC medium + 5 % HI-HS of 37 °C/well and maintain the cells under culture conditions.

5. Refresh the culture medium once a day.

3.2 Production of Vesicular Stomatitis Virus G Protein (VSV-G)-Pseudotyped SIN-LVs

3.2.1 Seeding of 293T cells in 175 cm² Cell Culture Flasks

1. Use 293T cells of low passage number for LV production. Check the 293T cells using an inverted phase-contrast microscope. When the cultures are ±80 % confluent, remove the medium and rinse each 175 cm² cell culture flask once with 10 ml PBS.

2. Add 1.5 ml of TrypLE Express/flask and after 1-min incubation gently rock the flask to facilitate cell detachment. Once the cells have detached from the plastic support collect them in DMEM-HG + 10 % FBS.

3. Generate a single-cell suspension by forcefully pipetting the cell suspension up and down 3–5 times to disrupt remaining cell clusters without destroying the cells (can be checked microscopically).

4. Suspend 7×10^7 viable 293T cells in 85 ml DMEM-HG + 10 % FBS and add 20 ml of the cell suspension to each 175 cm² cell culture flask (i.e., 1.65×10^7 cells/flask). Make sure that the cells become evenly spread over the plastic support (*see* **Note 3**).

3.2.2 Transfection of 293T Cells

1. The day after, before starting with the transfection make sure that the cells are ±60–70 % confluent, evenly spread over the surface of the culture flasks and viable (*see* **Note 4**).

2. Prepare the DNA/NaCl solution by diluting 154 µg of plasmid DNA (molar ratio SIN-LV shuttle plasmid:psPAX2:pLP/VSV-G is 8:5:5) in 4.4 ml of 150 mM NaCl in a sterile 50 ml polypropylene screw-cap tube (*see* **Note 5**). Add the DNA to the NaCl solution instead of the other way around and mix by gentle vortexing. The cell type(s) in which and the level at which the transgene is expressed will be determined by the promoter that drives its expression (Fig. 1).

3. Prepare the PEI/NaCl solution by adding 504 µl of 1 mg/ml PEI solution to 4296 µl of 150 mM NaCl in a sterile 50 ml polypropylene screw-cap tube. Mix the PEI with the NaCl solution by gentle vortexing.

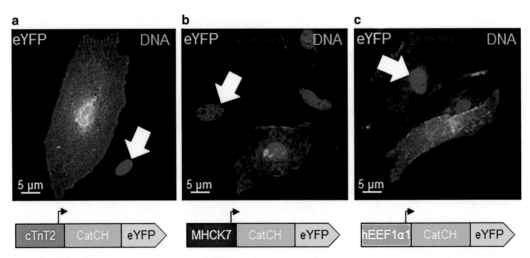

Fig. 1 Immunocytological confirmation of aCMC-specific transgene expression. aCMC cultures were incubated with CatCh~eYFP-encoding SIN-LVs in which transgene expression was driven by (**a**) the striated muscle-specific MHCK7 promoter [20], (**b**) the cardiomyocyte-specific chicken Tnnt2 promoter [21], or (**c**) the ubiquitous human EEF1A1 promoter. The MHCK7 and Tnnt2 promoter give rise to transgene expression in α-actinin-positive cells (i.e., cardiomyocytes) only, while the promoter of the housekeeping gene EEF1A1 directs transgene expression in α-actinin-positive cells as well as α-actinin-negative cells (i.e., cardiac fibroblasts [*white arrows*]). The *blue* fluorescence corresponds to cell nuclei stained with Hoechst 33342

4. Add 4.4 ml of the PEI/NaCl solution to the DNA/NaCl solution in a dropwise fashion. Gently rock/swirl the tubes during the addition of the PEI solution. Homogenize the content of the tubes by vortexing for 10 s (*see* **Note 6**).

5. Incubate the tubes for 15 min at room temperature (RT) to allow the formation of DNA/PEI complexes, and then add 2 ml of this transfection mixture to each flask. Move the flasks to completely mix the transfection mixture and culture medium.

3.2.3   Replacement of the Transfection Medium by DMEM-HG + 5 % FBS + 25 mM HEPES-NaOH (pH 7.4)

1. One day after the addition of the DNA/PEI complexes, replace the transfection medium in each of the 175 cm² cell culture flasks by 15 ml fresh DMEM-HG + 5 % FBS + 25 mM HEPES-NaOH (pH 7.4) (*see* **Note 7**).

3.2.4   Harvesting of the Culture Supernatants and Concentration/Purification of the SIN-LV Particles

1. Before harvesting the LV particle-containing culture supernatants, ±36–48 h after transfection, check the transfection efficiency of the 293T cells using an inverted fluorescence microscope (Fig. 2).

2. Transfer the culture medium of each pair of two 175 cm² cell culture flasks to a sterile 50 ml polypropylene screw-cap tube.

3. Centrifuge the tubes for 10 min at 3000 × *g* and RT in a tabletop centrifuge.

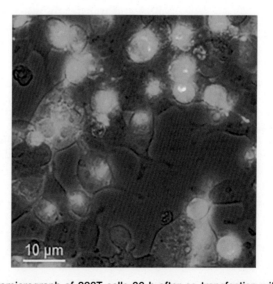

Fig. 2 Fluoromicrograph of 293T cells 36 h after co-transfection with psPAX2, pLP/VSV-G, and pLV.MHCK7.CatCh~eYFP.WHVPRE. LV yield strongly depends on the transfection efficiency, which should be >90 %. The *rounded shape* of the cells is largely due to the cytotoxic effects associated with the accumulation and aggregation (brightly fluorescent perinuclear dots) of the CatCh~eYFP fusion protein in the LV producer cells

4. Push the cleared culture media through 0.45 μm pore size polyethersulfone syringe filters and collect the filtrate in autoclaved 38.5 ml polyallomer ultracentrifuge tubes.

5. Carefully move a 5 ml stripette containing 7 ml of 20 % sucrose in PBS through the cell culture medium to the bottom of the ultracentrifuge tube and slowly release 5 ml of the sucrose solution to underlay the culture medium with a sucrose cushion.

6. After taring with DMEM-HG + 5 % FBS + 25 mM HEPES-NaOH (pH 7.4) spin the ultracentrifuge tubes for 2 h with slow acceleration and without braking at 15,000 revolutions/min ($\pm$ 40,000 × g_{max}) and 4 °C.

7. Aspirate the supernatant and place the ultracentrifuge tube upside down on a piece of sterile filter paper to absorb the remaining supernatant.

8. Add 400 μl of ice-cold PBS-1 % BSA to the LV particle-containing pellet in each ultracentrifuge tube.

9. Place each ultracentrifuge tube in upright position in a 50 ml polypropylene screw-cap tube and incubate overnight at 4 °C in the cold room while shaking gently and with the cap closed.

3.2.5 Aliquoting of the Concentrated SIN-LV Suspensions

1. Collect the SIN-LV suspensions in one of the ultracentrifuge tubes.

2. Wash each ultracentrifuge tube with 100 μl of ice-cold PBS-1 % BSA and transfer the wash solution to the collection tube.

3. Divide the supernatant on ice in 50–100 μl aliquots using precooled 0.5 ml microtubes for storage at –80 °C (*see* **Note 8**).

3.3 Transduction of aCMCs

3.3.1 Transduction

1. Four days after aCMC isolation thaw the SIN-LV stock on ice and add the desired amount of SIN-LV particles to prewarmed CMC medium + 5 % HI-HS.

2. Gently mix to ensure homogenous distribution of SIN-LV particles and replace the culture medium with 400 μl of inoculum/well of a 24-well cell culture plate (*see* **Note 9**).

3.3.2 Medium Refreshment

1. Approximately 24 h after transduction aspirate the inoculum, wash the cells once with PBS, and add 1 ml of prewarmed CMC medium + 5 % HI-HS/well of a 24-well cell culture plate.

3.4 Light Stimulation and Optical Voltage Mapping

3.4.1 Preparation of the Optical Mapping System

1. Place the plano-convex lens on top of the 470 nm LED.

2. Align the LED assembly with the center of the MiCAM ULTIMA-L camera.

3. Connect the lens-LED complex to the stimulus generator.

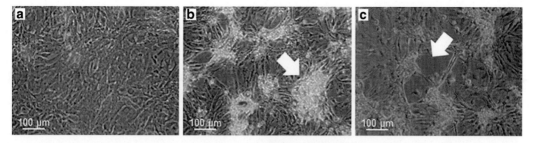

Fig. 3 Morphology of aCMC monolayers at day 9 after isolation. (**a**) Non-interrupted, homogenous aCMC monolayer well suited form optical mapping studies. (**b**) Inhomogeneous aCMC monolayer containing star-shaped cell aggregates (*white arrow*). (**c**) Noncontinuous aCMC monolayer due to the presence of acellular areas (*white arrow*). Cultures (**b**) and (**c**) do not allow reliable acquisition of whole culture optical mapping data

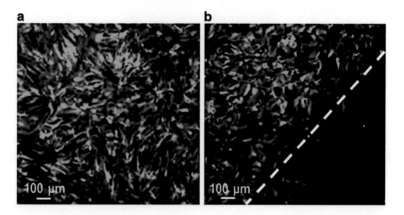

Fig. 4 Fluoromicrograph of confluent aCMC cultures 3 days after transduction with LV.MHCK7.CatCh~eYFP.WHVPRE. (**a**) Homogeneously transduced aCMC culture. (**b**) Heterogeneously transduced aCMC culture. Only cultures showing (near-)quantitative transduction and homogenous transgene expression should be used for optical mapping experiments

3.4.2 Preparation of aCMC Cultures

1. Check the aCMC monolayers for structural homogeneity, ±72 h after transduction, using an inverted phase-contrast microscope (*see* **Note 10**) (Fig. 3).

2. Check the aCMC monolayers for homogeneity of transgene expression by visualizing the enhanced yellow fluorescent protein tag fused to the channelrhodopsin with the aid of an inverted fluorescence microscope (*see* **Note 11**) (Fig. 4).

3. Prepare an 8 μM di-4-ANEPPS solution in prewarmed DMEM/F12.

4. Replace the CMC medium by 500 μl of the potentiometric dye solution and incubate for 10 min under culture conditions.

5. Aspirate the di-4-ANEPPS solution and add 500 μl of DMEM/F12.

6. Place the cells under the MiCAM ULTIMA-L camera.

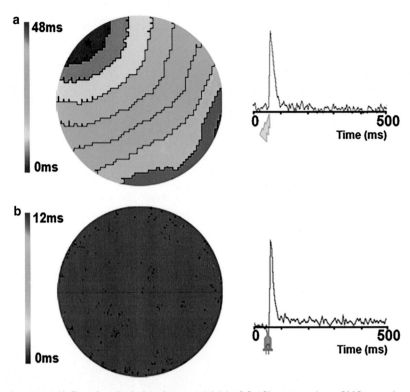

Fig. 5 Activation maps (*left*) and optical signal traces (*right*) of CatCh-expressing aCMC monolayers following (**a**) electrical or (**b**) optical induction of APs. The activation map in (**a**) shows uniform convex AP propagation initiated at the bipolar pacing electrode (*star*). The activation map in (**b**) shows synchronous induction of APs in the entire monolayer following 10-ms exposure to 470 nm LED light (0.08 mW/mm² irradiance)

3.4.3 Optical Voltage Mapping

1. Focus the MiCAM ULTIMA-L camera and check for functional homogeneity by electrical point stimulation at a frequency of 1 Hz using 10-ms rectangular pulses of 8 V (*see* **Note 12**).

2. While recording pace the cultures at a frequency of 1 Hz using 10-ms blue light pulses at the maximum current tolerated by the LED (i.e., 700 mA) (Fig. 5).

4 Notes

1. Selective removal of the atria can be accomplished by placing an opened scissor tightly around the ventricles and gently moving the scissor upwards until the atria will fall on top of the blades of the scissor and can be cut off.

2. Standardization of the chopping procedure is complicated due to subtle differences in the technique used by individual researchers. Experience has taught us that both insufficient and excessive cutting strongly reduces cardiomyocyte yields.

3. Differences in 293T cell density within each culture flask or between culture flasks should be avoided by frequent homogenization of the cell suspension during the seeding procedure and by ensuring that the shelves of the CO_2 incubator are perfectly horizontal (check with a spirit level).

4. For LV production, only use 293T cells of low passage number (i.e., <60). Furthermore, do not use 293T cells that have been seeded at too low a density, "over-trypsinized", or allowed to grow overconfluent.

5. The quality of the packaging/helper constructs and of the LV shuttle plasmid strongly influences LV yields. To achieve a high transfection efficiency, the plasmid DNA preparations should be ≥95 % supercoiled as checked by agarose gel electrophoresis, have a low endotoxin content, and contain minimal amounts of (in)organic contaminants.

6. The PEI/NaCl solution should be added to the DNA/NaCl solution instead of the other way around to ensure a high transfection efficiency [17].

7. 293T cells should be transfected at the right cell density (i.e., subconfluency). If the cell cultures are already ±60–70 % confluent in the morning of the day following seeding, the 293T cells should be transfected at that moment and refreshment of the transfection medium should be done in the late afternoon (i.e., after 6–8 h).

8. VSV-G-pseudotyped LV stocks can be repeatedly thawed and frozen without significant loss of functional titer provided that they are kept on ice/at 4 °C during inoculum preparation [18].

9. The transduction efficiency of target cells by LVs is not only depending on the vector dose and infection period but also on the inoculum volume/vector concentration. To achieve the highest possible transduction efficiency, the inoculum volume should be kept as small as possible [19]. However, the target cells should still be covered with sufficient culture medium to keep them healthy.

10. The aCMC monolayers should not show acellular areas and/or star-shaped aggregated cell clusters.

11. The aCMC monolayers should show homogenous transgene expression.

12. The aCMC monolayers should not show spontaneous activity and areas of conduction slowing or block. Moreover, the monolayers should not show areas of significant optical signal prolongation.

Acknowledgement

This work was supported by a VIDI grant (91714336) from the Dutch Organization for Scientific Research (NWO) to Daniël Pijnappels. Antoine de Vries is a recipient of a Chinese Exchange Programme grant (10CDP007) from the Royal Netherlands Academy of Arts and Sciences (KNAW) and received additional support by ICIN-Netherlands Heart Institute.

References

1. Boyden ES, Zhang F, Bamberg E, Nagel G, Deisseroth K (2005) Millisecond-timescale, genetically targeted optical control of neural activity. Nat Neurosci 8:1263–1268

2. Entcheva E (2013) Cardiac optogenetics. Am J Physiol Heart Circ Physiol 304:1179–1191

3. Bruegmann T, Malan D, Hesse M, Beiert T, Fuegemann CJ, Fleischmann BK, Sasse P (2010) Optogenetic control of heart muscle in vitro and in vivo. Nat Methods 7:897–900

4. Nagel G, Szellas T, Huhn W, Kateriya S, Adeishvili N, Berthold P, Ollig D, Hegemann P, Bamberg E (2003) Channelrhodopsin-2, a directly light-gated cation-selective membrane channel. Proc Natl Acad Sci U S A 100:13940–13945

5. Zhang F, Wang LP, Brauner M, Liewald JF, Kay K, Watzke N, Wood PG, Bamberg E, Nagel G, Gottschalk A, Deisseroth K (2007) Multimodal fast optical interrogation of neural circuitry. Nature 446:633–639

6. Chow BY, Han X, Dobry AS, Qian X, Chuong AS, Li M, Henninger MA, Belfort GM, Lin Y, Monahan PE, Boyden ES (2010) High-performance genetically targetable optical neural silencing by light-driven proton pumps. Nature 463:98–102

7. Cosentino C, Alberio L, Gazzarrini S, Aquila M, Romano E, Cermenati S, Zuccolini P, Petersen J, Beltrame M, Van Etten JL, Christie JM, Thiel G, Moroni A (2015) Optogenetics. Engineering of a light-gated potassium channel. Science 348:707–710

8. Wietek J, Wiegert JS, Adeishvili N, Schneider F, Watanabe H, Tsunoda SP, Vogt A, Elstner M, Oertner TG, Hegemann P (2014) Conversion of channelrhodopsin into a light-gated chloride channel. Science 344:409–412

9. Berndt A, Lee SY, Ramakrishnan C, Deisseroth K (2014) Structure-guided transformation of channelrhodopsin into a light-activated chloride channel. Science 344:420–424

10. Müller K, Naumann S, Weber W, Zurbriggen MD (2015) Optogenetics for gene expression in mammalian cells. Biol Chem 396:145–152

11. Beyer HM, Naumann S, Weber W, Radziwill G (2015) Optogenetic control of signaling in mammalian cells. Biotechnol J 10:273–283

12. Zhang K, Cui B (2015) Optogenetic control of intracellular signaling pathways. Trends Biotechnol 33:92–100

13. Kleinlogel S, Feldbauer K, Dempski RE, Fotis H, Wood PG, Bamann C, Bamberg E (2011) Ultra light-sensitive and fast neuronal activation with the Ca^{2+}-permeable channelrhodopsin CatCh. Nat Neurosci 14:513–518

14. Bingen BO, Engels MC, Schalij MJ, Jangsangthong W, Neshati Z, Feola I, Ypey DL, Askar SF, Panfilov AV, Pijnappels DA, de Vries AA (2014) Light-induced termination of spiral wave arrhythmias by optogenetic engineering of atrial cardiomyocytes. Cardiovasc Res 104:194–205

15. DuBridge RB, Tang P, Hsia HC, Leong PM, Miller JH, Calos MP (1987) Analysis of mutation in human cells by using an Epstein-Barr virus shuttle system. Mol Cell Biol 7:379–387

16. Askar SF, Ramkisoensing AA, Schalij MJ, Bingen BO, Swildens J, van der Laarse A, de Vries AA, Ypey DL, Pijnappels DA (2011) Antiproliferative treatment of myofibroblasts prevents arrhythmias in vitro by limiting myofibroblast-induced depolarization. Cardiovasc Res 90:295–304

17. Boussif O, Lezoualc'h F, Zanta MA, Mergny MD, Scherman D, Demeneix B, Behr JP (1995) A versatile vector for gene and oligonucleotide transfer into cells in culture and in vivo: polyethylenimine. Proc Natl Acad Sci U S A 92:7297–7301

18. Higashikawa F, Chang L (2001) Kinetic analyses of stability of simple and complex retroviral vectors. Virology 280:124–131

19. Zhang B, Metharom P, Jullie H, Ellem KA, Cleghorn G, West MJ, Wei MQ (2004) The significance of controlled conditions in lentiviral vector titration and in the use of multiplicity of infection (MOI) for predicting gene transfer events. Genet Vaccines Ther 2:6

20. Salva MZ, Himeda CL, Tai PW, Nishiuchi E, Gregorevic P, Allen JM, Finn EE, Nguyen QG, Blankinship MJ, Meuse L, Chamberlain JS, Hauschka SD (2007) Design of tissue-specific regulatory cassettes for high-level rAAV-mediated expression in skeletal and cardiac muscle. Mol Ther 15:320–329

21. Prasad KM, Xu Y, Yang Z, Acton ST, French BA (2011) Robust cardiomyocyte-specific gene expression following systemic injection of AAV: in vivo gene delivery follows a Poisson distribution. Gene Ther 18:43–52

Chapter 23

A Multichannel Recording System with Optical Stimulation for Closed-Loop Optogenetic Experiments

Carmen Bartic, Francesco P. Battaglia, Ling Wang, Thoa T. Nguyen, Henrique Cabral, and Zaneta Navratilova

Abstract

Selective perturbation of the activity of specific cell types in the brain tissue is essential in understanding the function of neuronal circuits involved in cognition and behavior and might also provide therapeutic neuromodulation strategies. Such selective neuronal addressing can be achieved through the optical activation of light-sensitive proteins called opsins that are expressed in specific cell populations through genetic methods—hence the name "optogenetics." In optogenetic experiments, the electrical activity of the targeted cell populations is optically triggered and monitored using arrays of microelectrodes. In closed-loop studies, the optical stimulation parameters are adjusted based on the recorded activity, ideally in real time. Here we describe the basic tools and the protocols allowing closed-loop optogenic experiments in vivo.

Key words Optical stimulation, Closed-loop, Multichannel recordings, Tetrode, Optic fibers, Spike-sorting

1 Introduction

Selective excitation of genetically expressed proteins called "opsins," capable of transforming a light signal into an electrical stimulus, is currently a powerful stimulation tool in neuroscience—called "optogenetics" [1, 2]. The light stimuli with variable intensity and temporal parameters are delivered usually through implanted multimode optic fibers coupled to LED or laser sources, while the evoked electrical activity might be detected by microelectrodes connected to low noise amplifiers. The microelectrodes are either twisted wires assembled into a miniaturized driving frame (i.e. microdrive) allowing depth positioning with accuracies in the micrometer range [3] or sometimes multielectrode arrays processed on silicon neuroprobes [4, 5]. Optogenetic neural perturbation and multichannel activity recordings can be performed either in open- or closed-loop. In a typical open-loop experiment, the stimulation parameters are set beforehand and the evoked responses are

Arash Kianianmomeni (ed.), *Optogenetics: Methods and Protocols*, Methods in Molecular Biology, vol. 1408,
DOI 10.1007/978-1-4939-3512-3_23, © Springer Science+Business Media New York 2016

detected, while in closed-loop operation, the measured responses are used to determine the next stimulus parameters with the purpose of enhancing or suppressing the activity of particular cells or networks. Closed-loop operation has been used in studying the mechanisms of pathologically synchronized firing in a demand-controlled way [6–9], altering the functional connectivity and motor behavior through single-electrode stimulation based on extracellular recordings in the primate cortex [10] or in regulating the firing activity of the medial temporal lobe neurons through visual regulation in humans [11]. Closed-loop electrical stimulation that was triggered by a hippocampal place cell activation during sleep has been shown to induce behavioral changes [12]. Also, closed-loop optogenetic stimulation has been applied in seizure suppression [13, 14] or in modifying the functional interactions between neuronal populations in an in vitro model [15].

Closed-loop operation is advantageous when neural selectivity is desired, since it allows feedback-controlled stimulus optimization. The stimulus signal, in the case of optogenetics delivered through implanted optic fibers, can be triggered online whenever the neural activity is threshold detected [16, 17]. However, the online spike-sorting often has to be sacrificed in order to meet the speed requirements for real time decoding and/or visualization (millisecond latency) [11, 15]. Although several algorithms suitable for online signal processing have been reported [18–20], only few have actually been implemented in a real-time system [21].

We describe the implementation of a system allowing electrical recordings and optical stimulation, interfaced with a microdrive containing at least 24 tetrodes and at least two optical fibers. This system with 32 recording channels has been described in [22], but here we also show how the system can be scaled up to 96 channels at least. The prototype is closed-loop operated by implementing the template matching as a spike-sorting algorithm. Our CPU implementation is fast enough to process signals from 8 integrated tetrodes (32 channels) and deliver optical stimuli with a controlling frequency of 8 ms, while the GPU-based spike detection [23] offers enhanced data processing speed and is suitable for systems containing up to 128 channels.

Such closed-loop opto-electronic systems are valuable tools in studying dynamic brain processes in neuroscience.

2   System Configuration and Assembly

The hardware configuration shown in Fig. 1 allows data acquisition from 96 parallel channels together with reliable and fast signal processing capability and controlled delivery of light pulses with selectable temporal features.

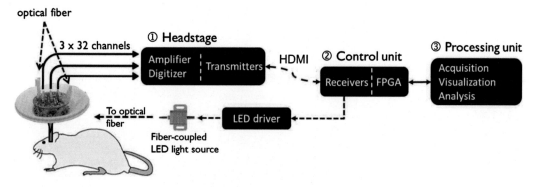

Fig. 1 Architecture of the 96-channel closed-loop platform

A custom-designed *microdrive* loaded with 24 tetrodes (96 channels) allows accurate positioning of the tetrodes in the brain and in situ electrical activity detection. This microdrive allows also the positioning of three individually controllable multimode or single-mode optic fibers. The data acquisition system consists of three blocks: (1) three head-mounted amplifier boards, (2) a control unit, and (3) a processing unit (*see* Fig. 1). The assembly and operation of each component are described below.

2.1 Microdrive

A microdrive allowing the placement of up to 24 independently moveable tetrodes was designed for targeting forebrain brain areas such as the medial prefrontal cortex or the dorsal hippocampus. The drive shown in Fig. 2 (adapted from a design described [24]) consists of a 3D printed plastic shell, with 24 screw-driven shuttles allowing adjustment of each tetrode depth.

Tetrodes are made out of four twisted strands of Teflon-insulated 13 μm diameter Nichrome (Sandvik, Halstahammar, Sweden) wire. The insulation is melt after twisting, so that the four strands stick to one another. Tetrodes are then successively inserted in polyimide tubes (inner diameter: 0.0071″, outer diameter: 0.0116″, High Performance Conductors, Inc., Inman, SC) that are further glued to the shuttles.

The microdrive can contact the brain by mean of an exit bundle, where the guide tubes that protect each tetrode can be arranged in one or more groups, targeting different brain areas.

After the drive assembly, the tetrode tips were freshly cut and gold electroplated to a final impedance of 300–500 kΩ, using a Nano-Z computer-controlled current source (Neuralynx, Bozeman, MT).

2.2 Electrode Interface Board (EIB)

A custom PCB designed and commercialized by Atlas Neuroengineering—http://www.atlasneuro.com/—ensures the tetrode connection to the amplifier headstage and the vertical, central positioning of the optic fibers. By using a similar driving system, as for tetrodes, the optic fibers can be individually moved up

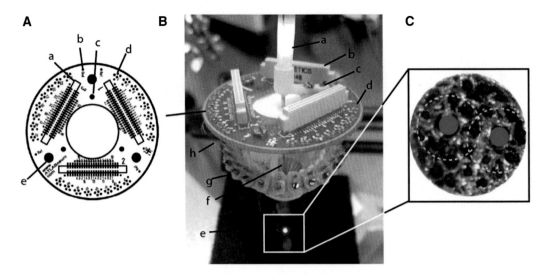

Fig. 2 Microdrive for 24 recording tetrodes and up to three optic fibers: (**a**) Electrode interface board (EIB) scheme with (a) Omnetics connector, (b) ground connection, (c) fiber microdrive screw hole, (d) tetrode pin, (e) attachment screw hole; (**b**) photograph of the microdrive with the tetrode-fiber array: (a) fiber connector sleeve, (b) Omnetics connector, (c) fiber microdrive, (d) EIB, (e) tetrode-fiber array, (f) fiber, (g) tetrode microdrive, (h) tetrode; (**c**) cross-sectional view through the tetrode bundle with highlighted tetrode (*red dots*) and fiber (*blue dots*) positions; *dashed white circles* highlight the ring arrangement of the tetrodes around the fibers

and down in small increments and with high precision. Three 44 pin tetrode connectors (nanoseries), provided by Omnetics Connector Corporation—http://www.omnetics.com—are triangularly arranged around the central hole allowing the insertion of the fibers. A more cost-effective alternative for the Omnetics connectors is provided by the SlimStack™ connectors from Molex—www.molex.com—although the Omnetics connectors, given their compatibility with commercial headstage (e.g. Neuralynx headstages), offer possibilities for additional validation using commercial data acquisition systems.

2.3 Controlled Light Delivery System

Both optic single mode (SM—core diameter size 5–10 μm) and multimode (MM—core diameter size 100–200 μm) fibers can be implanted in the brain for light delivery depending on the desired stimulation volume. Numerical simulations based on the optical properties of rodent brain tissues have shown that fibers with higher numerical apertures (NA) and large core diameters generate larger and more uniform irradiance distribution volumes in the brain. On the other hand, for highly localized stimulation of particular cell groups, SM fibers might be more suitable.

Depending on the used fiber and desired power levels, light sources for optogenetics experiments include lasers, LEDs, and incandescent lamps. Some advantages and disadvantages of using different light sources in optogenetics have been discussed in [1]. Here, we point out that only coherent light can be coupled into SM fibers with sufficient efficiency. For the closed-loop system, one can use either DPSS lasers and as well as LEDs, respectively, for different experimental purposes. The DPSS lasers provide powerful and high coherent light beams, while the LEDs are simple and inexpensive. Both light sources can be electrically controlled to deliver pulses or continuous illumination over long periods.

The controlled illumination system (*see* Fig. 3) incorporates a DPSS laser (VM-TIM, DPSS-V473-150F) and an acousto-optic modulator (AOM) (AA-SA, MTS110-A3-VIS) to modulate the light intensity. AOMs, also called Bragg cells, use the acousto-optic effect to diffract and shift the frequency of light using sound waves (usually at radio-frequency). When one of the high-order outputs is coupled into the optical fiber, the emitted light power at the other fiber end changes linearly to the amplitude of the radio-frequency signals applied on the AOM device. Therefore, by combining the AOM device and the DPSS laser, light patterns with changeable intensity, flexible duty cycle, and stable and accurate temporal resolution can be obtained. The maximum light intensity modulation can reach >100 kHz (sufficient for optogenetics experiments). In total, more than 25 % of the laser light can be coupled into a common SM fiber (10 μm core diameter) for optical communication. This efficiency increases to 60–70 % for a MM fiber with a core diameter larger than 50 μm.

Light-emitting diodes (LEDs) are incoherent light sources, only suitable for MM fibers. The output power of LEDs depends on the electric current. A simple home-made, high-power LED driver can be used to support the pulse modulation in a cost-effective way. Additionally a commercial LED driver can be used (e.g. Thorlabs, DC2100).

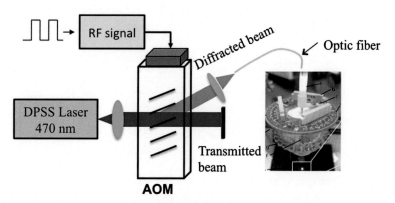

Fig. 3 Scheme of the controlled illumination system based on a DPSS laser and an AOM

2.4 Amplifier Boards for 96 Channels Data Acquisition

Three head-mounted boards (i.e. headstages), each of 32 channels sharing the same architecture (*see* Fig. 4), interface with the tetrode-containing microdrive through a Slimstack Molex/Omnetic connectors, and are connected with the control unit via a custom cable. Each board includes an Intan amplifier (32 channels) and an ADC. Each headstage communicates to the control unit through seven digital signal lines, which are: (1) the digital input channel selector, commanding the amplifier to sample the next channel when this line is activated; (2) digital channel reset, which commands the amplifier to return to the first input channel;

a

b

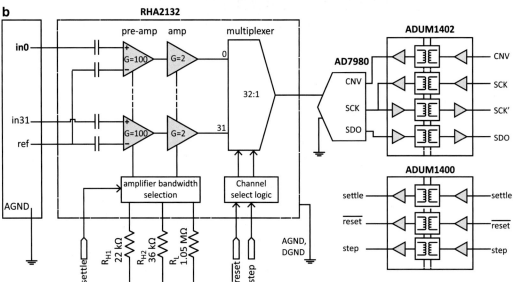

Fig. 4 (a) Photographs of one headstage (top and bottom views) (PCB size: 29.5 mm × 43.4 mm), containing a miniature connector, a packaged amplifier, an ADC, and digital isolators; **(b)** schematic illustrations of the signal paths. Pre-amplification = 100×, amplification = 2×, filter bandwidth = 0.2–5000 Hz, multiplexing ratio = 32:1, 16-bit ADC and digital isolators

(3) digital settle, for rapidly discharging all capacitors in the front end amplifiers to ground in the event of the amplifier saturation; (4) ADC signal conversion, which initiates the conversion and outputs the most significant bits on the falling edge when the conversion is complete; (5) ADC clock signal, which outputs the remaining data bits at subsequent falling edges at the acquisition phase; (6) serial digital output of the ADC, returning the converted acquired signal; and (7) the return ADC clock signal, which ensures accurate signal acquisition at the computer's side.

The 32-channel amplifier is the RHA2132 chip (Intan technologies html). Neural signals are recorded against a common reference electrode. The ground is connected to the drive ground. Although the amplifier gain is fixed at 200×, the signal bandwidth can be set by external resistors between 0.2 and 5 kHz. The amplifier is dc-coupled, with a gain of 1 at 0 Hz. The multiplexed signals are digitized with 16-bit resolution (AD7980) before sending them to the digital I/O board.

The integrated multiplexers permit sampling speeds of up to 31.25 kSps (kilosamples per second) per channel. When a DAQ device (NI, PCI-6259) with a maximum aggregate sampling rate of 10 MHz is used for the 32-channel system, the sampling rate can reach up to 400 kSps ($=1/2500$ ns) or 12.5 kSps per channel. The reduced value results from the TTL minimum pulse width generated by the DAQ card clock of 100 ns ($=1/10$ MHz). Considering the requirements of the ADC conversion and acquisition time, we choose to set a fixed cycle time of 2500 ns (i.e. 900 ns of conversion and 16×100 ns of acquisition of 16-bit data samples), The high digital transmission speed of 10 MHz can cause several side effects, such as ringing, crosstalk, reflections, and ground bounce. To ensure the signal integrity, impedance bridging source termination and line drivers are implemented.

For the 96-channel system, the data acquisition is developed from a multifunction reconfigurable I/O board (NI, PCIe-7842R) armed with a programmable field-programmable gate array (FPGA) chip. The cycle time can thus be further reduced to 1540 ns (i.e. 900 ns of conversion and 16×40 ns of acquisition of 16-bit data samples), resulting in the sampling rate of 649 or 20.3 kSps per channel. Tests with shorter clock periods (<40 ns) show distortion in the TTL pulse. Data acquisition on 96 channels is achieved when the three head-stages perform in parallel on the FPGA device. The extension potential of the acquisition channels depends on the hardware resources in the system.

The amplifier board operates with two battery-supplied power sources: a regulated +3 V and a +5 V source, respectively. A 2.5 V reference voltage (ADP1710) is used for the ADC conversion. The digital isolators (ADUM1400 series) are used to decouple the ground of the amplifier board from the noisy computer ground and eliminate ground loops. These digital isolators are included in a home-made signal collection board (SCB), which is connected to the FPGA card (National Instrument, PCIe-7842R) via a commercial high-performance shielded cable (National Instrument, SHC68-68-RMIO). The signal lines from three headstages meet at this SCB before communicating to the control unit. Since, for animal experimenting purposes, the cables between the headstages and the SCB must be flexible and robust, light weight, and with low cross-talk at high frequencies, we customized cables that contain 8 pairs of twisted insulated wires. The diameter of the wires is 130 μm, and the total diameter of the cable is ~1.5 mm. These cables have been tested to work at frequencies up to 40 MHz with a length of 3 m.

3 Software

Custom software controlling the acquisition visualization and signal analysis consists of two parts: (1) the hardware-timed acquisition and (2) the online signal processing.

3.1 Hardware-Timed Acquisition

The real-time data acquisition is timed by the FPGA clock (an embedded FPGA chip for the DAQ device PCI-6259). The software is developed on a Windows 7 64-bit platform running Labview 2012. Within the Labview environment, programming and accessing the FPGA occurs at two sides, device and host. The device side, programmed within the Labview FPGA module, serves for the TTL pulse generation and data buffering, interfacing the PCB boards and the host computer. FIFO memory buffers were designed to balance the hardware acquisition and reading the values by the host program. The programming codes are multiplicated to interface the three headstages in parallel. Real-time digital signal filtering can also be optionally implemented in the FPGA. Finally, the programming codes for the device side are compiled and written to the FPGA chip as a bit file. At the host side, the programs mainly include three functions in parallel loops: (1) to invoke the device program and read the values from the FIFO buffer, (2) perform the signal processing, and (3) store the data. The synchronization among these three functions depends on the processing and storing speeds. For example, some signals might have to be skipped when the spike-sorting program takes more time than the data acquisition. Therefore, for a large number of recording channels (e.g. 96 channels) we implemented the spike-sorting on a GPU platform.

3.2 Online Signal Processing Based on the GPU

The signal-processing algorithm consists of filtering, spike detection, and classification steps. The GPU implementation is schematically illustrated in Fig. 5. The data read from the FIFO memory buffer are firstly transferred from the host computer memory to

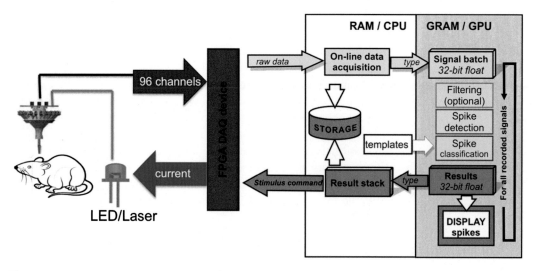

Fig. 5 Architecture of the closed-loop system with GPU spike-sorting

the GPU memory and converted into 32-bit floating-point type (single precision). If the signals have not been processed by a filter operator (e.g. a simpler finite impulse response filter) on the FPGA, a zero-phase bandpass filter is applied before the spike detection. Spikes can be detected by applying a negative amplitude threshold calculated based on the standard deviation of the background noise using the formula proposed by Quiroga et al. in [25]. Spike templates were extracted from a baseline-recording period at the beginning of the session. For the template definition, we have used a standard mixture-of-Gaussians expectation maximization (EM) algorithm [26]. Online spike-sorting is then performed using the simplest form of template matching, i.e. using the correlation between single spike waveforms and each template. The normalized correlation of the waveform with all templates is taken, and the spike assigned to the template resulting in the maximum correlation score. With the convolution theorem, a fast Fourier transformation algorithm was used to implement the correlation instead of the convolution in order to reduce the computation time. Finally, the results of the spike-sorting are sent back to the host memory. The system generates then the required stimulation strategy (depending on the experimental goals) immediately and sends a command to the light source. Figure 4 shows the architecture of the closed-loop system and the data flow.

Both programming and computation in this work can be implemented with a workstation hosting an NVIDIA GeForce GTX 680 GPU (PCIe 3.0 16× interface, 1536 cores at 1.006 GHz, 2.0 GB graphics memory), an Intel i7-3770 CPU (3.9 GHz) and a 16 GB host memory. In contrast to programming codes directly through NVIDIA's Compute Unified Device Architecture (CUDA), one can employ the commercial software library, ArrayFire (Version 2.0, AccelerEyes, GA, USA) that makes GPU programming simpler by packaging most basic CUDA codes into array-based functions. This relatively simple scheme for GPU programming and GPU-memory access allows for faster and more flexible adaptation of the existing signal processing routines. Since we use Labview as a coding tool for data acquisition and system control, while the GPU-based spike-sorting codes are performed in VC++ (Visual Studio 2008, Professional Edition), we used a shared library compiled from VC++ scripts to implement the spike-sorting. Thus the VC++ scripts are first compiled to generate a dynamic link library (DLL) available for program platforms such as LabView.

4 Animal Surgeries

The protocol used by our team for the genetic targeting of a particular depolarizing opsin (i.e. ChR2) is described here. Other opsins or protocols may be selected depending of the experimental goals. Animal experiments should be always performed in accordance to the national and international regulations (in our case the national Belgian and European Union regulations and with the approval of the ethical committee of KU Leuven).

In a first surgical step, adeno-associated based viral vectors (AAV), kindly provided by the laboratory of Prof. V. Baekelandt and the viral vector core facility of the KU Leuven, were stereotaxically injected to allow the expression of the opsin mChR2 based on the alphaCaMKII promoter in the target structures (medial prefrontal cortex: AP 3.5–4.5, ML 0.5, DV 1.8–3.5; dorsal hippocampus: AP –3.5, ML 2, DV 1.5–2.5), by means of a glass micro-pipette controlled by a micro-injector. The small craniotomies were then filled with a biocompatible silicon based elastomer, and the skin sutured.

At least 2 weeks are required for obtaining sufficient opsin expression levels. After this time, the microdrive implantation was performed on rats anesthetized with isoflurane (4 % induction, 1.5–2 % maintenance) that have also received subcutaneous carprofen for pain management. Craniotomies and dura dissection were performed above the target structure. The microdrive was then positioned and anchored to the skull by means of bone screws and dental cement. During recovery from surgery, the tetrodes were slowly advanced towards their target structure. Local field potential markers, such as hippocampal sharp waves, cortical delta waves, and spindles were used in order to guide the positioning process. After the tetrodes reached the desired location, electrophysiological recording were performed using the system described above, during quiet wakefulness or sleep. Various stimulation protocols can be applied according to both open- and closed-loop protocols, depending on the experimental goals.

Acknowledgements

The authors thank Prof. Veerle Baekelandt and Dr. Chris Van den Haute for providing the viral vectors used in this work. We thank Mr. Valentijn Tuts for helping with the PCB assembly and custom cables. We thank Dr. Fabian Kloosterman at NERF for assistance and suggestions on the design of the tetrode micro-drive.

These methods and protocols have been developed within the frame of the FP7 EC project ENLIGHTENMENT. The project ENLIGHTENMENT acknowledges the financial support of the Future and Emerging Technologies (FET) programme within the Seventh Framework Programme for Research of the European Commission

References

1. Yizhar O, Fenno LE, Davidson TJ, Mogri M, Deisseroth K (2011) Optogenetics in neural systems. Neuron 71(1):9–34

2. Knopfel T, Boyden ES (2012) Optogenetics: tools for controlling and monitoring neuronal activity. Prog Brain Res 196:2–278

3. Kloosterman F, Davidson TJ, Gomperts SN, Layton SP, Hale G, Nguyen DP, Wilson MA (2009) Micro-drive array for chronic in vivo recording: drive fabrication. J Vis Exp (26):e1094. doi:10.3791/1094

4. Anikeeva P, Andalman AS, Witten I, Warden M, Goshen I, Grosenick L, Gunaydin LA, Frank LM, Deisseroth K (2012) Optetrode: a multi-channel readout for optogenetic control in freely moving mice. Nat Neurosci 15:163–170

5. Wu F, Stark E, Im M, Cho I-J, Yoon E-S, Buzsáki G, Wise KD, Yoon E (2013) An implantable neural probe with monolithically integrated dielectric waveguide and recording electrodes for optogenetics applications. J Neural Eng 10(5):056012

6. Tass PA, Klosterkötter J, Schneider F, Lenartz D, Koulousakis A, Sturm V (2003) Obsessive-compulsive disorder: development of demand-controlled deep brain stimulation with methods from stochastic phase resetting. Neuropsychopharmacology 28(Suppl 1): S27–S34

7. Tass PA (2002) Effective desynchronization with bipolar double-pulse stimulation. Phys Rev E Stat Nonlin Soft Matter Phys 66:036226

8. Tass PA (2003) A model of desynchronizing deep brain stimulation with a demand-controlled coordinated reset of neural subpopulations. Biol Cybern 89(2):81–88

9. Hauptmann C, Popovych O, Tass PA (2007) Desynchronizing the abnormally synchronized neural activity in the subthalamic nucleus: a modeling study. Expert Rev Med Devices 4(5):633–650

10. Jackson A, Mavoori J, Fetz EE (2006) Long-term motor cortex plasticity induced by an electronic neural implant. Nature 444:56–60

11. Cerf M, Thiruvengadam N, Mormann F, Kraskov A, Quiroga RQ, Koch C, Fried I (2010) On-line, voluntary control of human temporal lobe neurons. Nature 467:1104–1108

12. de Lavilléon G, Lacroix MM, Rondi-Reig L, Benchenane K (2015) Explicit memory creation during sleep demonstrates a causal role of place cells in navigation. Nat Neurosci 18:493–495

13. Paz JP, Davidson TJ, Frechette ES, Delord B, Parada I, Peng K, Deisseroth K, Huguenard JR (2013) Closed-loop optogenetic control of thalamus as a tool for interrupting seizures after cortical injury. Nat Neurosci 16:64–70

14. Armstrong C, Krook-Magnuson E, Oijala M, Soltesz I (2013) Closed-loop optogenetic intervention in mice. Nat Protoc 8:1475–1493

15. Zrenner C, Eytan D, Wallach A, Their P, Marom S (2010) A generic framework for real-

time multi-channel neuronal signal analysis, telemetry control, and submillisecond latency feedback generation. Front Neurosci 4:173

16. Venkatraman S, Elkabany K, Long JD, Yao Y, Carmena JM (2009) A system for neural recordings and closed-loop intracortical microstimulation in awake rodents. IEEE Trans Biomed Eng 56:15–22

17. Sahin M, Durand DM, Haxhiu MA (2000) Closed-loop stimulation of hypoglossal nerve in a dog model of upper airway obstruction. IEEE Trans Biomed Eng 479(7):919–925

18. Rutishauser U, Schuman EM, Marmelak AN (2006) Online detection and sorting of extracellularly recorded action potentials in human medial temporal lobe recordings, in vivo. J Neurosci Methods 154:204–224

19. Takahashi S, Anzai Y, Sakurai Y (2003) A new approach to spikes sorting for multineuronal activities recorded with a tetrode - how ICA can be practical. Neurosci Res 46:265–272

20. Franke F, Natora M, Boucsein C, Munk MH, Obermayer K (2010) An online spike detection and spike classification algorithm capable of instantaneous resolution of overlapping spikes. J Comput Neurosci 29:127–148

21. Guger C, Gener T, Pennartz CMA, Brotons-Mas JR, Edlinger G, Bermúdez i Badia S, Verschure P, Schaffelhofer S, Sanchez-Vives MV (2011) Real-time position reconstruction with hippocampal place cells. Front Neuroprosthetics 5:85

22. Nguyen TK, Navratilova Z, Cabral H, Wang L, Gielen G, Battaglia FP, Bartic C (2014) Closed-loop optical neural stimulation based on a 32-channel low-noise recording system with online spike sorting. J Neural Eng 11(4):046005

23. Wang L, Nguyen TK, Cabral H, Gysbrechts B, Battaglia FP, Bartic C (2014) Closed-loop optical stimulation and recording system with GPU-based real-time spike sorting. In: Popp J, Tuchin V, Matthews D, Pavone F (eds) Biophotonics: photonic solutions for better health care IV: vol. 9129 (91293U-1)

24. Nguyen DP, Layton SP, Hale G, Gomperts SN, Davidson TJ, Kloosterman F, Wilson MA (2009) Micro-drive array for chronic in vivo recording: tetrode assembly. J Vis Exp 26:pii: 1098

25. Quiroga QR (2004) Unsupervised spike detection and sorting with wavelets and superparamagnetic clustering. Neural Comput 16(8):1661–1687

26. Harris KD, Henze DA, Csicsvari J, Hirase H, Buzsáki G (2000) Accuracy of tetrode spike separation as determined by simultaneous intracellular and extracellular measurements. J Neurophysiol 84(1):401–414

Chapter 24

Optogenetic Control of Fibroblast Growth Factor Receptor Signaling

Nury Kim, Jin Man Kim, and Won Do Heo

Abstract

FGFR1 is a member of the fibroblast growth factor family, which controls diverse cellular functions such as cell proliferation, migration, and differentiation. OptoFGFR1, an optogenetic method to modulate the FGFR signaling pathway with light by utilizing PHR domain of cryptochrome2 and cytoplasmic region of FGFR1, enabled light-guided activation of FGFR to study its effects on downstream signaling pathway and during diverse biological processes such as cell migration. Here, we describe about optogenetic and microscopic methods to spatiotemporally manipulate FGFR signaling in a single cell or group of cells using confocal microscope and LED array.

Key words FGFR, optoFGFR1, Optogenetics, Spatiotemporal regulation, Signaling pathway, Migration, Live cell imaging

1 Introduction

Biological processes are regulated by signal relays with high spatio-temporal control, and analysis of a biological phenomenon requires regulation of a specific protein—the functional component of signals [1]. Therefore, creating a signaling module offers new opportunities to study biological systems of interest [2]. In particular, RTK (receptor tyrosine kinase) which acts as the antenna for extracellular stimuli has been a research topic of interest, as these proteins are involved in many biological processes such as development and cancer [3, 4]. Synthetic modules that induce dimerization upon the addition of chemicals, i.e., CID (chemical inducer of dimerization) [5], have been introduced to mimic activation of RTKs via ligand-dependent dimerization for generating synthetic RTKs such as PDGFR [6], EGFR [7], and FGFR [8]. However, the use of chemicals has limited spatiotemporal resolution and application as they require delivery and long diffusion times to the region of interest and cause side effects by binding to unexpected targets. These limitations led to the discovery of light-responsive

Arash Kianianmomeni (ed.), *Optogenetics: Methods and Protocols*, Methods in Molecular Biology, vol. 1408,
DOI 10.1007/978-1-4939-3512-3_24, © Springer Science+Business Media New York 2016

interacting modules and development of new tools that can be functionally modulated by light, and facilitated studies requiring high spatiotemporal resolution of signal regulation [9].

Along with other optogenetic components that induce changes in protein interactions [10], the PHR domain of cryptochrome 2 (CRY2PHR) is a blue-light responsive homo-interacting module for optical regulation of protein [11]. Parallel to synthetic RTKs using CID, we generated synthetic RTKs including optoFGFR and optoTrk by combining the cytoplasmic region of FGFR and Trk with the PHR domain, respectively, which have a precise spatio-temporal resolution sufficient to stimulate subcellular regions at the time of interest, reversibly and repetitively [12, 13]. Especially, we have shown efficient intracellular signal modulation and functional outcomes of optoFGFR1, such as cell polarization and light-induced cell migration mimicking phototaxis with blue light.

In this article, three different methods using optoFGFR1 are described: (A) activation of canonical signaling pathways of FGFR1 and (B) inducing cell migration by light under a confocal microscope. As the confocal microscope is effective for single cell analysis but not for activating or analyzing a group of cells, methods for (C) light activation of cells grown on a 6-well plate using the LED array, obtaining protein samples and western blot are also explained. We have attempted to cover every detail from cell plating to analysis of results, to enable use of optoFGFR1, not only by researchers interested in the FGFR and RTK but for broader applications.

2   Materials

2.1   Plasmids

1. OptoFGFR1: OptoFGFR1 is constructed by combining the myristoylation signal peptide derived from the human LYN proto-oncogene (amino acids 1-11, NM_002350), cytoplasmic region of human FGFR1 (amino acids 397-822, NM_023110), CRY2PHR from A. thaliana cryptochrome 2 (amino acids 1–498, NM_100320) [14], and mCitrine (A gift from Robert Campbell & Michael Davidson & Oliver Griesbeck & Roger Tsien, #54594, Addgene) (*see* **Note 1**). CRY2PHR is codon-optimized for expression in human cell lines (*see* **Note 2**).

2. For analyzing canonical downstream pathways: R-GECO1 was a gift from Robert Campbell (#32444, Addgene) [15], dTomato-PH$_{Akt1}$ (PH domain of mouse Akt1 attached to the C-terminus of dTomato: amino acids 2–147, NM_009652), Erk1-dTomato (full-length Erk1 from mouse attached to N-terminus of dTomato, NM_011952). dTomato is obtained from pThy1-Brainbow-1.0L, a gift from Joshua Sanes (#18725, Addgene) [16].

3. For quantification of cell migration: FusionRed-NLS (NLS from pDsRed2-Nuc fused to pmCherry-C1, Clontech), and iRFP682-Lifeact (Lifeact [17] in iRFP670-N1, a gift from Vladislav Verkhusha (#45457, Addgene) [18]).

2.2 Cell Culture	1. Plates: T75 flask (70075, SPL), 96-well plate (89626, Ibidi), 6-well plate (30006, SPL), plastic tape cover for well plate (Tape Pads, 1018104, QIAGEN).

1. Plates: T75 flask (70075, SPL), 96-well plate (89626, Ibidi), 6-well plate (30006, SPL), plastic tape cover for well plate (Tape Pads, 1018104, QIAGEN).

2. HeLa cells: Dulbecco's Modified Eagle's Medium (DMEM) containing 10 % fetal bovine serum, humidified incubator at 37 °C with 10 % CO_2.

3. Human umbilical vein endothelial cells (HUVEC, C-003-5C, Gibco) (*see* **Note 3**): Medium 200 (M-200-500, Gibco) supplemented with 2 % Low Serum Growth Supplement (S-003-10, Gibco), humidified incubator at 37 °C with 5 % CO_2.

4. Serum starvation medium: serum-free DMEM for HeLa cells, Human Endothelial Serum Free Medium (11111-044, Gibco) containing 0.1 % bovine serum albumin for HUVECs.

5. Transfection: Lipofectamine LTX (15338-100, Invitrogen), Neon transfection system (MPK5000, Invitrogen).

6. Collagen, DPBS, Trypsin-EDTA, 0.25 %.

2.3 Microscope

1. Confocal microscope: Nikon A1R.

2. Lens: Nikon CFI Apo 60× Oil λS (NA 1.4, WD 0.14 mm), Nikon CFI Plan Apochromat λ 20× (NA 0.75, WD 1.00 mm).

3. Immersion oil: Immersol-518F (444962-0000-000, Carl Zeiss).

4. Imaging and analysis: NIS-element AR (Ver. 4.13, Laboratory Imaging), MetaMorph software (Ver. 7.7, Molecular Devices), Microsoft Excel® (Ver. 2010)

5. Laser: Nikon Multi-line Ar Laser, 488 nm/40 mW, Coherent Sapphire Solid Laser 561 nm/20 mW, Coherent CUBE diode Laser, 640 nm/40 mW (*see* **Note 4**). Photoactivation is conducted through galvano scanner incorporated in hybrid confocal scan head with high-speed hyper selector in A1R.

6. Cell chamber: Chamlide WP (includes stage-top incubator & cover, lens heater, temperature controller, gas flow rate controller for CO_2 and humidifier, Live Cell Instrument), CO_2 supply (gas cylinder).

7. Optical power meter (8230E, ADCMT).

8. Lens wipes: Cleaning wiper "Dusper" (E703, Nikon).

2.4 Western Blot

1. Cell harvest and lysis: Scraper, PBS, ProPREP solution (17081, iNtron Biotechnology, Korea).

2. 4× SDS sample buffer: 0.08 % bromophenol blue, 4 % β-mercaptoethanol, 40 % glycerol, 8 % sodium dodecyl sulfate, 0.25 M Tris-Cl (pH 6.8). To make 10 ml of stock solution, add 2 ml of 1 M Tris-Cl (pH 6.8), 0.8 g of SDS, 4 ml of 100 % glycerol, 1 ml of 0.5 M EDTA, and 8 mg of bromophenol blue (*see* **Note 5**).

3. Measuring protein concentration: Nanodrop 1000 (Thermo Scientific).

4. Electrophoresis and transfer: 4–12 % gradient polyacrylamide gel, iBlot gel transfer device (IB1001, Invitrogen).

5. Antibody and related reagents: phospho-ERK1/2 (1/2000 diluted, 9101S, Cell signaling), ERK1/2 (1/2000 diluted, 4696, Cell signaling), anti-mouse IR Dye® 800CW (1/3000 diluted, 926-32210, LI-COR), anti-rabbit IR Dye® 600CW (1/3000 diluted, 926-68071, LI-COR), Odyssey® Blocking Buffer (927-40000, LI-COR), PBST (PBS supplemented with 0.1 % of Tween-20). Antibodies were diluted in blocking buffer (*see* **Note 6**).

6. Ponceau S solution.

7. Stripping buffer: BlotFresh™ Western blot stripping buffer (#SL100324, SignaGen Laboratories).

8. Imaging: ODYSSEY CLx (P/N 9140-WP, LI-COR).

2.5 Miscellaneous

1. Centrifuge: Eppendorf 5415D, Eppendorf 5415R.

2. LED: six blue LEDs (cat. no. PP465-8L63-ESESBI; Photron, Korea) mounted on a customized array, each positioned at the center of each well of a 6-well plate (Live Cell Instrument, Inc., Korea), with a power of 5.5 μW, measured 1 cm from LED (Fig. 1).

3. Inhibitors: PD0325901 (50 μM stock solution in DMSO), LY294002 (50 mM stock solution in DMSO), SU5402 (20 mM stock solution in DMSO), DMSO.

4. Swing arm table lamp, with red bulb (1 W).

5. Dark room, with red light (20 W).

3 Methods

Three specified experimental situations for light-induced activation of optoFGFR1 are discussed in detail, specifically, (A) observation of PLCγ, PI3K, and MAPK signaling pathway using genetically encoded biosensors: R-GECO1, dTomato-PH$_{Akt1}$, and Erk1-dTomato in HeLa cells, (B) inducing directed cell migration of HUVECs, and (C) verifying ERK phosphorylation via western blot in HeLa cells. Six wells in a 96-well plate are used for (A) and

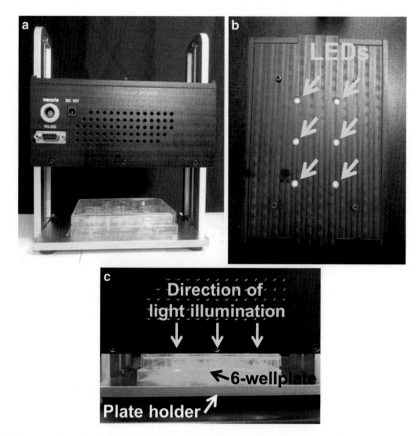

Fig. 1 LED device for activation of cells grown on a 6-well plate. (**a**) Full image of the LED device. (**b**) LED array. Each LED is located at the center of individual wells of a 6-well plate, indicated by *yellow arrows*. (**c**) Stimulation mode of the LED device. The 6-well plate (*black arrow*) is tightly embedded in the device on the plate holder (*white arrow*), and light from LED is irradiated from the upside of the well (*yellow arrow*)

(B), and seven plates of 6-well plate for (C). Transfection was conducted with Lipofectamine LTX in (A) and (C), or Neon electroporation system in (B). All cells were cultured in a T75 flask at a confluence of 80 %. All reagents and consumables should be sterile. The Nikon A1R confocal microscope is utilized for imaging experiments.

(A) Observation of FGFR1 downstream signal activities using biosensors in HeLa cells expressing optoFGFR1, with blue light illumination.

3.1 Cell Handling

1. Prepare a 96-well plate and sterilize wells with UV on the clean bench for at least 20 min.

2. Trypsinize and harvest HeLa cells in a 75T dish and count cells. Prepare 1.5 ml of cells at a concentration of 6×10^4 cells/ml in a microcentrifuge tube. Transfer 200 μl of cells per one well of a 96-well plate; three wells in rows and two wells in columns (1.2×10^4 cells/well) (*see* **Note 7**).

3. Wait for 5 min to allow cells to attach (*see* **Note 8**), and place in the incubator for at least 12 h.

3.2 Transfection

1. Aliquot 0.5 ml of Opti-MEM in a microcentrifuge tube and prewarm at 37 °C.

2. Prepare four microcentrifuge tubes. Transfer 250 ng of optoF-GFR1 DNA and 250 ng of biosensor DNA to three tubes: R-GECO1, dTomato-PH$_{Akt1}$, and Erk1-dTomato. Add 62.5 µl of prewarmed Opti-MEM and add 0.5 µl of Plus reagent in each tubes (*see* **Note 9**). Mix by pipetting 2–3 times.

3. Put 4.8 µl of Lipofectamine LTX into a remaining empty tube. Add 200 µl of prewarmed Opti-MEM and mix by pipetting five times carefully to avoid formation of bubbles. Incubate at R.T. for 5 min.

4. Transfer 62.5 µl of Lipofectamine LTX mixture to each tube with the DNA mixture, and pipette 8–10 times with care. Incubate at R.T. for 20 min.

5. Add 50 µl of Lipofectamine LTX and DNA mixture to wells of 96-well plate, drop wise. Place in the incubator for at least 18 h.

3.3 Microscopic Observation and Light Stimulation

1. Start serum starvation of cells at least 6 h before the experiment (*see* **Note 10**). Warm two microcentrifuge tubes containing 1.5 ml of serum starvation medium to 37 °C. Remove the culture medium in 96-well plate and wash with 200 µl of serum starvation medium, twice. Add 200 µl of serum starvation medium, wrap the 96-well plate with aluminum foil to protect from light and place in the incubator.

2. Dilute inhibitors to 2× concentration in serum starvation medium in a microcentrifuge tube: 100 nM of PD0325901, 100 µM of LY294002, 40 µM of SU5402 (*see* **Note 11**). Turn off the clean bench and room lights, switch on the red table lamp, then add inhibitors to wells and re-wrap with foil. Place in the incubator for 30 min.

3. At least 10 min before the experiment, warm up the microscope lens, stage-top incubator, and cover (*see* **Note 12**). Clean the surface of the lens initially with a lens wipes dipped in 100 % ethanol and subsequently with a dry lens wiper to remove traces of ethanol (*see* **Note 13**).

4. Turn off the light in the room, drop immersion oil onto the lens, and place 96-well plate in the plate chamber. Cover the chamber and find focus with red fluorescence (*see* **Note 14**). Adjust the laser power of 561 nm to 2 (*see* **Note 15**) to take an image of red fluorescence.

5. Draw stimulation ROI on cells to stimulate, using the Bezier shape in NIS software (Fig. 2) (*see* **Note 16**). Adjust power,

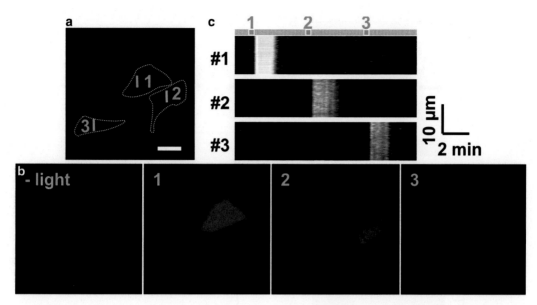

Fig. 2 ROI drawings for sequential stimulation. Three independent HeLa cells expressing optoFGFR1 and R-GECO1 were sequentially stimulated to show spatiotemporal specificity and reversibility. (**a**) Images of three independent ROIs for stimulation (white dotted lines, numbered in blue). (**b**) Changes in R-GECO1 intensity via sequential stimulation from areas 1 to 3 in (**a**). (**c**) Kymographs from *blue vertical lines* in each cell in (**a**) are presented. Light is illuminated at the indicated time (*blue-square*) on *top* of (**c**) to the corresponding region of (**a**). Scale bar in (**a**) = 20 μm

scan speed, and scan number (*see* **Note 17**). Adjust the ND sequence program to start imaging process (*see* **Note 18**). After completion of the imaging process, take a yellow channel image to validate the optoFGFR1 expression level (*see* **Note 19**). Repeat this step for all wells.

3.4 Quantification

1. The NIS program can be applied to quantify signaling activities. For R-GECO1, draw ROI with the Bezier line to cover nearly the whole cell area. For dTomato-PH$_{Akt1}$, draw ROI with the Bezier line covering the cytoplasmic area only. For Erk1-dTomato, draw two ROIs with a circle or ellipse, one covering the nucleus and one the cytoplasm, and divide nuclear intensity by cytoplasmic intensity (Fig. 3). "Time measurement" will give the change in intensity during time-lapse imaging (*see* **Note 20**).

2. For calculating T1/2 with the four-parameter logistic curve, use the "Solver" program in Microsoft Excel®. To calculate, prepare intensity data with the specified time, arrayed vertically. Prepare five empty cells in the empty area of the Excel® sheet, denoted as R1C1, R2C2, R3C3, R4C4, and R5C5 where R is the row and C the column of data. These cells represent the minimum, maximum, T1/2, Hill coefficient, and

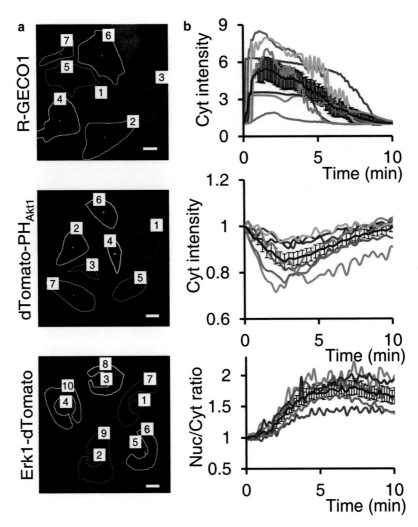

Fig. 3 ROI drawings for intensity analysis. (**a**) Examples of ROI for R-GECO1, dTomato-PH$_{Akt1}$, and Erk1-dTomato. (**b**) Quantified results of cytosolic (Cyt) intensity of R-GECO1 and dTomato-PHAkt1, and ratio between nuclear (Nuc) and cytosolic intensities of Erk1-dTomato are plotted versus time. *Gray lines* indicate individual cells, and the *red line* signifies average intensity. Error bars indicate SEM. Scale bars = 10 μm

sum of squares of differences values in four-parameter logistic curve fitting, respectively. Enter the initial intensity value in R1C1, maximum intensity value in R2C2, expected T1/2 value or time when change of intensity looks significant in R3C3, and a number around 10 in R4C4 (*see* **Note 21**). Add as "=$R1$C1+($R2$C2−$R1$C1)/(1+(Rt$Ct/$R3$C3)^(−$R4$C4))" on the right side of intensity data, RtCt is the time of measured (*see* **Notes 22** and **23**). In R5C5, calculate sum of squares of differences using SUMXMY2 function between real and expected data. In the solver program, put target as R5C5 with option to be minimum, R1C1 to R4C4 as variables, GRC as nonlinear and do solve. Calculated T1/2 is presented in the R3C3.

(B) Inducing phototaxis movement in HUVECs expressing optoFGFR1, with blue light illumination.

3.5 Cell Handling and Transfection

1. Sterilize a 96-well plate with UV on the clean-bench for at least 20 min.

2. Add 100 μl of 300 μg/ml bovine collagen solution to wells. Place in the culture incubator for 1 h. Aspirate out afterwards. Wash coated wells with 200 μl of PBS, twice. Add 100 μl of culture medium to wells for the experiment and place in the incubator.

3. Prewarm culture reagents, Neon buffer E, Neon buffer R at 37 °C.

4. Prepare two microcentrifuge tubes. Add 300 ng of optoF-GFR1, 150 ng of FusionRed-NLS, and 150 ng of iRFP682-Lifeact DNA to one tube, and add 1.5 ml of prewarmed growth medium in the other one (*see* **Note 24**). Add 3.5 ml of prewarmed buffer E in the Neon tube. Preset the Neon system to: 1350 mV, 30 ms, 1 pulse for HUVECs.

5. Trypsinize and harvest HUVECs in a 75T dish and count cells. Aliquot 1.5×10^5 cells in a microcentrifuge tube. Harvest via centrifugation at $4500 \times g$ for 30 s in an Eppendorf 5415D centrifuge. Pipette out medium, and add 1 ml of prewarmed PBS without disturbing the pellet. Centrifuge once more at $4500 \times g$ for 30 s, and pipette out PBS thoroughly. Add 11 μl of prewarmed Neon buffer R and pipette to resuspend cells.

6. Using the electric pipette, carefully fill the tip with cells resuspended in buffer R. Transfer to the tube with DNA, and pipette five times with extreme care not to form any bubbles. Transfer the whole pipette to Neon, electroporate, and release cells in the tube with growth medium (*see* **Note 25**).

7. Carefully resuspend cells by pipetting and transfer 50–100 μl or 0.5–1×10^4 cells into the 96-well plate (*see* **Note 26**) Wait 5 min to allow cells to attach, and place in the incubator for at least 18 h.

3.6 Microscopic Observation and Light Stimulation

1. At least 6 h before the experiment, warm microcentrifuge tubes with 1.5 ml of serum starvation medium to 37 °C. Wash wells in 96-well plate with 200 μl of serum starvation medium, once, and add 200 μl of serum starvation medium. Wrap the 96-well plate with aluminum foil to protect from light and place in the incubator.

2. At least 10 min before the experiment, warm up the microscope lens, plate chamber, plate cover, and CO_2 mixer (*see* **Note 27**).

3. Turn off the light in the room, place the 96-well plate in the plate chamber and find focus with red fluorescence. Adjust the laser power of 561 and 640 nm to 1–3 and 10–15 (*see* **Note 15**), depending on the expression level of FusionRed-NLS and iRFP682-Lifeact, respectively. Find a place where the center of the imaging field is empty with several surrounding cells (*see* **Note 28**). Obtain images of red and infrared fluorescence.

4. Draw a circular stimulation ROI of 320 μm (half of the imaging width) in diameter on cells for stimulation, located in the center of the imaging area. Adjust power, scan speed, and scan number (*see* **Note 29**). Use the ND sequence program to image in the red and infrared channel (*see* **Note 30**). Take a yellow channel image to validate the optoFGFR1 expression level before and after the migration experiment (laser power has no influence on migration speed (Fig. 4)).

3.7 Quantification

1. Parameters for migrating HUVECs are quantified using MM (Metamorph software). Before tracking the migration paths of HUVECs, all images are subjected to median filtering for decoding noise signals. Nuclear location is traced by "multidimensional motion analysis" application to measure the movement of cells (*see* **Note 31**). Calculate the distance of movement for cells in every frame using Pythagoras' theorem. As the distance is shown in "pixels" in MM, this should be corrected to the appropriate unit—μm in this case (*see* **Note 32**).

2. The sum of the distance is the migration path, and distance from the starting point to endpoint should yield the displacement. Dividing displacement by migration path should provide the directionality. Measure the distance from the starting point and endpoint to the center of the image (*see* **Note 33**),

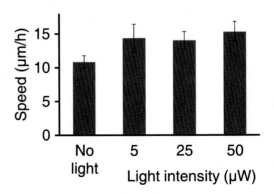

Fig. 4 Effect of laser intensity on optoFGFR1-induced phototaxis in HUVECs. Migrating speed of HUVECs stimulated by 488 nm laser with distinct power (5, 25, and 50 μW) was analyzed. Error bars indicate SEM. Reproduced from [12] with permission from Elsevier

and use the law of cosines to calculate the migration angle. Use acos function in Excel® to calculate the arc-cosine value (*see* **Note 34**).

(C) Western blot

3.8 Cell Handling

1. Prepare seven 6-well plates and sterilize wells with UV on the clean-bench for at least 20 min.

2. Trypsinize and harvest HeLa cells in two 75T dishes and count cells. Prepare 15 ml of cells at a concentration of 1×10^5 cells/ml in a 50 ml tube. Transfer cells to two wells of 6-well plate, 2 ml each (2×10^5 cells/well), agitate horizontally to disperse cells, and place in the incubator for at least 18 h.

3.9 Transfection

1. Prewarm Opti-MEM at 37 °C.

2. Prepare a 50 ml tube. Transfer 30 μg of optoFGFR1 DNA to the tube followed by 7.5 ml of prewarmed Opti-MEM and 30 μl of Plus reagent (*see* **Note 35**). Mix by pipetting with a 5 ml pipette for 4–5 times and incubate at R.T. for 5 min.

3. Add 90 μl of Lipofectamine LTX to mixture and mix by pipetting with a 5 ml pipette 8–10 times carefully to avoid the formation of bubbles. Incubate at R.T. for 30 min.

4. Add 500 μl of mixture dropwise with a 1000P pipette to wells of a 6-well plate. Place in the incubator for at least 18 h.

3.10 Light Stimulation, Protein Sample Preparation, and Western Blot

1. At least 6 h before the experiment, warm serum starvation medium to 37 °C. Wash wells in 6-well plate with 1 ml of serum starvation medium, and add 1.5 ml of serum starvation medium. Wrap each plate with aluminum foil separately to protect from light and place plates in the incubator.

2. Prepare a bucket of ice. Place 14 microcentrifuge tubes and one 50 ml tube with 30 ml of PBS on the ice (ice-cold PBS used for cell harvest). Prepare two 1000P pipettes adjusted to 1000 μl and 500 μl, one 200P pipette, tips, scrapers, LED array of 488 nm wavelength, a waste bucket, a tissue bed with 5–6 layers of paper towel, and pre-cool centrifuge (Eppendorf 5415R) to 4 °C in the dark room.

3. In the dark room with red light, place one plate without cover under LED. Adjust the height so that the LED part touches the top of the plate and turn on the light. After 30 min, turn off the light, discard medium by pouring into waste bucket, and remove the remaining medium within and around the rim of wells by tapping gently 2–3 times on the tissue bed. Add 1 ml of ice-cold PBS to each well and harvest cells with a scraper. Transfer cells to microcentrifuge tubes, add 0.5 ml of ice-cold PBS to each well and harvest once more. Repeat this step with exposing light to cells for 15, 10, 5, 3, and 1 min,

and simply harvest the last plate as a negative control without light (*see* **Note 36**).

4. Harvest cells in the centrifuge at 4 °C for 1 min at maximum speed. Discard the supernatant and spin for 3–4 s to pull down the remaining PBS on the tube wall. Completely remove PBS with a 200P pipette. Add lysis buffer to approximately 5 volumes of pellet (*see* **Note 37**). Resuspend pellet by pipetting carefully until the buffer becomes clear, and incubate on the ice for 30 min. Prepare a thermoblock at 95 °C during incubation. Prepare new seven microcentrifuge tubes on ice.

5. Centrifuge lysates at 4 °C for 30 min at maximum speed. Carefully transfer the supernatant to new tubes (two supernatants from the same sample into one tube) without disturbing the pellet. Measure protein concentrations with Nanodrop: absorbance at 280 nm.

6. Add 1/4 volume of 5× SDS sample buffer. Incubate in the 95 °C thermoblock for 5 min and cool on ice for 1 min. Centrifuge at 4 °C, 30 s at maximum speed (*see* **Note 38**).

7. Load 30 μg of total protein onto each well of polyacrylamide gel. Perform electrophoresis at 120 V until the blue line touches the bottom of the gel. Place gel in the iBlot gel transfer device and transfer the protein sample onto nitrocellulose membrane. Slice membrane with a razor blade to retrieve sample portion. Confirm the transfer with Ponceau S staining and wash the membrane with PBST until membrane becomes clear.

8. Block membrane in blocking buffer at R.T. with rocking at 30 rpm for 1 h. Discard blocking buffer and add phosphor-ERK1/2 and ERK1/2 antibody, diluted to 1/2000 in blocking buffer. Incubate at 4 °C with rocking at 30 rpm for 12 h.

9. Discard buffer and wash membrane five times with PBST in R.T. with rocking at 30 rpm for 5 min.

10. Discard PBST and add anti-mouse and anti-rabbit IR dye, diluted to 1/3000 in blocking buffer. Incubate at R.T. with rocking at 30 rpm, for 30 min.

11. Discard buffer and wash membrane five times with PBST at R.T. with rocking at 30 rpm for 5 min.

12. Obtain image of the membrane using ODYSSEY CLx. We simultaneously used phosphor-ERK1/2 and ERK1/2 antibodies, since the secondary antibody is tagged with dye, not HRP. If your secondary antibody is tagged with HRP, perform western blotting with phosphor-ERK1/2, strip the membrane, and repeat western blot steps with ERK1/2. For stripping, incubate the membrane in stripping buffer at R.T. for 20 min. Wash three times with PBST for 5 min at R.T., and start from the blocking step (Method (C) 3.3, **step 8**).

4 Notes

1. The detailed construction scheme is as follows: CRY2PHR is inserted into mCitrine-N1 via single digestion with *AgeI* to yield PHR-mCitrine. Lyn myristoylation sequence, followed by *EcoRI*, *XhoI*, and *BamHI* enzyme sites, is inserted into PHR-mCitrine with *NheI* and partial digested *AgeI*, to yield Lyn-PHR-mCitrine. Finally, the cytoplasmic region of human FGFR1 (cyFGFR1) is inserted into the *EcoRI* and *BamHI* sites to yield Lyn-cyFGFR1-PHR-mCitrine, i.e., optoFGFR1.

2. Codon optimization can be performed on the GenScript website (http://www.genscript.com/cgi-bin/tools/rare_codon_analysis). Codon-optimized CRY2PHR expression is nearly twice as high as non-optimized PHR when red fluorescence from codon optimized and non-optimized DsRed-tagged CRY2PHR is compared.

3. There are three major distributors of HUVECs: Gibco, Lonza, and ATCC. We found that HUVECs from Gibco grown in their medium are suitable for Neon transfection. HUVECs between passages 5 and 9 are used for experiments.

4. Wavelengths longer than 561 nm have no effect on optoFGFR1, with single time stimulation. Results from testing 405, 488, and 514 nm lasers show that all these lasers are able to activate optoFGFR1, with 488 and 514 nm being the most effective (Fig. 5).

5. SDS sample buffer is stored at −20 °C, which may cause precipitation of SDS. Thaw thoroughly by vortexing occasionally.

6. The list of other antibodies used for optoFGFR1 experiments is as follows: phospho-FGFR1 Tyr653/654 (1/4000 diluted, 3476S, Cell signaling), FGFR1 (1/4000 diluted, 9740S, Cell signaling), and beta-actin (1/4000 diluted, A5316, Sigma).

7. A 96-well plate can be covered with plastic tape and cut open with a sterile razor blade so that unused wells can be preserved for later use without contamination.

8. Do not move immediately or try to agitate horizontally, as cells will congregate on the edge of the wells.

9. For transfection with Lipofectamine LTX, a mixture of 200 ng of DNA with 0.2 μl of Plus reagent in 25 μl of Opti-MEM and 0.6 μl of lipofectamine LTX in 25 μl of Opti-MEM is used per well of a 96-well plate. In case of co-transfection of optoFGFR with biosensors, use 100 ng of optoFGFR and 100 ng of biosensor DNA. If cells other than HeLa are used, refer to the manual provided by Invitrogen (http://www.lifetechnologies.com/order/catalog/product/15338100).

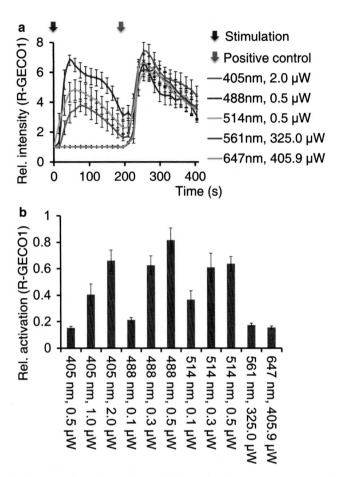

Fig. 5 Relative efficiencies of optoFGFR1 activation by lasers with different wavelengths. Light was illuminated onto optoFGFR1 and R-GECO1 expressing HeLa cells twice: first with the indicated wavelength (405–647 nm) and power (0.5–405.9 μW) and second at 488 nm and 3 μW. Images were obtained every 15 s. Initial stimulation was conducted at 0 s (*red arrow*), and the second stimulation at 180 s (*blue arrow*). (**a**) Quantified intensities of R-GECO1 from time-lapse images. (**b**) Relative activation levels calculated by dividing maximum intensity induced by first stimulation with that induced by second stimulation are presented. Error bars indicate SEM

10. For PH_{Akt1}, more than 10 h is necessary to observe dramatic changes in localization. HeLa cells in our lab display relatively high membrane localization of PH_{Akt1} compared to other cell lines such as NIH3T3 fibroblasts.

11. Final concentration of PD0325901, LY294002, and SU5402 will be 50 nM, 50 μM, and 20 μM, respectively. Additionally, prepare serum starvation medium for negative controls with the same amount of DMSO.

12. Differences in temperature between microscope and well plate may cause focus drift.

13. A 60× lens is used for observation of biosensors.

14. Any light may trigger activation of optoFGFR1. All lights, including microscopic lamps, should be turned off. If light from the monitor of the computer reaches the plate, tilt the position and use a dark background to minimize the light. If lamp for brightfield should be used, try with other wells away from the experimental wells, then clean and replace the immersion oil. In case of accidental illumination, start the experiment at least 30 min later.

15. In Nikon confocal microscopes, there is an indicator showing relative power of laser. In our system, power 1 corresponded to 0.5 μW or 32.6 μW/mm² when using the 60× lens.

16. NIS software can classify stimulation ROI up to three groups so that three groups of cells or areas are independently stimulated. For sequential activation of different cells, draw ROI smaller than a cell, as light may reach out of the ROI depending on the laser power. A circular ROI larger than 5 μm in diameter in any area of cells is sufficient to induce activation of the signaling pathway.

17. Our usual conditions for signal activation are 488 nm laser power at 5 (see **Note 15**), and scan speed at 1 s/frame (1 μs/pixel), with scanning two times. Laser power influences activation kinetics, but not amplitude (Fig. 6a, b).

18. The time required for changes in biosensor intensity or localization differs significantly. Usually, R-GECO1 is observed with the red channel every 3–15 s, and 0.5–3 min for dTomato-PH_{Akt1} and Erk1-dTomato, respectively.

19. Expression level of optoFGFR1 also influences the kinetics of signal, not intensity (Fig. 6c, d).

20. The R-GECO1 signal increases sharply within 1 min ($T1/2 \approx 0.2$ min), and dTomato-PH_{Akt1} will decreases in about 5 min ($T1/2 \approx 3.1$ min). Nuclear Erk1-dTomato intensity increases in about 10 min ($T1/2 \approx 8.3$ min) in HeLa cells. T1/2 values are reproduced from [12] with permission from Elsevier.

21. Initial values are important using the solver program. Different T1/2 values or Hill coefficient should be attempted in case the graph does not fit well.

22. This formula represents the four-parameter logistic curve, expressed as "minimum value + (maximum value − minimum value)/(1 + x/T1/2)^(−Hill's coefficient)", where x is the time for calculating T1/2 and concentration for IC_{50}. The actual address of Excel® in the datasheet is presented in column and row sequences, e.g., A3 denote the third row first column. Therefore R3$C3 should be entered as $A3, not A$3, so that

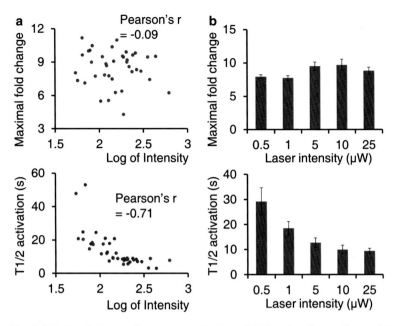

Fig. 6 Effects of diverse factors on activation of PLCγ signaling pathway. (**a**) Effects of optoFGFR1 expression levels on the signal amplitude of PLCγ pathway. (**b**) Effects of laser intensity on the activation kinetics of PLCγ pathway. OptoFGFR1 and R-GECO1 expressing HeLa cells were stimulated with different intensities of 488 nm laser. Maximal fold change and half maximal time for activation were compared with (**a**) optoFGFR1 expression measured based on intensity of yellow fluorescence (*see* **Note 39**) or (**b**) laser intensity. Error bars indicate SEM. Reproduced from [12] with permission from Elsevier

time variables can be differentially calculated for each frame. "$" fixes the location of a cell used for the calculation of Excel®.

23. For PH_{Akt1}, the intensity will decrease as PH_{Akt1} translocates to the membrane. Therefore, the formula "=$R2$C2 – ($R2$C2 – $R1C1)/(1+(Rt$Ct/$R3$C3)^(–$R4$C4))" should be used instead.

24. For transfection with Neon, use 600 ng of DNA in total per 10 μl reaction. FusionRed-NLS is used for tracking migration path and iRFP682-Lifeact for observing cell morphology.

25. This step should be performed with extreme care, since a small bubble in the electroporation tip may cause a "spark," which drastically decreases transfection rate and cell survival.

26. Cells should be plated at a low density as sparse plating leads to better response and easier observation of migration.

27. A use 20× lens is used for the migration experiment.

28. If cells are plated at high density, their migration paths may converge and interfere with each other. Imaging fields with 10–30 cells (in case of HUVECs, mean cell width ≈ 50 μm) were good for migration experiments in our case.

29. Our usual activation conditions for cell migration are 488 nm laser power at 1–10 (*see* **Note 15**), with scan speed at 1.

30. Our usual imaging interval is 5–10 min. As cell migration takes tens of hours, add 300 µl of serum free medium before starting experiment to minimize the effects caused by evaporation of medium.

31. The nucleus tracking conditions are as follows: XY diameter (pixels): 10–50, local intensity above background: 50, maximum travel between frames (pixels): 20.

32. In our system, the imaging field is 512×512 pixels, which covers 640×640 µm, so that 1 pixel calculated using MM corresponds to 1.25 µm.

33. As specified in Methods (B) 3.4, **step 5**, the image center is the center of the illuminated area.

34. If the A is the displacement, B the distance from start point to the center of illumination, and C the distance from the endpoint and to the center of illumination, the angle θ is calculated as "=acos$((A^2 + B^2 - C^2)/2 \times A \times B)$" in Excel®.

35. For transfection in a 6-well plate with Lipofectamine LTX, a mixture of 2 µg of DNA with 2 µl of Plus reagent and 6 µl of Lipofectamine LTX in 500 µl of Opti-MEM is used per well. We used the one-tube protocol of LTX (https://tools.lifetechnologies.com/downloads/Lipofectamine_LTX_pps.pdf).

36. To stimulate repeatedly, use the digital time switch (TR611 top2, Theben).

37. Typically, 30 µl of lysis buffer is added.

38. If any aggregate precipitates, transfer the supernatant to a new tube.

39. Pearson's r is calculated using the "CORREL" function in EXCEL®. The value is slightly different from our published data, as we used log of intensity, not raw intensity values, which resulted in a smaller correlation value between maximal fold change and intensity, and conversely, an increase in negative correlation between T1/2 activation time and intensity.

References

1. Pryciak PM (2009) Designing new cellular signaling pathways. Chem Biol 16:249–254. doi:10.1016/j.chembiol.2009.01.011

2. Lim WA (2010) Designing customized cell signalling circuits. Nat Rev Mol Cell Biol 11:393–403. doi:10.1038/nrm2904

3. Lemmon MA, Schlessinger J (2010) Cell signaling by receptor tyrosine kinases. Cell 141(7):1117–1134. doi:10.1016/j.cell.2010.06.011

4. Hanahan D, Weinberg RA (2011) Hallmarks of cancer: the next generation. Cell 144:646–674. doi:10.1016/j.yane.2012.02.046

5. Klemm JD, Schreiber SL, Crabtree GR (1998) Dimerization as a regulatory mechanism in signal transduction. Annu Rev Immunol 16:569–592. doi:10.1146/annurev.immunol.16.1.569

6. Van Stry M, Kazlauskas A, Schreiber SL, Symes K (2005) Distinct effectors of platelet-derived

growth factor receptor-α signaling are required for cell survival during embryogenesis. Proc Natl Acad Sci U S A 102:8233–8238

7. Hofman EG, Bader AN, Voortman J et al (2010) Ligand-induced EGF receptor oligomerization is kinase-dependent and enhances internalization. J Biol Chem 285:39481–39489. doi:10.1074/jbc.M110.164731

8. Welm BE, Freeman KW, Chen M et al (2002) Inducible dimerization of FGFR1: development of a mouse model to analyze progressive transformation of the mammary gland. J Cell Biol 157:703–714. doi:10.1083/jcb.200107119

9. Toettcher JE, Weiner OD, Lim WA (2013) Using optogenetics to interrogate the dynamic control of signal transmission by the Ras/Erk module. Cell 155:1422–1434. doi:10.1016/j.cell.2013.11.004

10. Tischer D, Weiner OD (2014) Illuminating cell signalling with optogenetic tools. Nat Rev Mol Cell Biol 15:551–558. doi:10.1038/nrm3837

11. Bugaj LJ, Choksi AT, Mesuda CK et al (2013) Optogenetic protein clustering and signaling activation in mammalian cells. Nat Methods 10:249–252. doi:10.1038/nmeth.2360

12. Kim N, Kim JM, Lee M et al (2014) Spatiotemporal control of fibroblast growth factor

receptor signals by blue light. Chem Biol 21:903–912. doi:10.1016/j.chembiol.2014.05.013

13. Chang K-Y, Woo D, Jung H et al (2014) Light-inducible receptor tyrosine kinases that regulate neurotrophin signalling. Nat Commun 5:4057. doi:10.1038/ncomms5057

14. Kennedy MJ, Hughes RM, Peteya LA et al (2010) Rapid blue-light-mediated induction of protein interactions in living cells. Nat Methods 7:973–975. doi:10.1038/nmeth.1524

15. Zhao Y, Araki S, Wu J et al (2011) An expanded palette of genetically encoded Ca^{2+} indicators. Science 333:1888–1891. doi:10.1126/science.1208592

16. Livet J, Weissman TA, Kang H et al (2007) Transgenic strategies for combinatorial expression of fluorescent proteins in the nervous system. Nature 450:56–62. doi:10.1038/nature06293

17. Riedl J, Crevenna AH, Kessenbrock K et al (2008) Lifeact: a versatile marker to visualize F-actin. Nat Methods 5:605–607. doi:10.1038/nmeth.1220

18. Shcherbakova DM, Verkhusha VV (2013) Near-infrared fluorescent proteins for multicolor in vivo imaging. Nat Methods 10:751–754. doi:10.1038/nmeth.2521

Chapter 25

Protein Inactivation by Optogenetic Trapping in Living Cells

Hyerim Park, Sangkyu Lee, and Won Do Heo

Abstract

Optogenetic modules that use genetically encoded elements to control protein function in response to light allow for precise spatiotemporal modulation of signaling pathways. As one of optical approaches, LARIAT (Light-Activated Reversible Inhibition by Assembled Trap) is a unique light-inducible inhibition system that reversibly sequesters target proteins into clusters, generated by multimeric proteins and a blue light-induced heterodimerization module. Here we present a method based on LARIAT for optical inhibition of targets in living mammalian cells. In the protocol, we focus on the inhibition of proteins that modulate cytoskeleton and cell cycle, and describe how to transfect, conduct a photo-stimulation, and analyze the data.

Key words LARIAT, Protein inhibition, Optogenetics, Cryptochrome 2 (CRY2), CIB1, Clustering, Cytoskeleton, Cell division

1 Introduction

Perturbation of specific protein activity in a complex signaling network is an important step for elucidating many cellular, developmental, and physiological processes. Even though genetic approaches such as knockout and knockdown have been widely used as powerful tools to identify specific functions of proteins, yet they typically require a relatively long time to see the effects or can have lethal effects on embryonic development [1, 2]. At protein level, conditional approaches such as small molecule-based inhibition or targeted degradation have been utilized to control protein activities posttranslationally [3–5]. However, these techniques suffer from intrinsic limitations, such as intricate design needs for each target protein, low reversibility, off-target effects, and poor spatial resolution. Development of optogenetic tools that use genetically encoded light-sensitive proteins to control protein activities have offered opportunities to overcome these drawbacks [6–9]. These tools allow to modulate signaling pathways with rapid response, good reversibility, and high spatiotemporal resolution in mammalian cells.

Arash Kianianmomeni (ed.), *Optogenetics: Methods and Protocols*, Methods in Molecular Biology, vol. 1408,
DOI 10.1007/978-1-4939-3512-3_25, © Springer Science+Business Media New York 2016

Unlike most optogenetic modules developed for protein activation, LARIAT (Light-Activated Reversible Inhibition by Assembled Trap) is a versatile method for inhibiting protein function by reversible sequestering target proteins into optically assembled clusters in living cells [10]. LARIAT utilizes multimeric proteins (MPs) and a light-mediated heterodimerization between Cryptochrome 2 (CRY2) and CIB1 [8]. The light-activated CRY2 proteins simultaneously oligomerize [9] and interconnect with CIB1-MPs through CRY2-CIB1 interaction, thereby inducing cluster formation. These clusters would serve as synthetic intracellular compartments to conditionally trap and consequently inactivate target proteins upon blue-light illumination (Fig. 1). In addition, utilizing a single-domain antibody (called "nanobody") against the intracellular green fluorescence protein (GFP) [11], we extend application of LARIAT system to inactivate various GFP-tagged proteins. Here we present a protocol to spatiotemporally control protein functions exemplified by inactivation of Vav2 and Tubulin which are involved in cell migration and mitosis, respectively. We divided this protocol into two categories: (1) inhibition of Vav2 involved in reorganization of actin cytoskeleton and (2) disruption of microtubule structure during mitosis.

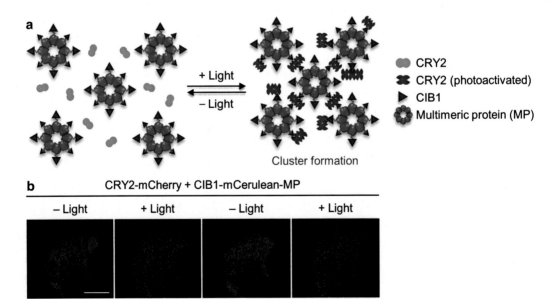

Fig. 1 Blue light-mediated cluster formation. (**a**) Schematic diagram of LARIAT. Photoactivated CRY2 binds to CIB1 conjugated with multimeric protein (MP) and induces the formation of large clusters by interconnecting CIB1-MPs. (**b**) Reversible cluster formation in a HeLa cell co-expressing CRY2-mCherry and CIB1-mCerulean-MP illuminated with 488 nm. Scale bar, 20 μm. It is reproduced with permission from *Nature Methods*

2   Materials

2.1   Plasmids (See Note 1)

1. LARIAT: Two strategies—direct conjugation of CRY2 to targets and use of CRY2-fused anti-GFP nanobody—can be applied to inhibit targets (Fig. 2).

(A) Components for inhibition of CRY2-fused target proteins

1. CIB1-mCerulean-MP (Addgene; plasmid 58366, *see* **Note 2**).
2. mCitrine-CRY2-Vav2 (*see* **Note 3**).

(B) Components for inhibition of GFP-labeled target proteins

1. CIB1-CLIP-MP (*see* **Note 4**).
2. SNAP-CRY2-V_HH(GFP) (Addgene; plasmid 58370).
3. Target proteins: EGFP-Vav2, EGFP-αTubulin (*see* **Note 5**).

(C) Fluorescence readout proteins to monitor inhibitory effect

1. mCherry-Lifeact to monitor reorganization of actin cytoskeleton [12].
2. mCherry-H2B to monitor cell cycle.

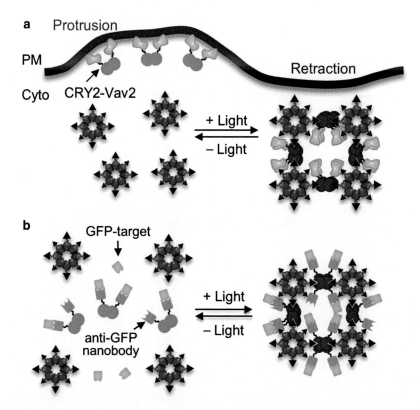

Fig. 2 Two strategies of LARIAT to inhibit target proteins. (**a**) Inhibition of CRY2-conjugated target proteins (CRY2-Vav2) by light-inducible trapping in clusters. (**b**) Inhibition of GFP-labeled target proteins by trapping via CRY2-conjugated anti-GFP nanobody (V_HH(GFP). It is reproduced with permission from *Nature Methods*

2.2 Cell Culture and Transfection

1. Plates: 96-well plastic bottom plates (ibidi), T75 flasks.

2. Tubes: 1.5-mL microcentrifuge tubes, 15-mL conical tubes.

3. Cells: HeLa cell, NIH3T3 cell.

4. Culture medium: Dulbecco's Modified Eagle's Medium (DMEM) supplemented with 10 % fetal bovine serum (FBS). Stored at 4 °C.

5. 0.25 % Trypsin-EDTA. Stored at 4 °C.

6. Dulbecco's Phosphate Buffered Saline (DPBS).

7. Opti-MEM (GIBCO). Stored at 4 °C.

8. 100 µg/mL poly-D-lysine solution (PDL, Sigma): dilute 5 mg of PDL in 50 mL deionized water. Stored at 4 °C.

9. Lipofectamine LTX with Plus Reagent (Invitrogen). Stored at 4 °C.

10. Neon Transfection System (Invitrogen): Neon Transfection Device, Neon Pipette, Neon Pipette Station.

11. Neon Transfection System 10 µL Kit (Invitrogen): 10 µL Neon Tips, Neon Tubes, Resuspension Buffer R, Electrolytic Buffer E. Store the remaining buffers at 4 °C.

12. Hemocytometer.

13. CO_2 incubator with 37 °C and 10 % CO_2.

2.3 Microscope

1. Nikon A1R confocal microscope mounted onto a Nikon Eclipse Ti body.

2. Objectives: 60× Plan Apochromat VC objective (Nikon), 40× Plan Fluor objective (Nikon).

3. Lasers: Nikon Multi-line Argon Laser for 457, 488, and 514 nm, Coherent Sapphire Solid Laser 561 nm (*see* **Note 6**).

4. Chamlide TC system (Live Cell Instrument): Stage-top incubator, lens warmer, humidifier, and controller placed on a microscope stage.

5. Immersion oil (Carl Zeiss).

6. Optical power meter (ADCMT).

7. Image analysis software: NIS-Elements (Nikon, ver 4.1), MetaMorph (Molecular Devices, ver 7.8), Microsoft Excel.

3 Methods

3.1 Reorganization of Actin Cytoskeleton by Inactivating Vav2

3.1.1 Cell Transfection with a Lipofectamine LTX

1. Add 100 µL of 100 µg/mL poly-D-lysine solution to wells of 96-well plate and incubate 30 min at 37 °C. Discard the remained solution and wash the wells three times with DPBS and allow the plate to dry at room temperature (RT).

2. Prepare the NIH3T3 cells that are ~80 % confluency in a T75 flask. Trypsinize and count the cells.

3. Plate the cells on the coated wells at a density of 1.5×10^4 cells per well in 200 µL of complete culture medium. Incubate overnight at 37 °C supplemented with 10 % CO_2.

4. Dilute 200 ng total amount of plasmid DNA into 20 µL of Opti-MEM in a 1.5-mL microcentrifuge tube (*see* **Note 7**). Add 0.2 µL Plus reagent in the tube (*see* **Note 8**). Mix well by pipetting and incubate at RT for 5 min.

5. Dilute 0.6 µL Lipofectamine LTX into the DNA solution (*see* **Note 9**). Mix well by pipetting and incubate at RT for 25 min.

6. Meanwhile, replace the media in the wells to 100 µL of new complete media.

7. Add 20 µL of DNA-Lipofectamine LTX mixture dropwise into the well and incubate overnight at 37 °C and 10 % CO_2.

3.1.2 Imaging and Photo-stimulation

1. At least 10 min before imaging, switch on a Chamlide TC system to maintain environmental condition in 37 °C and 10 % CO_2. Just before imaging, replace the medium with 200 µL of prewarmed Opti-MEM. Place the 96-well plate on the plate chamber.

2. Select appropriate channels for imaging specific fluorescence signals and find the cells via observing mCherry signals with a 60× oil-immersion objective (*see* **Note 10**).

3. For photo-stimulation, acquire a mCherry image first, and draw a ROI area where the photo-stimulation is to be applied. Select this ROI as "Used as Stimulation ROI" to designate a ROI area as the photo-activation area.

4. Designate 488-nm laser as stimulating light source and select laser power and illumination time (*see* **Note 11**).

5. Press the "Photo Activation" button and set the photo-activation experiment sequence to suit specific imaging needs such as Fig. 3 (*see* **Note 12**).

6. Click the "Apply Stimulation Settings" button to send the photo-activation area information and click the "Run now" button to start monitoring.

7. After the imaging is finished, take the images with other fluorescence channels to validate expression levels of CRY2- and CIB1-fused proteins.

3.1.3 Image Analysis

1. For analyzing membrane dynamics, threshold the image based on the mCherry-Lifeact signal to mask the whole cell area using "Automated Measurement" tool in the NIS-Element software and apply the threshold on all frames of the movie (*see* **Note 13**).

2. Measure the area of binary images using "ROI statistics" tool. Export raw data to Microsoft Excel software and plot the area change curve (Fig. 4a, b).

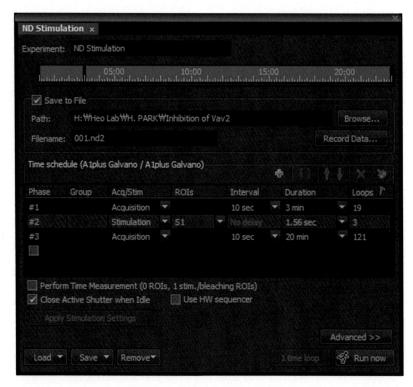

Fig. 3 Photo-activation experiment sequence for inhibition of Vav2. Interface of NIS-element AR software to set up a condition (time intervals and duration) of photo-stimulation and imaging

3. For generating kymograph, select the "Create Kymograph by Line" and draw a line across the cell body (in direction from cytoplasm to edge of the plasma membrane) (Fig. 4c).

4. For protrusion and retraction analysis, threshold the image based on the mCherry-Lifeact signal to cover the whole cell area using "Threshold Image" tool on "Measure" tap in the MetaMorph software. Select "Binary Operations" on "Process" tap and convert the mCherry images of before and after light stimulation to binary images.

5. To obtain retraction image, subtract the binary image of "after light illumination" from that of "before light illumination" using "Arithmetic" tool on "Process" tap.

6. To obtain protrusion image, subtract the binary image of "before light illumination" from that of "after light illumination" using "Arithmetic" tool.

7. Operate the "Logical AND" with the two binary images to generate image showing areas that overlapped right before stimulation.

8. Using "Color Combine" tool, combine the images presenting each regions to generate protrusion and retraction map.

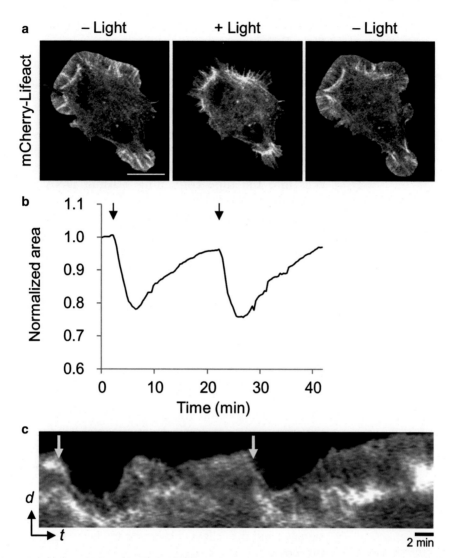

Fig. 4 Inactivation of CRY2-conjugated Vav2 by LARIAT. (**a**) Fluorescence images of a retraction of NIH3T3 cell co-expressing mCherry-Lifeact, mCitrine-CRY2-Vav2, and CIB1-mCerulean-MP upon repeated light illumination with 20-min interval. (**b**) Time-lapse measurements of cell areas. (**c**) Kymograph of Lifeact corresponding to the *red line* in panel **a**. *d*, distance; *t*, time. Scale bars, 20 μm. It is reproduced with permission from *Nature Methods*

3.2 Inhibition of Cell Cycle

3.2.1 Cell Transfection with a Neon Transfection System

1. Fill the Neon Tube with 3 mL of prewarmed Buffer E and insert into the Neon Pipette Station until having a click sound. Set the desired pulse conditions as 980 V for Voltage, 35 ms for Width, and 2 pulses for Pulses, in case of HeLa cell.

2. Prepare two 1.5-mL microcentrifuge tubes. Transfer 600 ng total amount of plasmid DNA (200 ng of SNAP-CRY2-V_HH(GFP), 200 ng of CIB1-CLIP-MP, 120 ng of EGFP-αTubulin, and 80 ng of mCherry-H2B) in one tube and 810 μL of complete culture medium in another tube.

3. Prepare the HeLa cells that are ~80 % confluency in a T75 flask. Trypsinize and count the cells.

4. Transfer 2×10^5 cells to a new 1.5-mL microcentrifuge tube and centrifuge at 7000 rpm for 0.5 min at RT. Discard the medium and wash the cell pellet with 100 μL of DPBS by centrifugation at $4500 \times g$ for 0.5 min at RT.

5. Discard the DPBS and resuspend the cell pellet with 11 μL of Resuspension Buffer R. Gently pipette the cells to avoid air bubbles (*see* **Note 14**). Transfer 10.8 μL of resuspended cells into the tube containing plasmid DNA and mix by pipetting.

6. Fill a 10 μL Neon Tip with the cell-DNA mixture. Insert the Neon Pipette with the sample into the Neon Tube placed in the Neon Pipette Station until having a click sound. Press Start button on the screen (*see* **Note 15**).

7. After "Completed" is displayed on the screen, transfer the samples from the Neon Tip into the prepared tube containing the complete culture medium.

8. Mix by pipetting and transfer 200 μL of the medium containing electroporated cells into each wells of a 96-well plate. Incubate the plate overnight at 37 °C and 10 % CO_2.

3.2.2 Imaging and Photo-stimulation

1. At least 10 min before imaging, switch on a Chamlide TC system was used to maintain environmental condition in 37 °C and 10 % CO_2. Just before imaging, replace the medium with 200 μL of prewarmed Opti-MEM. Place the 96-well plate on the plate chamber.

2. Select the 488-nm laser for observation of EGFP signals and photo-stimulation and the 561-nm laser for imaging mCherry signals. Find the cells via observing mCherry signals with a 40× objective.

3. For multipoint and time-lapse acquisition, add the positions which the acquisition is to be applied.

4. Set time-lapse parameters such as intervals and duration (*see* **Note 16**).

5. Set laser power of 488 nm to 65 μW/mm² for photo-stimulation and adjust laser power of 561 nm for monitoring mCherry-H2B.

6. Simultaneously, acquire images with both 488 and 561 nm by clicking the "Run now" button (Fig. 5) (*see* **Note 17**).

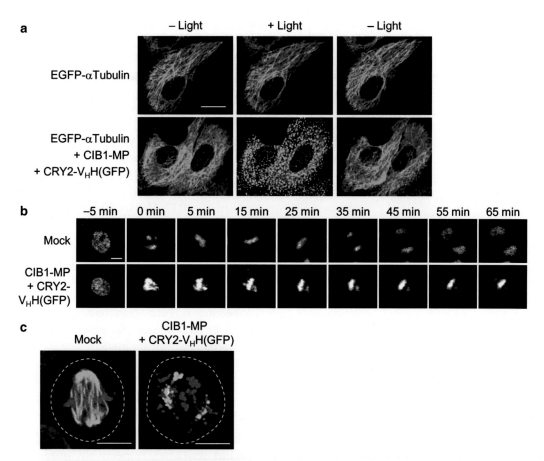

Fig. 5 Perturbation of cell cycle by trapping tubulins into clusters. (**a**) Trapping GFP-labeled tubulins into clusters by blue-light illumination. (**b**) Time-lapse images of mCherry-H2B to monitor cell division of HeLa cells co-expressing EGFP-tubulin, mCherry-H2B and each indicated fusion proteins. (**c**) Structure of mitotic spindles (*cyan*) and arrangement of chromosomes (*magenta*) in HeLa cells co-expressing EGFP-tubulin, mCherry-H2B and each indicated fusion proteins. Scale bars, 20 μm (**a**) and 10 μm (**b, c**). It is reproduced with permission from *Nature Methods*

4   Notes

1. Plasmids: The PCR-amplified sequence encoding association domain of CaMKIIα (amino acids 315–478) [13], indicated as MP, was cloned into pmCerulean-C1 vector (Clontech) between EcoRI and BamHI, which resulted in a plasmid encoding mCerulean-MP. In order to generate a CIB1-mCerulean-MP vector, sequence encoding CIB1 (amino acids 1–147) was inserted into the mCerulean-MP plasmid between NheI and AgeI sites. Expression plasmid for CRY2-mCherry was generated by inserting sequence encoding CRY2 into NheI and AgeI sites of pmCherry-C1 (Clontech). In order to

generate pmCitrine-CRY2-Vav2, PCR-amplified mCitrine flanked by NheI and AgeI were inserted into pmCherry-C1, original mCherry was replaced with PCR-amplified sequence encoding CRY2 at AgeI and BsrGI sites, and PCR-amplified sequence encoding Vav2 (amino acids 167–541) was excised by BspEI and XhoI and inserted into pmCitrine-CRY2.

To generate SNAP-CRY2-V_HH(GFP) vector, sequence encoding V_HH(GFP) from pcDNA3_NSlmb-vhhGFP4 (Addgene; plasmid 35579) was PCR-amplified and cloned into pmCitrine-CRY2 at XhoI and EcoRI sites, and sequence encoding SNAP (New England BioLabs) was PCR-amplified and inserted into pmCitrine-CRY2-V_HH(GFP) at NheI and AgeI sites, replacing sequence for mCitrine.

2. We used an association domain of Ca^{2+}/calmodulin-dependent protein kinase IIα (CaMKIIα), which self-assembles into multimeric protein (MP) with 12 identical subunits in LARIAT method [14]. However, you can choose other MPs to induce light-mediated interconnection. It is important to note that since labeling MPs with dimeric or weak dimeric fluorescence protein such as EGFP or EYFP variants causes spontaneous cluster formation regardless of light illumination, monomeric fluorescence proteins (mRFP, mCherry, or mCerulean) should be used [15].

3. For efficient mammalian expression, codon-optimized sequence encoding CRY2 (amino acids 1–498) was used.

4. The plasmids for CIB1 and CRY2 are conjugated with CLIP and SNAP (New England BioLabs), respectively, instead of fluorescence proteins to avoid trapping by anti-GFP nanobody. To label the CLIP or SNAP tags, we used CLIP-Cell TMR-Star (New England BioLabs) or SNAP-Cell Oregon Green (New England BioLabs), respectively. Dilute a 0.5 μL SNAP or CLIP-tag substrate in 100 μL culture medium. Replace the cell medium to the substrate-containing medium and incubate at 37 °C and 10 % CO_2 for 30 min. Wash the cells three times with complete culture medium.

5. An anti-GFP nanobody can recognize not only GFP but also other GFP variants such as CFP and YFP in living cells (Fig. 6).

6. Although we usually use the 488-nm laser for photo-activation, illumination with 405- and 457-nm light also activates LARIAT system. However, illumination with light longer than 514 nm could not induce the clustering (Fig. 7).

7. You can choose the scheme to inhibit the cytoskeletal protein among (option A) direct conjugation of target to CRY2 and (option B) use of CRY2-fused anti-GFP nanobody. In our optimal transfection condition, for option A, we usually use each 80 ng of CIB1-mCerulean-MP and mCherry-CRY2-Vav2, and 40 ng of mCherry-Lifeact. For option B, we transfect with each

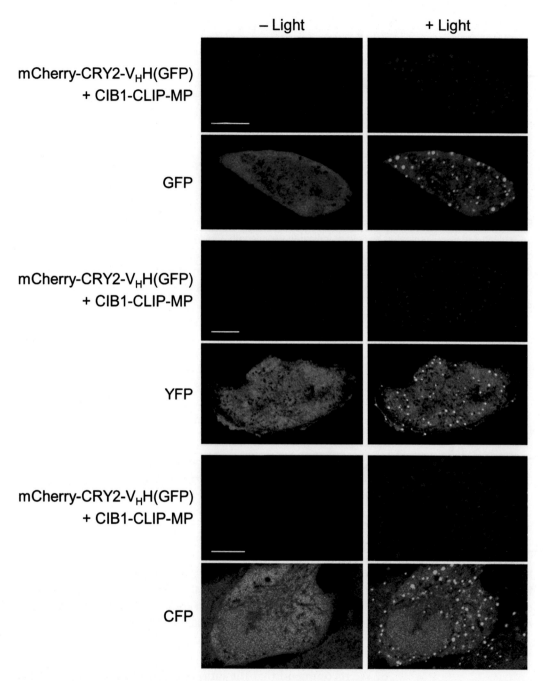

Fig. 6 Binding of anti-GFP nanobody to GFP variants. HeLa cells were co-transfected with expression plasmids as indicated. Fluorescence images were captured before and after light illumination. Scale bars, 10 μm

70 ng of CIB1-CLIP-MP and SNAP-CRY2-V_HH(GFP), 40 ng of EGFP-Vav2, and 20 ng of mCherry-Lifeact.

When we compared the morphological effects of EGFP-Vav2 and CRY2-Vav2, trapping EGFP-Vav2 with an anti-GFP nanobody showed a more effective membrane retraction than trapping CRY2-Vav2 (Fig. 8).

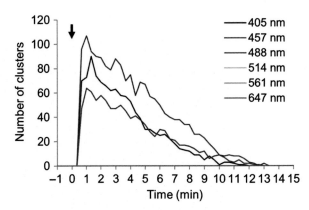

Fig. 7 Cluster formation with various wavelengths of light. A HeLa cell co-expressing CRY2-mCherry and CIB1-mCerulean-MP was illuminated by light of indicated wavelength. Cluster formation was monitored by fluorescent imaging of CRY2-mCherry in every 20 s. For quantification, clusters were defined as discrete puncta of fluorescence with criteria of fluorescence intensity (1500–4095 arbitrary units), size (>0.2 μm^2) and circularity (0.5–1.0 arbitrary units). *Arrow* indicates time point of irradiation. It is a representative data of two trials for each condition

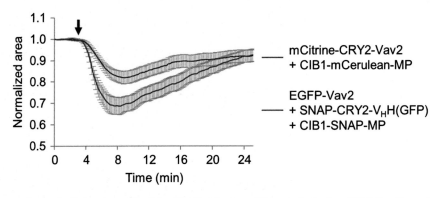

Fig. 8 Comparison of efficiencies of Vav2 inactivation by two different strategies. NIH3T3 cells co-transfected with expression plasmids as indicated, were illuminated by blue light (488 nm, 0.76 mW/mm^2). Morphological changes were monitored by fluorescent imaging of mCherry-Lifeact in every 10 s ($n = 11$; Error bars, s.e.m.) *Arrow* indicates time point of irradiation. It is reproduced with permission from *Nature Methods*

8. We recommend that the ratio of DNA to Plus reagent should be 1:1.

9. We recommend the 1:3 ratio of DNA to LTX reagent to achieve high transfection efficiency.

10. Excitation of CFP, GFP, or YFP can induce CRY2-CIB1 heterodimerization and cluster formation. So, you should find the cells by monitoring red fluorescence signals to avoid perturbation of the target before photo-stimulation.

11. In most LARIAT experiments, we use a 488-nm laser for photo-activation on whole cell area. In our case, illumination for 1.5 s at 0.3 mW/mm^2 of power density was enough to induce cluster formation. For photo-stimulation on partial region of cell, adjust the laser power to stimulated area.

12. Before stimulation, it is important to monitor basal change of cell morphology for at least 1 min to convince the effect of light-inducible protein inactivation. After light illumination, at least 20 min is required to recover cell morphology to the basal state.

13. If there is any unfilled spaces on the thresholded image, especially nearby nucleus due to difference of intensity between nucleus and cytoplasm, fill the spaces by clicking "Fill holes" button to avoid loss of cell area.

14. After resuspension with Resuspension Buffer R, go to next step as soon as possible to avoid any harmful effect on cell viability and consequent reduction of transfection efficiency.

15. Keep watching the Neon Tip during electric shock to see if there is any spark that can be caused by bubbles in the tip.

16. To monitor cell division, we usually acquire images at every 5 min for 24 h. Illumination by blue light (for excitation of EGFP) at 5 min intervals is enough to maintain clusters.

17. When both CIB1-CLIP-MP and SNAP-CRY2-V$_H$H(GFP) are coexpressed on a cell, cluster formation can be determined by observing EGFP signals. So, clustering EGFP signal reflects that CIB1- and CRY2-fused proteins are expressed and participate in cluster formation.

Acknowledgements

This work was supported by the Institute for Basic Science (no. IBS-R001-G1), Republic of Korea. Figures 1, 2, 4, 5, and 8 are reproduced with permission from *Nature Methods*.

References

1. Doupe DP, Perrimon N (2014) Visualizing and manipulating temporal signaling dynamics with fluorescence-based tools. Sci Signal 7(319):1

2. Turgeon B, Meloche S (2009) Interpreting neonatal lethal phenotypes in mouse mutants: insights into gene function and human diseases. Physiol Rev 89(1):1–26

3. Stockwell BR (2004) Exploring biology with small organic molecules. Nature 432(7019): 846–854

4. Zhou P (2005) Targeted protein degradation. Curr Opin Chem Biol 9(1):51–55

5. Banaszynski LA, Wandless TJ (2006) Conditional control of protein function. Chem Biol 13(1):11–21

6. Wu YI, Frey D, Lungu OI, Jaehrig A, Schlichting I, Kuhlman B, Hahn KM (2009) A genetically encoded photoactivatable Rac controls the motility of living cells. Nature 461(7260):104–108

7. Levskaya A, Weiner OD, Lim WA, Voigt CA (2009) Spatiotemporal control of cell signalling using a light-switchable protein interaction. Nature 461(7266):997–1001

8. Kennedy MJ, Hughes RM, Peteya LA, Schwartz JW, Ehlers MD, Tucker CL (2010) Rapid blue-light-mediated induction of protein interactions in living cells. Nat Methods 7(12):973–975

9. Bugaj LJ, Choksi AT, Mesuda CK, Kane RS, Schaffer DV (2013) Optogenetic protein clustering and signaling activation in mammalian cells. Nat Methods 10(3):249–252

10. Lee S, Park H, Kyung T, Kim NY, Kim S, Kim J, Heo WD (2014) Reversible protein inactivation by optogenetic trapping in cells. Nat Methods 11(6):633–636

11. Rothbauer U, Zolghadr K, Tillib S, Nowak D, Schermelleh L, Gahl A, Backmann N, Conrath K, Muyldermans S, Cardoso MC, Leonhardt H (2006) Targeting and tracing antigens in live cells with fluorescent nanobodies. Nat Methods 3(11):887–889

12. Riedl J, Crevenna AH, Kessenbrock K, Yu JH, Neukirchen D, Bista M, Bradke F, Jenne D, Holak TA, Werb Z, Sixt M, Wedlich-Soldner R (2008) Lifeact: a versatile marker to visualize F-actin. Nat Methods 5(7):605–607

13. Shen K, Meyer T (1998) In vivo and in vitro characterization of the sequence requirement for oligomer formation of Ca^{2+}/calmodulin-dependent protein kinase IIalpha. J Neurochem 70(1):96–104

14. Rosenberg OS, Deindl S, Sung RJ, Nairn AC, Kuriyan J (2005) Structure of the autoinhibited kinase domain of CaMKII and SAXS analysis of the holoenzyme. Cell 123(5):849–860

15. Shaner NC, Steinbach PA, Tsien RY (2005) A guide to choosing fluorescent proteins. Nat Methods 2(12):905–909

Chapter 26

Optogenetic Manipulation of Selective Neural Activity in Free-Moving *Drosophila* Adults

Po-Yen Hsiao, Ming-Chin Wu, Yen-Yin Lin, Chein-Chung Fu, and Ann-Shyn Chiang

Abstract

Activating selected neurons elicits specific behaviors in *Drosophila* adults. By combining optogenetics and laser-tracking techniques, we have recently developed an automated laser-tracking and optogenetic manipulation system (ALTOMS) for studying how brain circuits orchestrate complex behaviors. The established ALTOMS can independently target three lasers (473-nm blue laser, 593.5-nm yellow laser, and 1064-nm infrared laser) on any specified body part of two freely moving flies. Triggering light-sensitive proteins in real time, the blue laser and yellow laser can respectively activate and inhibit target neurons in artificial transgenic flies. Since infrared light is invisible to flies, we use the 1064-nm laser as an aversive stimulus in operant learning without perturbing visual inputs. Herein, we provide a detailed protocol for the construction of ALTOMS and optogenetic manipulation of target neurons in *Drosophila* adults during social interactions.

Key words Optogenetics, Channelrhodopsin, Halorhodopsin, Optical stimulation, Neural circuit, Freely walking, Behavior, *Drosophila*

1 Introduction

The function of the human brain arises from the coordinated activities of nearly 100 billion neurons. Unraveling information flows among these neurons is a formidable challenge for understanding the emergent properties of the brain [1]. The small brains of insects, which exhibit complex behaviors, provide an opportunity to solve this problem with new technologies before application in human brain studies. Although the fly brain has only around 135,000 neurons, these neuron form a complex network and exhibit a hierarchical structure with small-world characteristics and rich-club organization [2]. The organization of this hierarchy is highly similar to that in the mammalian brain [3], suggesting the presence of common basic operation principles in multiple

Arash Kianianmomeni (ed.), *Optogenetics: Methods and Protocols*, Methods in Molecular Biology, vol. 1408, DOI 10.1007/978-1-4939-3512-3_26, © Springer Science+Business Media New York 2016

functional modules that arise from the coordinated activity of large numbers of neurons and parallel neural circuits.

Recent advances in optogenetics using microbial opsins expressed in target neurons have allowed light-controllable manipulation of neuronal activities in specific brain circuits [4–6]. High spatiotemporal precision of light irradiation is particularly useful for resolving the causality of the information flows among different neurons within the same functional circuit, such as sensory perception [7, 8], parallel pathway shunting [9], information computation [2], decision making [10], and locomotion control [11, 12]. However, optogenetic methods may be challenging because most biological tissues are opaque, compromising the capacity for deep-tissue irradiation. Since light scattering is inversely proportional to the radiation wavelength, biochemists have recently developed several long-wavelength opsins that can be triggered by orange-red light for deep-tissue activation [13, 14] and/or inhibition [15–17].

As a proof-of-concept exercise, neuroscientists are now able to manipulate physiological activities and gene expression of almost any target neuron at any time in the genetically tractable model organism *Drosophila melanogaster* [18]. The most widely used light sources for optogenetic manipulation are light-emitting diodes (LEDs) and mercury lamps, which provides sufficient light intensity to activate the actuators in *Drosophila* adults [9, 19–21]. Alternatively, laser light with stronger intensity and higher spatiotemporal precision is used to irradiate target neurons in specified body parts. Combined with functional imaging of a restrained adult fly, optogenetic manipulation is gradually revealing the logic of information flow in the brain [22, 23]. To study how neural activities orchestrate complex behaviors, we have recently established an automated laser-tracking and optogenetic manipulation system (ALTOMS) for real-time manipulation of target neurons in two free-moving flies [11, 24]. Based on behavioral interactions computed with a high-speed image analysis system, ALTOMS can target three lasers (473-nm blue laser, 593.5-nm yellow laser, and 1064-nm infrared [IR] laser) independently on any specified body part of two freely moving *Drosophila* adults. Given its capacity for optogenetic manipulation of selected neurons through transient and independent activation/inactivation, ALTOMS offers opportunities to systematically construct a neuron-behavior map in *Drosophila* adults.

2 Materials

2.1 *ALTOMS Training Arena*

1. Upper and supporting base of the arena (customized design) (*see* **Notes 1** and **2**).

2. Cover glass for the arena (customized design) (*see* **Note 2**).

3. Fluon (*see* **Note 2**).

4. Water repellent (*see* **Note 2**).

2.2 Image Capture Module (ICM)

1. Charge-coupled device (CCD) camera.

2. CCD camera lens.

3. 1394 PCI 3-port card.

4. CCD camera adaptor for positioning the camera on the stand.

5. Copy stand.

6. CCD focusing stage.

7. Notch filter to block the blue spot that interferes with image analysis (473 ± 10 nm).

8. Notch filter to block the yellow spot that interferes with image analysis (593.5 ± 10 nm).

9. Notch filter holder for positioning of the notch filter on the front of the lens.

10. LED.

11. Circuit white-light LED.

12. LED controller.

13. Diffuser for making the LED light uniform.

2.3 Laser-Scanning Module (LSM) and Alignment Tools

1. Three-wavelength laser mixer and three lasers (customized design).

2. Infrared polarizer.

3. Blue/yellow laser polarizer for adjusting the intensity of the blue/yellow laser.

4. IR laser polarizer for adjusting the intensity of the IR laser.

5. Polarizer mount.

6. Beam shutter.

7. Adapter cable.

8. Open frame driver controller.

9. Power supply.

10. Dichroic mirror/beam splitter.

11. Round protected silver mirrors.

12. Square protected silver mirror.

13. Dichroic mirror/round protected silver mirror mount.

14. Square protected silver mirror mount for reflection of the laser beam into the behavioral apparatus (customized design).

15. Galvanometer scanner to guide the laser spot onto the moving fly.

16. 5-mm XY mirror set for the galvanometer scanner.

17. 5-mm XY mount set for the galvanomotor scanner.

18. Galvanomotor driver.

19. Galvanomotor power supply.

20. Beam profiler for measuring the spot size and laser profile.

21. IR viewer (350–1550 nm).

22. C-mount adaptor.

23. Close-up lens.

24. Visible sheet polarizers.

25. Linear film polarizers.

2.4 Intelligent Central Control Module (ICCM)

1. Software developed using LabVIEW 2010 and vision module.

2. Data acquisition device.

3. 68-conductor ribbon cable.

4. I/O connector block P/N.

5. Computer.

Processor: >3.60 GHz; Memory: >8 GB DDR-3 1600; Hard drive: 750 GB SATA; Video card: GeForce GT640-2GD3 PCIE; 1394B Adaptor Card; Operating system: Windows 7 Professional 64 bit.

2.5 All-trans-Retinal for Driving Optogenetic Proteins

All-*trans*-retinal powder (*see* **Note 3**).

3 Methods

The ventral-irradiated ALTOMS contains four modules (Fig. 1): (I) the observation apparatus, (II) the ICM, (III) the LSM, and (IV) the ICCM, as described below.

3.1 Loading Optogenetic Flies to the Behavior Arena

The observation apparatus made from acrylic materials consisted of an upper arena (20 mm in diameter and 3 mm in height) and a supporting base (Fig. 2).

1. Maintain optogenetic *Drosophila* adult flies in glass vials containing standard fly food at 25 ± 1 °C and 70 % relative humidity under a 12:12-h light:dark cycle.

2. Transfer flies 1–2 days after pupation into a new vial containing standard food with 100 μM all-*trans*-retinal. Approximately 15 individuals may be kept in the same vial, provided the flies are of the same genotype and sex (*see* **Note 4**).

3. Incubate the vial at 25 ± 1 °C and 70 % relative humidity in darkness for 7 days prior to the experiment.

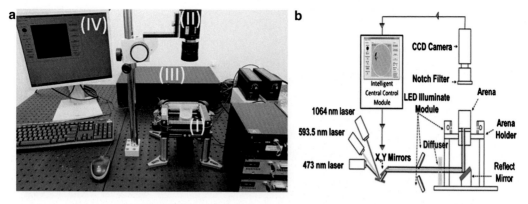

Fig. 1 The ventral-irradiated ALTOMS. (**a**) Photograph of the system. (**b**) Schematic design. ALTOMS contains four modules: (*I*) the observation apparatus, (*II*) the ICM, (*III*) the LSM, and (*IV*) the ICCM

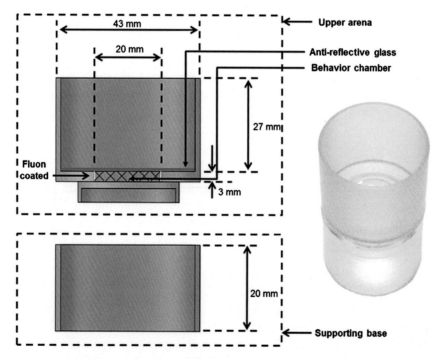

Fig. 2 The observation apparatus. The observation apparatus, which is made of acrylic materials, has two parts: an upper arena and a supporting base. A coarse outer surface is used for even light illumination

4. Anesthetize the flies on a narcotic pad connected to 100 % CO_2 and transfer them with a watercolor brush pen to the behavior arena.

5. Place the covered behavior arena on the observation apparatus.

6. Wait for the flies to wake up and acclimate for at least 2 min.

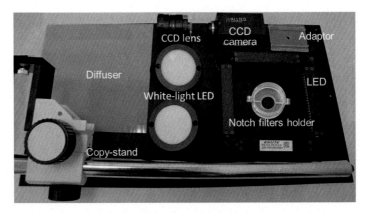

Fig. 3 The ICM setup. The ICM consists of a CCD camera, a set of white LED lights for illumination of the surrounding area and bottom, a diffuser plate for homogenizing the illumination, and a notch filter for blocking stray laser light reflected from the arena

7. Score the behavior response of the target fly when the laser is switched on and/or off to manipulate neuron activity.

3.2 The ICM Setup

1. The ICM consisted of two parts: a CCD camera and a set of white LEDs comprising the illumination module (Fig. 3).

2. The CCD camera integrated two notch filters (473 ± 10 and 593.5 ± 10 nm) in front of the CCD camera lens to block stray laser light reflected from the arena.

3. For recording, we set the CCD aperture to F5.6 for 500×480 pixels/frame resolution and 208 frames/s (FPS) speed. The arena diameter, designating the active area for the flies, is 20 mm, and the field of view of the CCD camera is positioned to fill this active area. We can obtain images of 25×24 pixels/mm^2 and provide suitable spatiotemporal resolution for fly behavior analysis.

4. Illuminate the arena with even LED light for high quality of image recording.

3.3 The LSM Setup

1. The LSM consisted of a pair of galvomotors with a 130-μs small-angle step response and three continuous wave diode-pumped solid-state lasers: a 473-nm blue laser with maximum intensity of 20 mW/mm^2, a 593.5-nm yellow laser with maximum intensity of 20 mW/mm^2, and a 1064-nm IR laser with maximum intensity of 300 mW/mm^2 (*see* **Note 5**).

2. A dichroic mirror (reflection band: 380–490 nm; transmission band: 520–700 nm), a broadband high-reflectivity dielectric mirror (reflection band: 400–872 nm; transmission band: 932–1300 nm), and Ag-coated surface mirrors were used to combine three different lasers (Fig. 4).

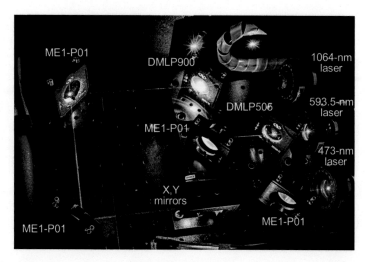

Fig. 4 The LSM setup. The LSM includes three lasers (473, 593.5, and 1064 nm) and three independent shutters. The polarizer allows adjustment of the laser intensity. The four high-reflection mirrors facilitate retention of laser intensity with minimum intensity loss, whereas the two dichroic mirrors allow transparency or reflection of specific wavelength ranges. A set of galvo scanners allows for rapid changes in the direction of the laser

3. Before starting an experiment, calibrating the spatial precision of laser irradiation is essential (Fig. 5a). To spot the location of laser irradiation, we placed a white paper at the bottom of the arena (Fig. 5b). In order to create a lookup table to calibrate the relative coordinates between the scanner and the specified locations on the arena, laser positioning was calibrated five times by matching the laser-induced fluorescence spots to five fixed points (the center point and the left, right, top, and bottom edges of the arena) on a white paper via manual input of different voltage signals (Fig. 5c).

4. Three mechanical shutters were activated within a 1-ms transfer time upon opening to modulate the laser stimulation by switching the laser on/off independently or simultaneously.

5. When the flies' positions were obtained from an image, the ALTOMS required a converter to convert the signals from digital to analog. In our system, we used a data acquisition (DAQ) device to change the digital signals (analysis results) to analog signals and drive the scanning motors.

3.4 The ICCM Setup

1. Ventral-irradiated ALTOMS enables the user to perform several optogenetic behavioral assays for *Drosophila* adults and larvae. Install LabVIEW 2010 and download the ALTOMS software from Brain Research Center, National Tsing Hua University (http://brc.life.nthu.edu.tw/resources.html). From the website, we provide four archive manuals for user

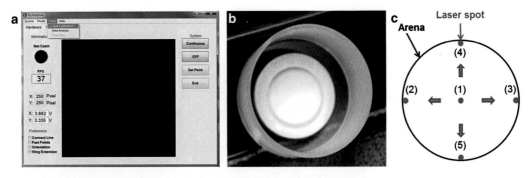

Fig. 5 Calibration of laser spot. (**a**) The graphical user interface (GUI) of laser spot calibration. (**b**) A "Calibrator" arena was specifically designed for laser spot calibration. The bottom of the arena, made of white acrylic material, was covered with a white paper. (**c**) The color threshold method for extracting the laser spot. The center was defined by theoretical approximation

download, including ALTOMS-Adult, ALTOMS-Larvae, ALTOMS-Shutter control, and ALTOMS-Total. ALTOMS-Adult and ALTOMS-Larvae are only for analysis of a single type, and ALTOMS-Shutter control provides an external trigger shutter switch (Continuous On/OFF or set switch period). The ALTOMS-Total integrates analysis and shutter control for both types (Adult/Larvae) of fly behavior assays (*see* **Note 6**).

2. Open the ALTOMS folder and double click the ALTOMS icon (Fly image, Fig. 6a).

3. The "Source" button in the graphical user interface (GUI) is used to select the type of analysis ("Image" button for off-line analysis, "CCD" button for online analysis (Fig. 6b)).

4. In off-line analysis, the user can select a location for all images for analysis in a single directory (Fig. 6c).

5. In online analysis, the user must set the channels of the DAQ device to convert the signal between scanning galvomotors and the analysis result (Fig. 6d).

6. The user selects the analysis type (Off-Line/On-Line analysis) and hardware setting (DAQ device) for the ALTOMS. The user can then click the "Display" button to change the display screen of the GUI (Fig. 6e).

7. ALTOMS allows the user to select either larva or adult behavior assays when performing the optogenetic experiment. The user chooses *Drosophila* larva/adult through the "Type" menu in the selection bar (Fig. 6f).

8. Before starting the analysis process, the user should press the "Start Acquisition" button to capture online images from the CCD camera or to upload off-line data from the prerecorded folder (Fig. 6g).

9. To start the experiment, the ALTOMS allows users to define the following parameters: switching period, number of shutter

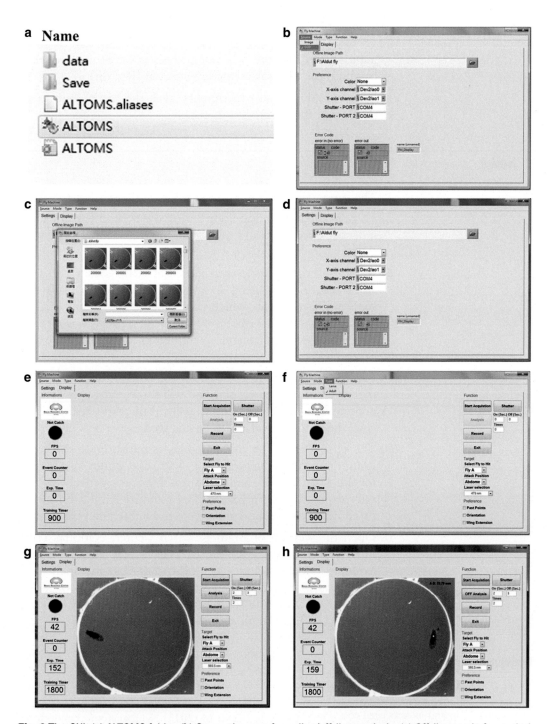

Fig. 6 The GUI. (**a**) ALTOMS folder. (**b**) Source images for online/off-line analysis. (**c**) Off-line mode for analysis of images from prerecorded video. (**d**) Hardware settings. (**e**) Software settings. (**f**) Experiment types: adult or larvae conditioning. (**g**) Software settings for optogenetic experiments. (**h**) Initiation of analysis

cycles, target region of the fly (head, thorax, or abdomen), and wavelength of the laser (1064, 593.5, or 473 nm; Fig. 6g).

10. Pressing the "Analysis" button will cause the system to automatically perform the behavior analysis and laser manipulation based on the user-selected configuration (Fig. 6h).

11. After completing all experiments, click the "Exit" button to shut down the system automatically.

4 Notes

1. The supporting base helps to provide even light illumination.

2. To keep the flies on the floor during the experiment, the anti-reflective cover glass and the wall of the arena were coated with Fluon.

3. All-*trans*-retinal powder was stored at −20 °C as a 100 mM stock solution dissolved in alcohol. Additionally, 100 μM sugar-retinal solution, which was prepared by heating and liquefying the sugar to mix evenly with the retinal, was directly added to glass vials covered in aluminum foil to prevent exposure to light.

4. Flies of the same age and sex should be compared to ensure that the flies are normal size and will show consistent optogenetic trigger behaviors. Additionally, if the experiment is specific, the number of flies can be changed.

5. The spot sizes of the collimated lasers measured in the arena using a beam profiler, were 1.1, 1.05, and 2 mm in diameter for the 473-, 593.5-, and 1064-nm lasers, respectively. Before each experiment, the laser power intensity should be monitored using visible sheet polarizers and linear film polarizers.

6. Open the website of Brain Research Center, National Tsing Hua University and click the ALTOMS icon, and the downloadable folders include ATOMS-Adult (file for analysis of adult flies only), ATOMS-Larvae (files for analysis of larvae only), ALTOMS-Shutter control (files for shutter control only), ALTOMS-Total, and ALTOMS-Manual.

References

1. Herculano-Houzel S (2012) The remarkable, yet not extraordinary, human brain as a scaled-up primate brain and its associated cost. Proc Natl Acad Sci U S A 109(Suppl 1):10661–10668

2. Shih CT, Sporns O, Yuan SL, Su TS, Lin YJ, Chuang CC, Wang TY, Lo CC, Greenspan RJ, Chiang AS (2015) Connectomics-based analysis of information flow in the *Drosophila* brain. Curr Biol 25:1249–1258

3. Kaiser M (2015) Neuroanatomy: connectome connects fly and Mammalian brain networks. Curr Biol 25:R416–R418

4. Fork RL (1971) Laser stimulation of nerve cells in *Aplysia*. Science 171:907–908

5. Nikolenko V, Poskanzer KE, Yuste R (2007) Two-photon photostimulation and imaging of neural circuits. Nat Methods 4:943–950

6. Lima SQ, Miesenbock G (2005) Remote control of behavior through genetically targeted photostimulation of neurons. Cell 121:141–152

7. Vosshall LB, Amrein H, Morozov PS, Rzhetsky A, Axel R (1999) A spatial map of olfactory receptor expression in the *Drosophila* antenna. Cell 96:725–736

8. Vosshall LB, Wong AM, Axel R (2000) An olfactory sensory map in the fly brain. Cell 102:147–159

9. Lin HH, Chu LA, Fu TF, Dickson BJ, Chiang AS (2013) Parallel neural pathways mediate CO_2 avoidance responses in *Drosophila*. Science 340:1338–1341

10. Yang CH, Belawat P, Hafen E, Jan LY, Jan YN (2008) *Drosophila* egg-laying site selection as a system to study simple decision-making processes. Science 319:1679–1683

11. Wu MC, Chu LA, Hsiao PY, Lin YY, Chi CC, Liu TH, Fu CC, Chiang AS (2014) Optogenetic control of selective neural activity in multiple freely moving *Drosophila* adults. Proc Natl Acad Sci U S A 111:5367–5372

12. Bath DE, Stowers JR, Hormann D, Poehlmann A, Dickson BJ, Straw AD (2014) FlyMAD: rapid thermogenetic control of neuronal activity in freely walking *Drosophila*. Nat Methods 11:756–762

13. Lin JY, Knutsen PM, Muller A, Kleinfeld D, Tsien RY (2013) ReaChR: a red-shifted variant of channelrhodopsin enables deep transcranial optogenetic excitation. Nat Neurosci 16:1499–1508

14. Klapoetke NC, Murata Y, Kim SS, Pulver SR, Birdsey-Benson A, Cho YK, Morimoto TK, Chuong AS, Carpenter EJ, Tian Z, Wang J, Xie Y, Yan Z, Zhang Y, Chow BY, Surek B, Melkonian M, Jayaraman V, Constantine-Paton M, Wong GK, Boyden ES (2014) Independent optical excitation of distinct neural populations. Nat Methods 11:338–346

15. Zhang F, Wang LP, Brauner M, Liewald JF, Kay K, Watzke N, Wood PG, Bamberg E, Nagel G, Gottschalk A, Deisseroth K (2007) Multimodal fast optical interrogation of neural circuitry. Nature 446:633–639

16. Gradinaru V, Thompson KR, Deisseroth K (2008) eNpHR: a Natronomonas halorhodopsin enhanced for optogenetic applications. Brain Cell Biol 36:129–139

17. Gradinaru V, Zhang F, Ramakrishnan C, Mattis J, Prakash R, Diester I, Goshen I, Thompson KR, Deisseroth K (2010) Molecular and cellular approaches for diversifying and extending optogenetics. Cell 141:154–165

18. Luo L, Callaway EM, Svoboda K (2008) Genetic dissection of neural circuits. Neuron 57:634–660

19. Suh GS, Ben-Tabou de Leon S, Tanimoto H, Fiala A, Benzer S, Anderson DJ (2007) Light activation of an innate olfactory avoidance response in *Drosophila*. Curr Biol 17:905–908

20. de Vries SE, Clandinin TR (2012) Loom-sensitive neurons link computation to action in the *Drosophila* visual system. Curr Biol 22:353–362

21. Inagaki HK, Jung Y, Hoopfer ED, Wong AM, Mishra N, Lin JY, Tsien RY, Anderson DJ (2014) Optogenetic control of *Drosophila* using a red-shifted channelrhodopsin reveals experience-dependent influences on courtship. Nat Methods 11:325–332

22. Keene AC, Masek P (2012) Optogenetic induction of aversive taste memory. Neuroscience 222:173–180

23. Haikala V, Joesch M, Borst A, Mauss AS (2013) Optogenetic control of fly optomotor responses. J Neurosci 33:13927–13934

24. Lin YY, Wu MC, Hsiao PY, Chu LA, Yang MM, Fu CC, Chiang AS (2015) Three-wavelength light control of freely moving *Drosophila Melanogaster* for less perturbation and efficient social-behavioral studies. Biomed Opt Express 6:514–523

Chapter 27

Guidelines for Photoreceptor Engineering

Thea Ziegler*, Charlotte Helene Schumacher*, and Andreas Möglich

Abstract

Sensory photoreceptors underpin optogenetics by mediating the noninvasive and reversible perturbation of living cells by light with unprecedented temporal and spatial resolution. Spurred by seminal optogenetic applications of natural photoreceptors, the engineering of photoreceptors has recently garnered wide interest and has led to the construction of a broad palette of novel light-regulated actuators. Photoreceptors are modularly built of photosensors that receive light signals, and of effectors that carry out specific cellular functions. These modules have to be precisely connected to allow efficient communication, such that light stimuli are relayed from photosensor to effector. The engineering of photoreceptors benefits from a thorough understanding of the underlying signaling mechanisms. This chapter gives a brief overview of key characteristics and signal-transduction mechanisms of sensory photoreceptors. Adaptation of these concepts in photoreceptor engineering has enabled the generation of novel optogenetic tools that greatly transcend the repertoire of natural photoreceptors.

Key words Light, Optogenetics, Protein engineering, Sensory photoreceptor, Signal transduction

1 Introduction

As genetically encoded, light-regulated actuators, photoreceptors provide the basis for optogenetics, the noninvasive, reversible, and spatiotemporally precise manipulation of biological processes by light. In signal transduction as in biology in general, researchers often tackle complex natural systems by disassembling them into simpler building blocks with more tractable attributes. For signal receptors such building blocks commonly correspond to sensor modules that receive environmental stimuli as input signals and effector modules that exert specific cellular functions in response to a given stimulus. These modules distribute into several classes with recurring structural and functional motifs as well as common principles of signal transduction. The modular nature of signaling proteins often allows the recombination of sensor and effector modules to accommodate new input or output modalities, or to vary functional parameters

---

*The first two authors contributed euqally to this chapter.

Arash Kianianmomeni (ed.), *Optogenetics: Methods and Protocols*, Methods in Molecular Biology, vol. 1408,
DOI 10.1007/978-1-4939-3512-3_27, © Springer Science+Business Media New York 2016

(e.g., light sensitivity, response kinetics, or dynamic range) of the composite sensor-effector system. In this chapter, we focus on the engineering of photoreceptors for which sensor and effector are organized in distinct protein domains or proteins [1–3]; by contrast, we only brush upon receptors in which these modules are integrated into a single domain, as for example within the large group of microbial rhodopsins acting as light-gated ion channels and pumps that kick-started optogenetics, reviewed elsewhere [4–6].

Intact signal transmission between photosensor and effector modules depends on diverse and dynamic allosteric coupling mechanisms. In many rational photoreceptor engineering approaches fundamental information on the structural and functional attributes of these modules is a prerequisite for the generation of functional sensor-effector combinations. To obtain this information, photoreceptors are often decomposed into isolated modules with a reduced number of parameters, and we organize this chapter in a corresponding manner by first introducing characteristic attributes of candidate photosensor (Subheading 2) and effector modules (Subheading 3), before continuing with the mechanistic principles of signal transduction (Subheading 4) that motivate the choice of the eventual design strategy (Subheading 5).

2   The Photosensor

To receive the environmental stimulus light, all photosensors harbor an organic chromophore (Subheading 2.1) with a conjugated π electron system that absorbs photons in the UV/visible range of the electromagnetic spectrum and transmits part of the absorbed energy to the protein scaffold [7, 8]. Light absorption by the dark-adapted state D initiates a so-called photocycle (Subheading 2.2), eventually leading to population of the signaling state S. This state then persists from milliseconds to many hours depending upon photosensor before it reverts to D in a thermally driven, spontaneous reaction, denoted "dark recovery". The kinetics for the reversion to D (Subheading 2.3) significantly affect the temporal resolution of optogenetic applications (*off* kinetics) and might effectively limit their reversibility on biologically relevant timescales.

2.1 Chromophore

The chromophore and the surrounding photosensor scaffold determine spectral sensitivity and photochemistry, based on which photoreceptors divide into several classes (Fig. 1) [7, 8]. The chromophore is embedded in the photosensor module, which mostly consists of a single protein domain but in case of phytochrome red-light sensors comprises three separate domains, denoted "photosensory core" (PSC) [9]. In view of eventual optogenetic applications, the choice of photosensor should be guided by at least two important considerations: chromophore availability in the target tissue; and wavelength used for stimulation.

Wavelength [nm]

| 400 | 500 | 600 | 700 | 800 |

Tissue penetration of light

Chromophore

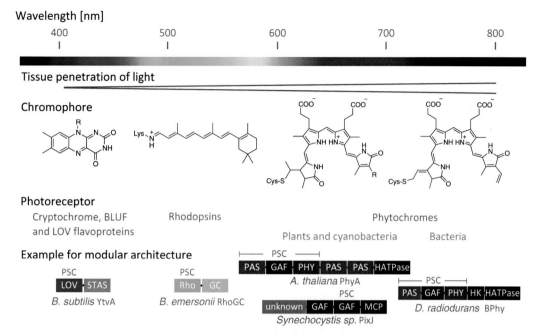

Photoreceptor

Cryptochrome, BLUF
and LOV flavoproteins

Rhodopsins

Phytochromes

Plants and cyanobacteria   Bacteria

Example for modular architecture

PSC

| LOV | STAS |

B. subtilis YtvA

PSC

| Rho | GC |

B. emersonii RhoGC

PSC

| PAS | GAF | PHY | PAS | PAS | HATPase |

A. thaliana PhyA

PSC

| unknown | GAF | GAF | MCP |

Synechocystis sp. PixJ

PSC

| PAS | GAF | PHY | HK | HATPase |

D. radiodurans BPhy

Fig. 1 Properties of selected chromophores and sensory photoreceptors. The penetration depth of light in mammalian tissue increases strongly with increasing wavelengths with a maximum in the near infrared denoted as "near infrared window". Blue-light sensitive cryptochromes, BLUF, and LOV flavoproteins incorporate flavin nucleotides, and rhodopsins use retinal as chromophore, all of which are naturally present in most mammalian tissues. Plant and cyanobacterial phytochromes (Phys) as well as cyanobacteriochromes (CBCRs) require exogenous supply of their cofactor phycocyanobilin or phytochromobilin for optogenetic applications in vivo, since these reduced bilin derivatives cannot be provided by most cells. In contrast, bacterial phytochromes (BPhys) utilize the oxidized linear tetrapyrrole biliverdin, which is a direct heme degradation product, meaning exogenous chromophore addition is not necessary in most tissues tested to date. Photoreceptors are often built modularly, consisting of at least a sensor and an effector module. Exemplary natural photoreceptors with modular architecture are the LOV protein YtvA from *Bacillus subtilis* [61], a rhodopsin guanylate cyclase from *Blastocladiella emersonii* [62], the plant phytochrome *Arabidopsis thaliana* PhyA [9], the *Synechocystis* sp. cyanobacteriochrome PixJ [63] and a bacterial phytochrome from *Deinococcus radiodurans* [64]

First, the chromophore must be available in sufficient amounts at the target site in situ to be autonomously incorporated into the functional photoreceptor. Plant UV-B receptors [10] employ intrinsic amino acids to absorb light but more commonly, photoreceptors use chromophore cofactors that derive from small metabolites. Specifically, LOV (light-oxygen-voltage), BLUF (blue-light sensors using flavin-adenine dinucleotide), and cryptochrome sensors employ flavin-nucleotide chromophores sensitive to blue light [11–13]; the rhodopsin family use retinal to respond to light from the UV to the red [4]; phytochromes use linear tetrapyrroles (bilins) to respond to red and near-infrared wavelengths [9], further extended to the entire visible spectrum in recently discovered algal phytochromes [14]; and cyanobacteriochromes also use linear tetrapyrroles and exhibit spectral sensitivity ranging from the UV to the near-infrared [15, 16]. Reduced tetrapyrroles, such as phycocyanobilin, that plant phytochromes and cyanobacteriochromes resort to, are not found in mammalian tissues which are frequent subjects of optogenetics. By contrast, the oxidized tetrapyrrole biliverdin, employed by bacterial phytochromes, retinal and flavin-nucleotide chromophores are apparently present in sufficient quantities in many mammalian tissues investigated to date [17–20].

Second, the wavelength used for photoreceptor activation determines the maximally achievable tissue penetration depth, phototoxicity, and potential combination of several optogenetic actuators and reporters. Limited tissue penetration of light complicates photon delivery to target sites within opaque tissues or deeper tissue layers. In particular, within the spectral region below 700 nm, penetration is substantially impeded by light scattering and absorption by lipids, hemoglobin, and other pigments. Mainly for longer wavelengths above ~700 nm, in a region denoted "near-infrared spectral window," so far only covered by members of the phytochrome family, high penetration depths are achieved. Especially at lower wavelengths, the absorbed light quanta can elicit inadvertent phototoxic effects, e.g., due to generation of reactive oxygen species. If photoreceptors are to be deployed in parallel and/or in combination with fluorescent reporters, the individual wavelengths used for photoreceptor activation should be spectrally separated such that activation of a selected process does not interfere with other ones; that is, stimulation of a given photoreceptor should be orthogonal without eliciting other responses.

2.2 Photocycle

The term photocycle refers to a series of structural and dynamic changes within the chromophore and the surrounding protein scaffold following light absorption. In addition to the dark-adapted state D and the signaling state S, the photocycle often encompasses short-lived intermediate states. Regardless of the presence of these intermediates, the photochemical reaction towards the signaling state S is generally completed within microseconds at most, which is much faster than many physiological responses; for the purpose of this guideline we hence disregard photocycle intermediates.

The absolute light sensitivity of a photoreceptor depends on the absorption coefficient at a given wavelength and on the intrinsic quantum efficiency for formation of the signaling state. Notably, natural photoreceptors are intrinsically optimized for sensitive light reception with suitably high quantum efficiencies, and absolute light sensitivity can usually not be enhanced to significant extent. Instead, to improve photoreceptor activation in optogenetic applications, light power can be ramped up but only to limited extent lest it causes severe biological damage. However, for optogenetic experiments conducted under constant illumination, a second route to optimizing photoreceptor activation is available. At photostationary conditions, an equilibrium is assumed between the dark-adapted and light-adapted states, which is not only determined by the kinetics of the light-driven forward reaction towards S but also by those of the thermally driven reverse reaction towards D (cf. Subheading 2.3). We denote the ratio of forward and reverse kinetics as the effective light sensitivity. For some photosensors, specifically LOV proteins and phytochromes, the dark recovery kinetics can be varied by many orders of magnitude via the introduction of mutations proximal to the chromophore, thereby offering an alternative way of modulating the effective light sensitivity [3].

2.3 Dark-Reversion Kinetics

The reversion from S to D occurs in a thermally driven reaction which can often be greatly accelerated by elevating temperature or changing solvent composition [21]. In addition to this spontaneous reaction, an alternative means of depleting S is offered in photochromic photoreceptors for which the signaling state S can actively be reverted to the dark-adapted state D by a subsequent light stimulus, typically of different color. The group of photochromic photoreceptors comprises phytochromes, cyanobacteriochromes, certain so-called "bistable" rhodopsins, and a re-engineered derivative of the photo-switchable fluorescent protein Dronpa [22]. The light-driven, bidirectional interconversion between D and S allows the regulation of downstream signaling events with superior temporal precision. Likewise, if activating and deactivating wavelengths are interleaved in space rather than time, superior spatial resolution can be obtained [3].

3   The Effector

The selection of a suitable effector module for photoreceptor engineering is largely determined by the desired output that should become subject to light control. The nature of the parental protein, from which the effector derives, governs a number of aspects that we discuss in turn: activity and dynamic range (Subheading 3.1); and availability of efficient activity assays (Subheading 3.2).

3.1 Activity and Dynamic Range

To elicit a suitable response in vivo, effector activity often has to be above certain threshold levels. Accordingly, activity of the photoreceptor in situ may have to be adjusted to match these levels, for example by varying overall expression levels of the photoreceptor and/or the specific activity of the effector. Another key consideration is the factor difference between the activities of an effector module in its low-activity and high-activity states, denoted as the "dynamic range" of the signal receptor. Notably, high dynamic ranges can only be achieved if the basal activity of the low-activity state is sufficiently low; for example, in light-activated receptors the dynamic range is often limited by residual dark activity. For engineered photoreceptors, a low dynamic range of the originally light-inert parental effector often limits the maximally attainable factor of light induction or repression. *Vice versa*, it is not guaranteed, that photoreceptors engineered on the basis of high-dynamic-range parental proteins will also yield strongly light-regulated derivatives. For example, the overall activity, the substrate affinities, and the maximal two-fold activation by light of *E. coli* dihydrofolate reductase (DHFR) fused to the *Avena sativa* phototropin 1 LOV2 (*As*LOV2) pale in comparison to the corresponding parameters of wild-type DHFR [23].

In at least certain cases, the dynamic range can be amplified via downstream cellular signaling pathways, e.g., those involving gene expression [24], second messengers [25] or signaling cascades like MAP kinase pathways [26].

3.2 Activity Assay   The engineering of photoreceptors often requires the testing of sizeable numbers of candidate constructs which is greatly aided by the availability of fast and convenient activity assays. In general, high-throughput approaches distribute into two groups: screening systems, often set up inside living cells, which rely on readily detectable reporter readout (e.g., fluorescence); and selection systems in which cell proliferation/survival under set selection settings (e.g., dark vs. light) is conditional on expression and activity of candidate photoreceptors.

An efficient in vivo screening setup can be established provided that the desired effector output is orthogonal to other cellular metabolic pathways; does not harm living cells; and generates a chromogenic, fluorogenic, or other easily detectable readout. High-throughput screening systems are particularly effective if the output of the engineered photoreceptor can be coupled to reporter gene expression [24], thus allowing the screening of large numbers of receptor variants, for example by fluorescence-activated cell sorting. In case of proteins that undergo light-regulated association reactions, several display techniques, i.e. phage, mRNA, or ribosome display, are well suited for screening [27]. Independently of the screening approach, iterative rounds of positive and negative screening under light and dark conditions are often necessary to optimize dynamic range. If high-throughput screening systems cannot be established, photoreceptor engineering can be facilitated by medium-throughput screening systems, e.g., assays that determine the presence of specific metabolites or enzymatic activities in crude or partially purified cell lysates (*see* Chapter 7 for a recent example).

Selection systems, allowing cell growth under either light or dark conditions, and conversely leading to cell death or growth arrest under the opposite condition, provide an alternative means of accelerating photoreceptor engineering [28]. However, such systems need to be carefully set up and calibrated which is often challenging, in particular when the initial activity difference between dark and light conditions is small [29].

4   Allosteric Mechanisms of Signal Transduction

The transduction of signals in receptors is achieved through allosteric coupling between sensor and effector modules. Regardless of the precise mechanism, in photoreceptors the reception of light generally leads to initial conformational and dynamic transitions within the chromophore-binding pocket and the surrounding photosensor scaffold. Signal transmission to the effector is often achieved through α-helical structures that serve as linkers between photosensor and effector modules. Allosteric coupling mechanisms widely differ across photoreceptors but usually involve conformational and dynamic transitions, such as local unfolding, refolding, domain rearrangement, association, or dissociation [3]. We arrange this section based on whether light absorption causes changes in oligomeric state of the photoreceptor (Associating photoreceptors; Subheading 4.1) or not (Non-associating photoreceptors; Subheadings 4.2.1 and 4.2.2) (Fig. 2).

4.1 Associating Photoreceptors

For this group of photoreceptors, the reception of light results in association/dissociation reactions, mostly dimerization, mediated by the uncovering or covering of interaction sites. We distinguish between homo- and hetero-oligomerization depending on whether association occurs between alike or different partners. Association processes can be tied to changes in biological activity in different manners, for example by assembly of proteins into their functional oligomeric form; by colocalization of interacting proteins; or by recruitment of proteins to subcellular compartments.

Examples of naturally occurring systems that homo-oligomerize upon light absorption are the blue light-sensing *Arabidopsis thaliana* cryptochrome 2 (*At*Cry2) [30] and the LOV photoreceptors Vivid from *Neurospora crassa* [31], aureochrome from stramenopiles [32], and EL222 from *Erythrobacter litoralis* [33]. By contrast, the homodimeric photoreceptor *At*UVR8 from *A. thaliana* dissociates into monomers upon UV-light exposure [34]. In case of hetero-associating systems, the most widely deployed representatives derive from higher plants, exemplified by *A. thaliana*: *At*Cry2 not only assembles into homo-oligomers upon light absorption but also forms a heterodimer with its interacting partner *At*CIB1 [35]; similarly, upon light-induced dissociation, *At*UVR8 forms a heterodimer with *At*COP1 [34]. The LOV protein *At*FKF1 interacts with its partner *At*GIGANTEA following blue-light absorption [36]; and the red/far-red sensing phytochromes A and B (*At*PhyA and *At*PhyB) associate in light-dependent manner with their interacting partners, of which *At*PIF3 and *At*PIF6 are the most popular in photoreceptor engineering (PIF, phytochrome interacting partner) [37, 38]. As we discuss in Subheading 5.1, light-regulated association/dissociation reactions have been utilized in numerous photoreceptor engineering studies.

Allosteric mechanisms (4.)

Changes in oligomeric state (4.1.)

Homo

Design strategies (5.)

Association / Dissociation (5.1.)

Photo-activatable dimerizer

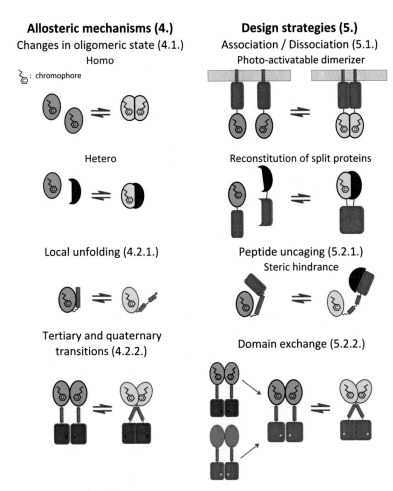

Hetero

Reconstitution of split proteins

Local unfolding (4.2.1.)

Peptide uncaging (5.2.1.)

Steric hindrance

Tertiary and quaternary transitions (4.2.2.)

Domain exchange (5.2.2.)

Fig. 2 Allosteric mechanisms of signal transduction in natural photoreceptors and representative strategies derived for photoreceptor engineering. Engineered photoreceptors undergoing light-induced changes in oligomeric state can act as photo-activatable dimerization modules, for example to mediate the assembly of functional oligomeric states or the reconstitution of split proteins. Light-directed local unfolding can be utilized to release steric hindrance with concomitant changes in effector oligomeric state and/or activity. Engineering by domain exchange allows other light-induced tertiary and quaternary transitions to control naturally light-insensitive proteins

4.2 Non-associating Photoreceptors

This category comprises a diverse group of photoreceptors for which signal transduction involves changes of tertiary and, in case of oligomeric receptors, quaternary structure but no change in oligomeric state. In contrast to the above cases, for non-associating photoreceptors the physical nature of the linker (sequence, length, structure, topology, dynamics) between photosensor and effector modules is of much greater importance, as the linker has to specifically interact with both photosensor and effector sites to enable signal propagation. Put another way, photosensor and effector have to be linked in a manner conducive to efficient thermodynamic coupling between these modules.

4.2.1 Uncaging of Peptide Epitopes/ Active Sites

As a paradigm of this class, *As*LOV2 exhibits light-triggered, local unfolding of its C-terminal Jα helix and concomitant dissociation from the LOV protein core [39]. In its original biological context within the multidomain receptor phototropin 1, Jα unfolding elicits subdomain rearrangements, but no apparent changes in the oligomeric state of the photoreceptor [40]. By contrast, in certain engineered photoreceptors, *As*LOV2 has been converted into an associating photoreceptor (cf. Subheading 5). Notably, light-regulated unfolding mechanisms are not restricted to *As*LOV2 but also contribute to signal transduction in other photoreceptors such as certain LOV domains (e.g., LOV2 from *A. thaliana* phototropin 1 [41], aureochrome 1a from *Phaeodactylum tricornutum* [42], *Rs*LOV from *Rhodobacter sphaeroides* [43]) and the photoactive yellow protein (PYP) from purple bacteria [44].

4.2.2 Tertiary and Quaternary Transitions

In this section, we treat a disparate class of photoreceptors in which signal transmission primarily depends neither on changes in oligomeric state nor on local unfolding but rather on other tertiary or quaternary structural transitions that are often transmitted between sensor and effector modules by helical elements. Many of these concepts are exemplified in two recent case studies.

First, the recent crystal structure of the monomeric LOV histidine kinase EL346 from *Erythrobacter litoralis HTCC2594* reveals long helices as mediators of photosensor-effector interdomain interactions [45]. These helices form an interface with the photosensory LOV core and maintain contact to the CA effector domain on the opposite side. The interdomain interactions stabilize the inhibited kinase form in the dark and weaken step-wise upon light induction, thereby causing a rearrangement of the CA domain that increases its catalytic activity.

Second, Takala et al. recently presented crystal and solution structures of dark- and light-adapted states of the PSC module of *Deinococcus radiodurans* bacterial phytochrome (cf. Fig. 1) [46]. The structures of this parallel homodimeric protein support a previously proposed toggle-model for photoconversion in phytochromes [47]. According to this model, signal-induced rotation of the D ring of the tetrapyrrole chromophore causes contact rearrangements of the GAF/PHY interface. These rearrangements are possibly transferred to the C-terminal effector module by causing a tug on the linker helix and a concomitant pivot motion of the effector modules.

5   Design Strategies

Having discussed the properties of natural photoreceptors, we regard how their signaling mechanisms have been co-opted (in some cases, even transcending the natural mechanism) in the engineering of novel photoreceptors [2, 3] (Table 1).

5.1   Association/ Dissociation

Our recent survey of photoreceptor engineering highlighted light-regulated association as a particularly versatile and promising design approach [3]. The prevalence of this approach is arguably explained by the frequent occurrence of oligomerization reactions in signal transduction and by less exigent demands on the linker connecting sensor and effector modules (cf. Subheading 4). Beyond providing a physical connection, requirements on the linker here are much less demanding, and linkers are often short, flexible, and predominantly hydrophilic. Association-based strategies are particularly well suited for effectors that are regulated by oligomerization reactions in their natural context, e.g., many transcription factors and transmembrane receptors. However, this approach is not restricted to naturally associating proteins but extends to proteins which are not originally regulated by oligomerization processes, in particular to split proteins.

For example, several recent studies described light-regulated variants of receptor tyrosine kinases in which activation is often based on ligand-induced receptor dimerization (RTKs) [26, 48, 49]. In all studies, control by ligand binding has been reprogrammed to control by light via fusion of the RTK to associating photoreceptors. Whereas Grusch et al. fused aureochrome LOV domains to the intracellular part of the membrane-bound RTK, the closely related studies by Chang et al. and Kim et al. accomplished the same via fusion to *At*Cry2.

Another set of studies employed associating photoreceptors to generate systems for light-induced expression of transgenes in eukaryotes [36, 50, 51]. For example, the *Neurospora crassa* Vivid LOV domain assembles into homodimers upon blue-light illumination; when linked to a truncated, nonfunctional monomeric form of the GAL4 transcription factor, this LOV domain mediated light-dependent association and concomitant reconstitution of the functional dimeric form of GAL4. Even earlier, in a pioneering application, Shimizu-Sato et al. exploited the light-regulated association of the *A. thaliana* phytochromes A and B with *At*PIF3 to furnish a transcription system that can be activated by red and inactivated by far-red light [38]. The system is based on the yeast two-hybrid approach, with the GAL4 DNA-binding domain fused to either full-length or the N-terminal PSC of *At*PhyA or *At*PhyB, and *At*PIF3 coupled to the GAL4 trans-activation domain. Conceptually similar, engineered photoreceptors are based on the *A. thaliana* cryptochrome *At*Cry2 and *At*CIB1 and mediate light-regulated transgene expression and MAP-kinase signaling [35, 52].

5.2   Other Strategies

5.2.1   Local Unfolding

Local unfolding reactions can be harnessed to alter the accessibility of active sites and surface epitopes in a light-dependent manner, and to thereby regulate the activity of effector modules and downstream metabolic pathways. A striking demonstration of this approach is provided by photoactivatable Rac1, a small GTPase involved in the regulation of cytoskeletal dynamics. Fusion to *As*LOV2 led to steric restriction of the active site of Rac1, which was relieved upon blue-light-induced unfolding of the *As*LOV2 Jα helix, cf. Subheading 4.2.1, [53]. Another example for light-dependent control of activity is the inhibition of potassium channels by peptide toxins, which were colocalized with the channels via fusion to membrane-tethered *As*LOV2 [54]. Upon blue-light-induced unfolding of the Jα helix, the increased mobility of the toxin led to a decrease in its local concentration and channel opening. In two conceptually similar studies, systems for light-induced protein degradation were generated on the basis of phototropin LOV2 domains [41, 55]. Degron peptide sequences were interleaved with the LOV Jα helix such that they were little accessible under dark conditions; only upon light-induced Jα unfolding and concomitant exposure of the degron sequences, proteasomal degradation of target constructs was greatly stimulated.

5.2.2   Domain Exchange

In case the above two engineering strategies do not apply, originally light-inert signal receptors can often be subjected under light control if their sensor modules are replaced by suitable (homologous) photosensor modules. For example, GAF (cGMP-specific phosphodiesterases, adenylate cyclases, and FhlA) domains could be exchanged by (bacterial) phytochrome photosensors that comprise structurally homologous GAF and PHY (phytochrome specific) domains; or LOV domains could replace PAS domains of which they are a subgroup. Often, the availability of three-dimensional structures allows the construction of structure-based alignments that guide the fusion between sensor and effector modules. When no suitable homologous relatives exist, domain exchange can still yield functional proteins but the lack of structure-based alignments complicates the planning of the fusion strategy. For exchange of sensor and effector modules linked by α-helical linkers (e.g., coiled coil linkers), an examination of the linker properties is helpful for the identification of the best fusion site. Linker helices of discrete length widely recur in natural signal receptors [56–59] and crucially determine activity and regulation by light in engineered photoreceptors as the below case studies illustrate.

As an example for a successful domain exchange, the engineered red-light-sensitive photoreceptor Cph8 connects the PCB-binding photosensor of the cyanobacterial phytochrome Cph1 from *Synechocystis* sp. to the effector module of the histidine kinase EnvZ from *E. coli* [60]. The light-activated cAMP/cGMP-specific phosphodiesterase LAPD represents another example for homologous domain exchange [18], in this case of two GAF domains of the human phosphodiesterase 2A against a biliverdin-binding PAS-GAF-PHY tandem of the *D. radiodurans* bacterial phytochrome. Notably, in both cases use of the correct linker length was crucial for obtaining light-regulated enzymes. Finally, the light-activated adenylate cyclase IlaC is based on heterologous domain exchange; specifically, the PAS-GAF-PHY PSC module of the *R. sphaeroides* bacterial phytochrome BphG1 was connected to the adenylate cyclase effector from *Nostoc* sp. CyaB1 and thereby replaced two regulatory PAS domains [19].

6   Summary

As discussed above, the properties of engineered photoreceptors strongly depend on the intrinsic characteristics of the constituent photosensor and effector modules (Table 1) [3]. Therefore, the attributes of both the photosensor (e.g., genetic encodability; light sensitivity; achievable temporal and spatial resolution) and the effector (e.g., specific activity; possibility of amplification and availability of a screening assay) should be carefully considered at the initial design stages. Additionally, the resultant photoreceptor needs

Table 1
Design aspects in photoreceptor engineering

Design aspect	Sensor/effector/receptor attributes
1. Genetic encoding	Delivery of photoreceptor DNA to target cells Chromophore availability in target organism
2. Light sensitivity	Chromophore type Photocycle kinetics
3. Temporal resolution	Photocycle kinetics Photochromic photoreceptors
4. Spatial resolution	Cell-type-specific promoters Intracellular trafficking signals Photochromic photoreceptors
5. Dynamic range	Effector specific activity Minimize dark activity Embedding in signaling networks for amplification
6. Allosteric mechanisms (Subheading 4) • Changes in oligomeric states (Subheading 4.1) • Uncaging of peptide epitopes/active sites (Subheading 4.2.1) • Tertiary/quaternary transitions (Subheading 4.2.2)	Design strategies (Subheading 5) • Association/dissociation (Subheading 5.1) • Local unfolding (Subheading 5.2.1) • Domain exchange (Subheading 5.2.2)

to be optimized regarding expression in situ, cell-type specific or subcellular targeting and dynamic range. Lastly, to engineer highly active and efficiently regulated photoreceptors, the signal transmission mechanisms of sensor and effector must be compatible.

To date, mainly three fundamental design strategies have proven successful in the engineering of photoreceptors, and they apply to different scenarios: (a) Association-based approaches, implementable for effectors whose activity is a function of their oligomeric state or subcellular location; (b) Approaches based on local unfolding that trigger uncaging of effector peptides or release of steric hindrance; and (c) Exchange of homologous or heterologous sensor and effector modules. We expect optogenetics and photoreceptor engineering to continue their rapid development and to thus grant light control over otherwise light-insensitive processes that were previously inaccessible to optogenetic intervention.

Acknowledgements

Research in our laboratory is generously supported through a Sofja-Kovalevskaya Award by the Alexander-von-Humboldt Foundation (to A.M.) and by the Deutsche Forschungsgemeinschaft within the Cluster of Excellence 'Unicat—Unifying Concepts in Catalysis'.

References

1. Möglich A, Moffat K (2010) Engineered photoreceptors as novel optogenetic tools. Photochem Photobiol Sci 9:1286–1300

2. Schmidt D, Cho YK (2015) Natural photoreceptors and their application to synthetic biology. Trends Biotechnol 33:80–91

3. Ziegler T, Möglich A (2015) Photoreceptor engineering. Front Mol Biosci 2:30

4. Ernst OP, Lodowski DT, Elstner M, Hegemann P, Brown LS, Kandori H (2014) Microbial and animal rhodopsins: structures, functions, and molecular mechanisms. Chem Rev 114: 126–163

5. Fenno L, Yizhar O, Deisseroth K (2011) The development and application of optogenetics. Annu Rev Neurosci 34:389–412

6. Schneider F, Grimm C, Hegemann P (2015) Biophysics of channelrhodopsin. Annu Rev Biophys 44:167–186

7. Hegemann P (2008) Algal sensory photoreceptors. Annu Rev Plant Biol 59:167–189

8. Möglich A, Yang X, Ayers RA, Moffat K (2010) Structure and function of plant photoreceptors. Annu Rev Plant Biol 61:21–47

9. Rockwell NC, Su Y-S, Lagarias JC (2006) Phytochrome structure and signaling mechanisms. Annu Rev Plant Biol 57:837–858

10. Brown BA, Cloix C, Jiang GH, Kaiserli E, Herzyk P, Kliebenstein DJ, Jenkins GI (2005) A UV-B-specific signaling component orchestrates plant UV protection. Proc Natl Acad Sci U S A 102:18225–18230

11. Conrad KS, Manahan CC, Crane BR (2014) Photochemistry of flavoprotein light sensors. Nat Chem Biol 10:801–809

12. Losi A, Mandalari C, Gärtner W (2015) The evolution and functional role of flavin-based prokaryotic photoreceptors. Photochem Photobiol 91:1021–1031. doi:10.1111/php.12489

13. Pudasaini A, El-Arab KK, Zoltowski BD (2015) LOV-based optogenetic devices: light-driven modules to impart photoregulated control of cellular signaling. Front Mol Biosci 2:18

14. Rockwell NC, Duanmu D, Martin SS, Bachy C, Price DC, Bhattacharya D, Worden AZ, Lagarias JC (2014) Eukaryotic algal phytochromes span the visible spectrum. Proc Natl Acad Sci U S A 111:3871–3876

15. Ikeuchi M, Ishizuka T (2008) Cyanobacteriochromes: a new superfamily of tetrapyrrole-binding photoreceptors in cyanobacteria. Photochem Photobiol Sci 7:1159–1167

16. Rockwell NC, Lagarias JC (2010) A brief history of phytochromes. Chemphyschem 11: 1172–1180

17. Filonov GS, Piatkevich KD, Ting L-M, Zhang J, Kim K, Verkhusha VV (2011) Bright and stable near-infrared fluorescent protein for in vivo imaging. Nat Biotechnol 29:757–761

18. Gasser C, Taiber S, Yeh C-M, Wittig CH, Hegemann P, Ryu S, Wunder F, Möglich A (2014) Engineering of a red-light-activated human cAMP/cGMP-specific phosphodiesterase. Proc Natl Acad Sci U S A 111:8803–8808

19. Ryu M-H, Kang I-H, Nelson MD, Jensen TM, Lyuksyutova AI, Siltberg-Liberles J, Raizen DM, Gomelsky M (2014) Engineering adenylate cyclases regulated by near-infrared window light. Proc Natl Acad Sci U S A 111:10167–10172

20. Shu X, Royant A, Lin MZ, Aguilera TA, Lev-Ram V, Steinbach PA, Tsien RY (2009) Mammalian expression of infrared fluorescent proteins engineered from a bacterial phytochrome. Science 324:804–807

21. Alexandre MT, Arents JC, van Grondelle R, Hellingwerf KJ, Kennis JT (2007) A base-catalyzed mechanism for dark state recovery in the Avena sativa phototropin-1 LOV2 domain. Biochemistry 46:3129–3137

22. Zhou XX, Chung HK, Lam AJ, Lin MZ (2012) Optical control of protein activity by fluorescent protein domains. Science 338:810–814

23. Lee J, Natarajan M, Nashine VC, Socolich M, Vo T, Russ WP, Benkovic SJ, Ranganathan R (2008) Surface sites for engineering allosteric control in proteins. Science 322:438–442

24. Ohlendorf R, Vidavski RR, Eldar A, Moffat K, Möglich A (2012) From dusk till dawn: one-plasmid systems for light-regulated gene expression. J Mol Biol 416:534–542

25. Jansen V, Alvarez L, Balbach M, Strünker T, Hegemann P, Kaupp UB, Wachten D (2015) Controlling fertilization and cAMP signaling in sperm by optogenetics. eLife 4:e05161

26. Grusch M, Schelch K, Riedler R, Reichhart E, Differ C, Berger W, Inglés-Prieto Á, Janovjak H (2014) Spatio-temporally precise activation of engineered receptor tyrosine kinases by light. EMBO J 33:1713–1726

27. Guntas G, Hallett RA, Zimmerman SP, Williams T, Yumerefendi H, Bear JE, Kuhlman B (2015) Engineering an improved light-induced dimer (iLID) for controlling the localization and activity of signaling proteins. Proc Natl Acad Sci U S A 112:112–117

28. Cosentino C, Alberio L, Gazzarrini S et al (2015) Optogenetics. Engineering of a light-gated potassium channel. Science 348:707–710

29. Goldsmith M, Tawfik DS (2012) Directed enzyme evolution: beyond the low-hanging fruit. Curr Opin Struct Biol 22:406–412

30. Bugaj LJ, Choksi AT, Mesuda CK, Kane RS, Schaffer DV (2013) Optogenetic protein clustering and signaling activation in mammalian cells. Nat Methods 10:249–252

31. Lamb JS, Zoltowski BD, Pabit SA, Crane BR, Pollack L (2008) Time-resolved dimerization of a PAS-LOV protein measured with photo-coupled small angle X-ray scattering. J Am Chem Soc 130:12226–12227

32. Takahashi F, Yamagata D, Ishikawa M, Fukamatsu Y, Ogura Y, Kasahara M, Kiyosue T, Kikuyama M, Wada M, Kataoka H (2007) AUREOCHROME, a photoreceptor required for photomorphogenesis in stramenopiles. Proc Natl Acad Sci U S A 104:19625–19630

33. Nash AI, McNulty R, Shillito ME, Swartz TE, Bogomolni RA, Luecke H, Gardner KH (2011) Structural basis of photosensitivity in a bacterial light-oxygen-voltage/helix-turn-helix (LOV-HTH) DNA-binding protein. Proc Natl Acad Sci U S A 108:9449–9454

34. Christie JM, Arvai AS, Baxter KJ et al (2012) Plant UVR8 photoreceptor senses UV-B by tryptophan-mediated disruption of cross-dimer salt bridges. Science 335:1492–1496

35. Kennedy MJ, Hughes RM, Peteya LA, Schwartz JW, Ehlers MD, Tucker CL (2010) Rapid blue-light-mediated induction of protein interactions in living cells. Nat Methods 7:973–975

36. Yazawa M, Sadaghiani AM, Hsueh B, Dolmetsch RE (2009) Induction of protein-protein interactions in live cells using light. Nat Biotechnol 27:941–945

37. Levskaya A, Weiner OD, Lim WA, Voigt CA (2009) Spatiotemporal control of cell signalling using a light-switchable protein interaction. Nature 461:997–1001

38. Shimizu-Sato S, Huq E, Tepperman JM, Quail PH (2002) A light-switchable gene promoter system. Nat Biotechnol 20:1041–1044

39. Harper SM, Neil LC, Gardner KH (2003) Structural basis of a phototropin light switch. Science 301:1541–1544

40. Christie JM, Blackwood L, Petersen J, Sullivan S (2015) Plant flavoprotein photoreceptors. Plant Cell Physiol 56:401–413

41. Renicke C, Schuster D, Usherenko S, Essen L-O, Taxis C (2013) A LOV2 domain-based optogenetic tool to control protein degradation and cellular function. Chem Biol 20:619–626

42. Herman E, Kottke T (2015) Allosterically regulated unfolding of the A′α helix exposes the dimerization site of the blue-light-sensing aureochrome-LOV domain. Biochemistry 54:1484–1492

43. Conrad KS, Bilwes AM, Crane BR (2013) Light-induced subunit dissociation by a light-oxygen-voltage domain photoreceptor from Rhodobacter sphaeroides. Biochemistry 52:378–391

44. Rubinstenn G, Vuister GW, Mulder FAA, Düx PE, Boelens R, Hellingwerf KJ, Kaptein R (1998) Structural and dynamic changes of photoactive yellow protein during its photocycle in solution. Nat Struct Mol Biol 5:568–570

45. Rivera-Cancel G, Ko W, Tomchick DR, Correa F, Gardner KH (2014) Full-length structure of a monomeric histidine kinase reveals basis for sensory regulation. Proc Natl Acad Sci U S A 111:17839–17844

46. Takala H, Björling A, Berntsson O et al (2014) Signal amplification and transduction in phytochrome photosensors. Nature 509:245–248

47. Anders K, Gutt A, Gärtner W, Essen L-O (2014) Phototransformation of the red light sensor cyanobacterial phytochrome 2 from Synechocystis species depends on its tongue motifs. J Biol Chem 289:25590–25600

48. Chang K-Y, Woo D, Jung H et al (2014) Light-inducible receptor tyrosine kinases that regulate neurotrophin signalling. Nat Commun 5:4057

49. Kim N, Kim JM, Lee M, Kim CY, Chang K-Y, Heo WD (2014) Spatiotemporal control of fibroblast growth factor receptor signals by blue light. Chem Biol 21:903–912

50. Nihongaki Y, Suzuki H, Kawano F, Sato M (2014) Genetically engineered photoinducible homodimerization system with improved dimer-forming efficiency. ACS Chem Biol 9:617–621

51. Wang X, Chen X, Yang Y (2012) Spatiotemporal control of gene expression by a light-switchable transgene system. Nat Methods 9:266–269

52. Aoki K, Kumagai Y, Sakurai A, Komatsu N, Fujita Y, Shionyu C, Matsuda M (2013) Stochastic ERK activation induced by noise and cell-to-cell propagation regulates cell density-dependent proliferation. Mol Cell 52:529–540

53. Wu YI, Frey D, Lungu OI, Jaehrig A, Schlichting I, Kuhlman B, Hahn KM (2009) A genetically encoded photoactivatable Rac controls the motility of living cells. Nature 461:104–108

54. Schmidt D, Tillberg PW, Chen F, Boyden ES (2014) A fully genetically encoded protein architecture for optical control of peptide ligand concentration. Nat Commun 5:3019

55. Bonger KM, Rakhit R, Payumo AY, Chen JK, Wandless TJ (2014) General method for regulating protein stability with light. ACS Chem Biol 9:111–115

56. Anantharaman V, Balaji S, Aravind L (2006) The signaling helix: a common functional theme in diverse signaling proteins. Biol Direct 1:25

57. Möglich A, Ayers RA, Moffat K (2009) Design and signaling mechanism of light-regulated histidine kinases. J Mol Biol 385:1433–1444

58. Möglich A, Ayers RA, Moffat K (2010) Addition at the molecular level: signal integration in designed Per-ARNT-Sim receptor proteins. J Mol Biol 400:477–486

59. Rockwell NC, Ohlendorf R, Möglich A (2013) Cyanobacteriochromes in full color and three dimensions. Proc Natl Acad Sci U S A 110:806–807

60. Levskaya A, Chevalier AA, Tabor JJ et al (2005) Synthetic biology: sngineering Escherichia coli to see light. Nature 438:441–442

61. Losi A, Polverini E, Quest B, Gärtner W (2002) First evidence for phototropin-related blue-light receptors in prokaryotes. Biophys J 82:2627–2634

62. Avelar GM, Schumacher RI, Zaini PA, Leonard G, Richards TA, Gomes SL (2014) A rhodopsin-guanylyl cyclase gene fusion functions in visual perception in a fungus. Curr Biol 24:1234–1240

63. Yoshihara S, Suzuki F, Fujita H, Geng XX, Ikeuchi M (2000) Novel putative photoreceptor and regulatory genes required for the positive phototactic movement of the unicellular motile Cyanobacterium Synechocystis sp. PCC 6803. Plant Cell Physiol 41:1299–1304

64. Davis SJ, Vener AV, Vierstra RD (1999) Bacteriophytochromes: phytochrome-like photoreceptors from nonphotosynthetic eubacteria. Science 286:2517–2520

INDEX

Arash Kianianmomeni (ed.), *Optogenetics: Methods and Protocols*, Methods in Molecular Biology, vol. 1408,
DOI 10.1007/978-1-4939-3512-3, © Springer Science+Business Media New York 2016

Printed in the United States
By Bookmasters